W9-CFF-549

LANDS AND PEOPLES

LANDS AND PEOPLES
ASIA · AUSTRALIA
NEW ZEALAND · OCEANIA

Volume 2

Lexicon Publications

COPYRIGHT © 1978 BY LEXICON PUBLICATIONS, INC.

COPYRIGHT © 1978, 1977, 1976, 1973, 1972 BY LEXICON PUBLICATIONS, INC.

Library of Congress Catalog Card Number: 78-60426
ISBN 0-7172-8006-3
*All rights reserved. No part of this book may be
reproduced without written permission from the publishers.*
Printed in the United States of America

CONTENTS
ASIA · AUSTRALIA
NEW ZEALAND · OCEANIA
Volume 2

ASIA

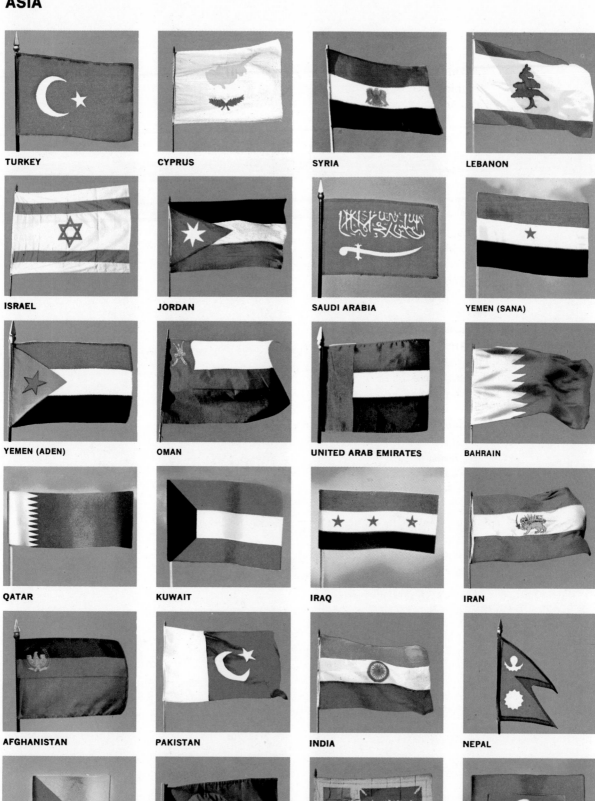

TURKEY

CYPRUS

SYRIA

LEBANON

ISRAEL

JORDAN

SAUDI ARABIA

YEMEN (SANA)

YEMEN (ADEN)

OMAN

UNITED ARAB EMIRATES

BAHRAIN

QATAR

KUWAIT

IRAQ

IRAN

AFGHANISTAN

PAKISTAN

INDIA

NEPAL

BHUTAN

BANGLADESH

CEYLON

MALDIVES

BURMA

THAILAND

LAOS

CAMBODIA (KAMPUCHEA)

VIETNAM

MALAYSIA

SINGAPORE

INDONESIA

BRUNEI

PHILIPPINES

MONGOLIA

CHINA, PEOPLE'S REPUBLIC OF

CHINA, REPUBLIC OF

HONG KONG

KOREA, DEMOCRATIC PEOPLE'S REPUBLIC OF (NORTH)

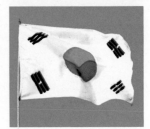

KOREA, REPUBLIC OF (SOUTH)

OCEANIA

JAPAN

AUSTRALIA

NEW ZEALAND

PAPUA NEW GUINEA

FIJI

NAURU

TONGA

WESTERN SAMOA

ASIA: AN INTRODUCTION

by Sirimavo R. D. BANDARANAIKE
Prime Minister of Sri Lanka

I am glad to contribute this introductory article on Asia for LANDS AND PEOPLES.

The conventional boundaries of Asia describe a vast area from the Ural Mountains, the Caspian and Black seas, the coasts of Asia Minor, the eastern Mediterranean and the Red Sea to the island chains of Indonesia, the Philippines, and Japan that extend up to the Kamchatka Peninsula. The ice-bound polar regions to the north and the tropical seas to the south make up the other frontiers of this part of the world. Asia encloses one third of the earth's land area, traverses the range of possible climates, embraces the most diverse conditions of human life, and includes the predominant number of mankind. Its heartland is an area of great magnitude and extent, the largest part of the largest landmass in the world, great parts of it thousands of miles removed from the sea. At the same time, the lands of Asia include islands like Sri Lanka, peninsulas like India—which extends over one thousand miles into the Indian Ocean—and archipelagos like Indonesia, where the sea forms a pervasive part of the human environment.

The peoples of Asia represent not only great numbers but also a bewildering kaleidoscope of cultures, some of them actually whole civilizations. The earliest beginnings of civilized life can be traced to Asia. The civilizations of the Tigris-Euphrates Valley, the Indus Valley cities of Mohenjo-Daro and Harappa, and the civilizations of China are only a few of the most ancient and illustrious of the achievements of man that had their origins in Asia. Two thousand five hundred years ago migrant peoples from North India, demonstrating a seafaring capability, arrived in Sri Lanka and began settlements which matured into a distinctive Asian civilization. The history of Asia is replete with such instances of cultural transmission and migration of peoples and the development of fresh centers of civilization which became intricate and distinctive variations of the original influence. Asian history, therefore, is a fabric of infinite variety and richness of texture.

The world religions, moreover, originated in Asia. The eastern Mediterranean area gave birth to Judaism and Christianity. From Arabia came Islam. Hinduism and Buddhism had their cradle in Asia. But some of these mighty religions did not remain in their homelands. Christianity spread to Europe. Islam spread from Arabia throughout North Africa, India, Southeast Asia, and many other parts of the world. Hinduism became the dominant religion of the Indian subcontinent. The message of the Compassionate Buddha, which was first uttered in India, was carried to Southeast Asia, China, and Japan. It was brought to Sri Lanka by Prince Mahinda at the time of Emperor Asoka in the 3rd century B.C.

and rapidly spread among its peoples. It has remained throughout the centuries, with unbroken continuity, the principal civilizing force and the nurturing spirit of the people of Sri Lanka. Despite geographical barriers of mountains, inaccessible country, and ocean expanse, and notwithstanding the immensity of space and the great periods of time involved, the fertilizing and civilizing influence of Asian centers of culture has spread to other parts of Asia and to other parts of the world.

Asia is not a uniform region. An area of such magnitude, embracing such a variety of peoples, cannot conceivably be one. It would be realistic, therefore, to speak of sub-regions of Asia, such as East Asia, Southeast Asia, Central Asia, North Asia, South Asia, Southwest Asia, as has increasingly been the trend in recent years. This accords with certain demographic and geographic patterns and helps to define areas where special influences have been dominant. The sub-regions do not, however, detract from the fact that throughout much of Asia an underlying layer of common cultural and folk elements exists. This has created an identity that gives some objective ground for regarding Asia and Asians as different from Europe or Europeans and from other non-Asians. There is, of course, no over-arching Asian culture, embracing all its diverse peoples and territories. It is easier to think of Asian identity by and large in terms of considerably diffused elements. The effect of centuries of evolution, expansion, and migration slowly came to create an identity that is segmental, unstructured, and decentralized, as is characteristic of basically peasant societies and traditional peoples. The center of Asian identity and Asian integrity lay in the villages and the many tiny social units of this otherwise vast geographical area. From time to time, of course, empires of great resplendence rose and fell. There were also identities built on race, tribe, language, and religion, but none implied an encompassing Asian unity or a sense of national consciousness.

It is predominantly the impact of Europe on Asia that led to the development of unified political and social structures. Within the frontiers demarcated by colonial powers in their scramble for colonies a territorial and administrative unity was created. European rule also stimulated within these boundaries feelings of nationalism, and as the tide of Western rule receded, the empires and colonies gave way to nation states. Nationalism brought about a cultural resurgence and renaissance. This reflected not only the pride and self-consciousness of individual nations but also their sense of identity as Asian countries and their awareness of common historical experiences which colonial rule had heightened.

Asia is emerging from its immediate post-colonial phase. The independent states in this part of the world reflect a vitality of their own. Major states have arisen in Asia representing major fulcrums of power and influence. Small states have tended to stabilize increasingly their independence and sovereignty. The future therefore is likely to be concerned less with the impact of Europe on Asia than with the Asian contribution to the world.

I am certain that this section on Asia will contribute much to a wider knowledge and understanding of an important part of the world.

ASIA

Although it is described as the largest continent, Asia is, in fact, the eastern section of an even greater landmass called Eurasia. The Ural Mountains are accepted as the dividing line between Europe and Asia. Such bodies of water as the Black Sea, the strait of Bosporus, and the Mediterranean Sea also separate parts of Asia from Europe, while the Suez Canal and the Red Sea separate Asia from Africa.

The word Asia was used by the ancient Greeks to apply to a rather limited area in what is now Turkey. The Romans took over the name, using it for one of their provinces situated in the same region, which is sometimes called Asia Minor. Much later, Europeans came to refer to all the lands in the distant east as Asia, but the inhabitants of these lands never heard of this name until modern times.

Marco Polo, who set out from Venice in 1271, left a fascinating account of his travels across the length of Asia and introduced Europe to a mysterious, unknown world. As an admirer of Marco Polo has written:

> He was the first Traveller to trace a route across the whole longitude of Asia, naming and describing kingdom after kingdom which he had seen with his own eyes; the Deserts of Persia [Iran], the flowering plateaux and wild gorges of Badakhshan [in Afghanistan], the jade-bearing rivers of Khotan [in China], the Mongolian steppes, cradle of the power that had so lately threatened to swallow up Christendom, the new and brilliant court that had been established at Cambaluc [Peking]: the first Traveller to reveal China in all its wealth and vastness, its mighty rivers, its huge cities, its rich manufactures, its swarming population, the inconceivably vast fleets that quickened its seas and its inland waters: to tell us of the nations on its borders with all their eccentricities of manners and worship; of Tibet . . .; of Burma with its golden pagodas and their tinkling crowns; of Laos, of Siam [Thailand], of Cochin China [part of South Vietnam], of Japan, with its rosy pearls and golden-roofed palaces; the first to speak of that Museum of Beauty and Wonder . . . the Indian Archipelago, source of those aromatics then so highly prized and whose origin was so dark; of Java, the Pearl of Islands; of Sumatra with its many kings, its strange costly products, and its cannibal races; of the naked savages of Nicobar and Andaman [in the Indian Ocean]; of Ceylon, the Isle of Gems with its Sacred Mountain and its Tomb of Adam; of India the Great . . . with its virtuous Brahmans . . . , its diamonds and strange tales of their acquisition, its seabeds of pearl, and its powerful sun; the first in mediaeval times to give any distinct account . . . of Siberia and the Arctic Ocean, of dog-sledges, white [polar] bears, and reindeer riding Tunguses [of Siberia].

These lines call to mind all that had been heard about the glamour, romance, and color of Asia and its strange and exotic peoples. Even the names of the places sound as if they come from poems. Marco Polo failed, however, to tell about other great wonders of Asia. Of these wonders, some were not along his way, and others were built after his journey. Now they may be seen by a traveler who heads eastward from the Mediterranean Sea.

ASIA

CANADA
120°
90°
60°
North Pole
Arctic Ocean
Greenland
0°
80°
West Spitsbergen
Novaya Zemlya
60°
30°
90°
70°
Barents Sea
Arctic Circle
Murmansk
Vorkuta
Atlantic Ocean
60°
Archangel
Europe Asia
S
50°
London
Leningrad
UNION OF
SOVIET
Baltic Sea
Perm
Ural Mountains
EUROPE
Moscow
Sverdlovsk
Berlin
Paris
Magnitogorsk
Kiev
40°
Madrid
Kharkov
Volgograd
(Stalingrad)
Aral Sea
30°
Odessa
Astrakhan
Rome
Europe Asia
Black Sea
Europe Asia
Bukhara
Samarkand
Istanbul
Caspian Sea
Baku
Ashkhabad
Mediterranean Sea
Smyrna
Ankara
TURKEY
Tabriz
Teheran
Meshed
AFGHANISTAN
Nicosia
Aleppo
CYPRUS
SYRIA
LEBANON
Beirut
Damascus
Baghdad
IRAN
ISRAEL
Jerusalem
Amman
IRAQ
Kerman
Kandahar
Suez Canal
JORDAN
Petra
Neutral Zone
Abadan
Persepolis
Cairo
The Neutral Zone was equally divided
between Kuwait and Saudi Arabia by
an agreement of December 18, 1969.
KUWAIT
Kuwait
Jahrum
AFRICA
BAHRAIN
Manama
QATAR
Doha
Abu
Dhabi
UNITED ARAB EMIRATES
Persian Gulf
Gulf of Oman
SAUDI
Muscat
Tropic of Cancer
Riyadh
Red Sea
Mecca
ARABIA
OMAN
20°
Sana
YEMEN
(Sana)
YEMEN (Aden)
Arabian
Madinat al-Shaab
Aden
Gulf of Aden
Socotra
10°

Map by J. Donovan

Alaska
(U.S.)

150°
180°
150°
120°

Aleutian Is.

International Date Line

Attu I.

Kamchatka Peninsula

Sakhalin

Kurile Islands

Pacific Ocean

S I B E R I A

SOCIALIST REPUBLICS

• Yakutsk

• Krasnoyarsk

• Tomsk

Lake Baikal

• Irkutsk

• Harbin

Hokkaido

• Vladivostok

N. KOREA

Pyongyang • Sendai

J A P A N

Tokyo

• Omsk

• Novosibirsk

Ulan Bator ★

Karakorum •

MONGOLIAN PEOPLE'S REP.

Shenyang
(Mukden) •

• Luta

Seoul ★

Kobe
Osaka

S. KOREA

Hiroshima •

• Karaganda

★ Peking

Tientsin •

Nagasaki •

L. Balkhash

• Urumchi

SINKIANG-UIGHUR
AUTONOMOUS REGION

Nanking •

Shanghai •

• Alma-Ata

PEOPLE'S REPUBLIC OF CHINA

Wuhan •

Wenchow •

Taipei
REPUBLIC OF CHINA
(Taiwan)

Okinawa (Japan)

Marianas Isls.

Guam
(U.S.)

• Tashkent

• Kashgar

• Chengtu

• Chungking

Foochow •

Amoy •

Tibet

Canton •

Victoria
HONG KONG
(Br.)

Bamian
(Bamiyan) •

• Srinagar

KASHMIR

★ Kabul

Islamabad •

• Jammu

HIMALAYAS Mt. Everest Lhasa •

Thimbu

BHUTAN

MACAO
(Port.)

Hainan

REPUBLIC OF THE
PHILIPPINES

Luzon

Quezon City •
Manila ★

Lahore •

PAKISTAN

Harappa •

• Amritsar

NEPAL ★
Katmandu

Gangtok ★ SIKKIM

Mandalay •

Hanoi •

Mindanao

• Sukkur

Mohenjo-Daro •

• Delhi

New Delhi •

• Agra

• Kanpur

BANGLADESH

Dacca ★

BURMA

Luang Prabang •

LAOS

Vientiane ★

Hue •

VIETNAM

South China
Sea

Calcutta •

THAILAND

Rangoon ★

• Hyderabad

• Karachi

• Ahmadabad

• Nagpur

INDIA

Bay of
Bengal

Bangkok ★

CAMBODIA

Pnompenh ★

Saigon •

Sabah

M A L A Y S I A

Bandar Seri Begawan ★

Sarawak BRUNEI

Celebes
(Sulawesi)

Sea

• Bombay

• Hyderabad

Bangalore •

• Madras

Andaman
Islands

Nicobar
Islands

Malaya

Kuala Lumpur ★

SINGAPORE ★

B o r n e o
(Kalimantan)

REPUBLIC OF INDONESIA

Strait of Malacca

• Cochin

SRI LANKA

Colombo ★

Indian Ocean

Equator

Sumatra

Java

Djakarta ★

MALDIVES

FACTS AND FIGURES

LOCATION: Mainland Asia extends from: **Latitude—** 1° 16′ N to 77° 41′ N. **Longitude—**26° 04′ E to 169° 40′ W.

AREA: Approximately 17,000,000 sq. mi. (44,000,000 sq. km.).

POPULATION: 2,000,000,000 (estimate).

CHIEF RIVERS: Yangtze, Yellow (Hwang Ho), Amur, Lena, Mekong, Yenisei, Ob, Indus, Irtysh, Brahmaputra, Salween, Euphrates, Amu Darya, Ganges, Kolyma, Syr Darya, Irrawaddy, Tarim, Si, Tigris.

HIGHEST POINT: Mount Everest, 29,028 ft. (8,848 m.).

LOWEST POINT: Dead Sea, about 1,300 ft. (400 m.) below sea level.

COUNTRIES AND TERRITORIES OF ASIA

COUNTRY	AREA (sq. mi.)	AREA (sq. km.)	POPULATION (estimate)	CAPITAL
Afghanistan	250,000	647,497	16,500,000	Kabul
Bahrain	230	596	200,000	Manama
Bangladesh	55,126	142,776	75,000,000	Dacca
Bhutan	18,147	47,000	800,000	Thimbu
Brunei	2,226	5,765	116,000	Bandar Seri Begawan
Burma	261,789	678,033	27,000,000	Rangoon
Cambodia (Kampuchea)	69,898	181,035	6,700,000	Pnompenh
China				
People's Republic of China	3,691,512	9,561,000	850,000,000	Peking
Republic of China (Taiwan)	13,885	35,961[1]	16,000,000	Taipei
Cyprus	3,572	9,251	630,000	Nicosia
East Timor	7,300	19,000	600,000	Dili
Hong Kong	398	1,031	4,000,000	Victoria
India	1,261,813	3,268,090	537,000,000	New Delhi
Indonesia	735,269	1,904,347	120,000,000	Djakarta
Iran	636,294	1,648,000	28,000,000	Teheran
Iraq	167,925	434,924	9,350,000	Baghdad
Israel	7,992	20,700	3,000,000	Jerusalem
Japan	145,737	377,459[2]	104,650,000	Tokyo
Jordan	37,738	97,740	2,145,000	Amman
Korea				
North Korea	46,540	120,538	13,300,000	Pyongyang
South Korea	38,022	98,477	31,000,000	Seoul
Kuwait	6,178	16,000	570,000	Kuwait
Laos	91,429	236,800	3,000,000	Vientiane
Lebanon	4,015	10,400	2,500,000	Beirut
Macao	6	16	260,000	Macao
Malaysia	128,430	332,633	10,700,000	Kuala Lumpur
Maldives	115	298	108,000	Male
Mongolia	604,248	1,565,000	1,250,000	Ulan Bator
Nepal	54,362	140,797	10,900,000	Katmandu
Oman	82,030	212,457	565,000	Muscat
Pakistan	310,402	803,940	55,000,000	Islamabad
Philippines	115,830	300,000	37,160,000	Manila
Qatar	4,000–8,500	10,000–22,000[3]	100,000	Doha
Saudi Arabia	829,997	2,149,690	7,200,000	Riyadh
Sikkim[4]	2,744	7,107	191,000	Gangtok
Singapore	224	581	2,000,000	Singapore
Sri Lanka	25,332	65,610	12,000,000	Colombo
Syria	71,498	185,180	5,900,000	Damascus
Thailand	198,456	514,000	34,700,000	Bangkok
Turkey	301,381	780,576	35,700,000	Ankara
United Arab Emirates	32,278	83,600	185,000	Abu Dhabi
Vietnam	128,402	332,559	39,300,000	Hanoi
Yemen (Aden)	111,075	287,683	1,200,000	Madinat al-Shaab
Yemen (Sana)	75,290	195,000	5,000,000	Sana

[1] Excludes islands of Quemoy and Matsu.
[2] Includes southern Kuriles and Okinawa.
[3] Area indefinite due to disputed boundaries.
[4] Now part of India.

WONDERS OF ASIA

Petra, in Jordan, has been described as "a rose-red city, half as old as time." It was built in a narrow gorge, between towering cliffs, in about the 2nd century B.C. Its builders were a relatively obscure people, the Nabataeans, traders who had settled in northern Arabia along the main caravan route between the Mediterranean Sea and the Arabian Peninsula. Elaborate temples and tombs were carved into the sheer faces of the red sandstone cliffs, and other structures were erected. Today, completely remote from the modern world, the site delights the imagination.

The church of **Hagia Sophia**, or "divine wisdom," in Turkey, was under construction at Constantinople (now Istanbul) from A.D. 532 to 537. Soon after its completion the dome collapsed and was replaced by a loftier one, with a span of over 100 feet (30 meters). Hagia Sophia covers an enormous area and is the outstanding monument of the Byzantine Empire and the Greek Orthodox Church. After the capture

Ancient ruins of tombs carved out of the rock at Petra in Jordan.

of Constantinople by the Turks in 1453, the church became a mosque, and its design inspired the architects of the great Turkish mosques. The building is now a museum.

Persepolis, situated in southern Iran, was the spiritual heart of the empire of the Achaemenid rulers of the Medes and the Persians, who combined to conquer much of Asia. Set upon a rock platform and rising from a vast plain are a series of gateways, audience halls, private palaces, storehouses for the treasures of empire, and headquarters for the military garrison. The major part of Persepolis was erected between the 6th and 5th centuries B.C. Especially noteworthy are the carved stone reliefs, which show the peoples from 23 lands of the empire bringing tribute—animals, weapons, vessels of metal and stone, and jewelry—on the occasion of the New Year festival. The buildings themselves stand in ruin after being burned by Alexander the Great in 330 B.C.

In the high mountains of Afghanistan is the site called **Bamian**. A vertical cliff of fairly soft stone was occupied there by Buddhist monks from the 2nd to the 9th century A.D. The cutrock rooms and temples are overshadowed by two standing figures of the Buddha carved into the

Bamian, in Afghanistan.

cliff. One figure rises to 115 feet (35 m.) and the other to 170 feet (50 m.); they were originally painted in bright colors and covered in gold leaf. As seen from across the valley, these statues make an unforgettable impression.

The **Badshahi Masjid**, or the Imperial Mosque, in Lahore, a very old city of Pakistan, is a monument of the Mogul period, when Muslims ruled India. The Imperial Mosque was erected by the Emperor Aurangzeb about 1673. Its great halls of red sandstone and its domes seem small when seen across the expanse of its open court—the largest court of any mosque in the world.

At Agra in India the 17th-century Mogul emperor Shah Jahan erected a mausoleum for his favorite wife, Mumtaz Mahal. Known as the **Taj Mahal**, it took about 20 years to build and is considered a supreme architectural gem. The tomb, constructed of white marble, has decorative patterns inlaid with semiprecious stones and is crowned by a soaring dome; the building takes on color from the atmosphere—rose-colored at dawn, stark white at midday, orange at dusk, and blue under a brilliant moon.

The Taj Mahal.

Angkor Wat.

Bangkok became the capital of Thailand near the end of the 18th century, and since that time successive rulers have enriched it with a series of Buddhist temples. These temples seen together are one of the wonders of Asia. Renowned among them is the **Wat Phra Keo**, the Temple of the Emerald Buddha, with its spires, or *chedis,* which house relics of the Buddha. Spires and temples alike gleam and glisten with bright colors and gold.

Another temple complex, **Angkor Wat**, lies within the ancient capital of Cambodia. About A.D. 1100 a number of great stone towers were erected and each was carved with figures and faces of Brahma and other Hindu gods. In later times these figures came to be mistaken for representations of the Buddha. The ruins of the temples were discovered in the 19th century and have since been cleared of the encroaching jungle. The magnificence of Angkor Wat lies in its great size and complexity.

Lhasa, the capital of Tibet, is the site of the **Potala**, which served at the same time as the palace of the Dalai Lamas, a monastery, and the administrative center of the country. It was probably constructed in the 17th century and rises in tier after tier on a hillside above the city. Hundreds of windows light the rooms of the golden-roofed building, which was long known as the Forbidden Palace.

The Great Wall of China.

The **Great Wall of China** was begun in the 3rd century B.C. and was rebuilt many times until it finally attained a length of some 1,500 miles (2,400 kilometers). The wall averages about 25 feet (8 m.) in height. This greatest engineering feat of Asia was undertaken to halt the movement of nomadic horsemen from Mongolia.

To the Western traveler, these and many other architectural wonders of Asia are visible reminders of the fascination, excitement, and strangeness of the East. But the people of Asia do not think of their surroundings as brimming with color, sounds, and light, or of themselves as picturesque and exotic. As in other parts of the globe, for century after century the ordinary people worked hard for a meager living, and there was little glamour in their lives. Much of their time was spent in working for the gratification and glorification of masters, overlords, princes, and other rulers, either as highly skilled craftsmen or as laborers who brought into being dams, canals, temples, and palaces. In many places in Asia highly developed cultures rose and flourished for long periods. Today we esteem the works of art left by the ancient kingdoms and empires of Asia and admire modern survivals of age-old handicrafts—brilliant handwoven silks, vessels of gleaming metal and fine pottery, jewelry encrusted with gems, and luminous paintings on silk.

Miles
0   250   500   750   1000   1250   1500

0   400   800   1200   1600   2000   2400
Kilometers

Lambert Azimuthal Equal-area Projection

☆ NATIONAL CAPITAL

★ PROVINCIAL CAPITAL

▢ TUNDRA   ▢ TROPICAL FOREST

▢ CONIFEROUS FOREST   ▢ STEPPE (SHORT GRASS)

▢ DECIDUOUS FOREST   ▢ DESERT

    ▢ MOUNTAIN VEGETATION

INDEX TO ASIA MAP

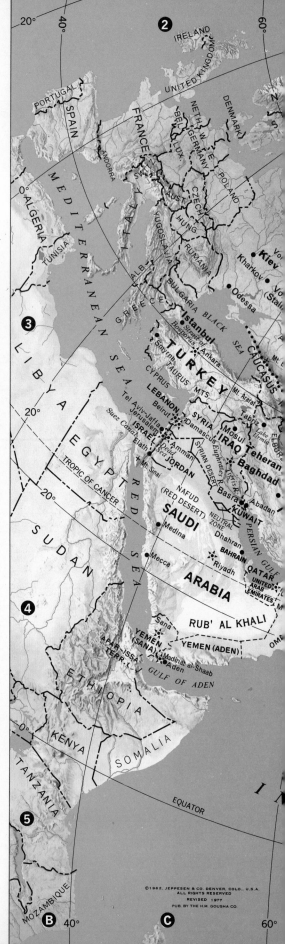

©1962, JEPPESEN & CO. DENVER, COLO., U.S.A.
ALL RIGHTS RESERVED
REVISED 1977
PUB. BY THE H.M. GOUSHA CO.

ASIA

ASIA AND THE WEST

It is said that the Western world symbolizes progress, and that Asia will advance only as it borrows and adapts the technical and scientific achievements of the West. And it is true that Asia has adopted Western methods of business, communications, transportation, and manufacturing. In some of these fields Japan, as one example, has not only copied from the West but has surpassed its achievements.

A number of explanations have been offered as to why Asia fell behind the surging, bustling Western world of the 18th and 19th centuries. One is that large areas of Asia were overcome by European military superiority to become colonies, which were then systematically exploited. That is, their raw materials were purchased cheaply and returned to them as much more expensive items, manufactured thousands of miles away. Another explanation is that Asians in general lacked the sense of political nationalism that causes people sharing a common habitat, or living area, to unite for the common good and to be aggressive in pursuit of their common interests—in other words, to form nations or states. It is fair to say that it was not until the Asians learned about the power that can be exerted by a nation through its organizations and institutions that they began to rebel against their colonial masters. And this knowledge came from observing the way foreigners drew up and enforced laws, collected taxes, ran state and local administrations, and established police forces. It came also from those young Asians who had gone to Europe to study and who became the leaders of forces that began to agitate and to fight for political freedom.

The tactics of nonviolent opposition to authority, so prevalent today, were developed in India by Mohandas K. Gandhi in the 1920's and 1930's as a weapon against the British presence. Guerrilla warfare was taken up as part of the struggle against foreign invaders, especially in World War II against the Japanese forces. Of course, many areas of Asia were never colonized by Europe. To name but a few, Japan, Thailand, Afghanistan, and Iran always maintained their independence. Other areas, such as Korea, were colonized by Asians rather than Europeans.

When confronted with a complex situation it is natural to search for elements of similarity and uniformity in order to draw meaningful generalizations. However, Asia does not lend itself to this kind of treatment. Asia is characterized by diversity, by extreme contrasts, and by a lack of easy classifications; and the problems of Asia reflect its diversities.

THE LAND OF ASIA

The continent of Asia is not as suitable for human life and activity as is, for example, the North American continent. Very large areas of Asia are desolate and arid, others are frozen by extreme cold, while others swelter under steaming rains. It has been estimated that less than 10 percent of the continent produces crops. And yet it contains over half the world's people—about 2,000,000,000 (billion) of them.

Asia, measured along the 40th parallel from Istanbul through Peking to a point north of Tokyo, stretches for over 6,000 miles (9,600 km.), and the distance is just about the same from its northernmost tip to the islands of Indonesia. The total area of the continent is about 17,000,000 square miles (44,000,000 square kilometers). Obviously no single topographical feature could dominate this vast area, but there are great

Part of the vast Himalayan mountain range.

mountain ranges that radiate from the knot of the high Pamirs of Central Asia and extend as far east as China and as far west as Turkey. Of these ranges the best-known is the Himalayan, with its group of towering peaks topped by Mount Everest, which at 29,028 feet (8,848 m.) is the highest mountain peak in the world. Other Asian ranges are the Hindu Kush, which extends to the west through Afghanistan; the Altai and Tien Shan, which thrust into China; and the Karakoram and the Altyn Tagh, near neighbors of the Himalayas. Associated with these ranges are the great plateaus of Tibet, Mongolia, and India.

The movement of peoples—migrating tribes, armies, nomads, travelers—has always been hampered by the need to go over or around mountain ranges. For example, Tibet, itself a land of great distances, served to shut off northern Asia from its southland. To the north, nature was rather hostile to man, displaying successive belts of desert, steppes, dark forests, and tundra. To the south the topography was less varied, and the climate hotter.

Of the many desert regions, the largest ones include the Thar in India and Pakistan; the Taklamakan in China; the Syrian and Rub' al Khali, which cover a large area of Southwest Asia; and the vast Gobi in Mongolia. The Gobi is well-known because of the discoveries there of the petrified eggs of dinosaurs and of the massive skeletons of these prehistoric creatures. Asia has great rivers. The most important include the Indus, Ganges, and Brahmaputra in Pakistan and India; the Mekong in Southeast Asia; and the Yangtze and Yellow (Hwang Ho) in China. But many of the rivers are not suitable for navigation. Also, Asia is quite short of lakes, and some of the major ones—the Aral, the Urmia, and the Caspian and Dead seas—are salt.

Over much of the continent the annual average rainfall is too slight to permit dry farming—the growing of crops without irrigation. Thus, irrigation works, including dams, canals, and devices to raise water from

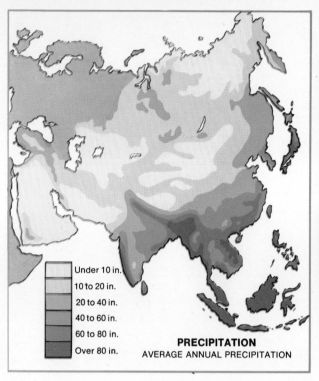

PRECIPITATION
AVERAGE ANNUAL PRECIPITATION

Under 10 in.
10 to 20 in.
20 to 40 in.
40 to 60 in.
60 to 80 in.
Over 80 in.

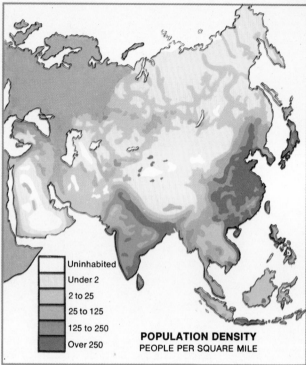

POPULATION DENSITY
PEOPLE PER SQUARE MILE

Uninhabited
Under 2
2 to 25
25 to 125
125 to 250
Over 250

rivers and wells, have been in constant use for centuries. Rainfall is abundant, however, along the southeastern shores of the continent, and it is there that population densities are the highest. It was there, too, that agricultural surpluses, particularly rice, were common until recently.

PEOPLES OF ASIA

Scholars do not always agree on terms or definitions, but one attempt at classification of the various peoples of Asia recognizes these types: Ainu, Dravidian, Indo-Aryan, Malay-Indonesian, Mongoloid, Negrito, Semitic, Sino-Mongoloid, and Turanian. The Ainu of Japan and the Dravidians of southern India are believed to be direct descendants of the original (called aboriginal) populations. The inhabitants of China, Korea, Japan (excluding the handful of Ainu), and Tibet are Sino-Mongoloid. Semitic peoples occupy much of Southwest Asia. Indo-Aryans are spread across much of India (particularly in the north), Pakistan, Afghanistan, and Iran. Negritos inhabit certain regions of south and southeast Asia and the Andaman Islands in the Indian Ocean. The word "Turanian" may be equated with "Turkic" and refers to peoples living in Central Asia and Turkey.

These terms, however, are remote from popular experience. A Chinese would never think of himself as a Sino-Mongoloid, or a Persian as an Indo-Aryan. Indeed, even today most Asians when asked what they are will not respond by saying "an Afghan," "an Iraqi," "an Indonesian," or any other national name. Instead, they will refer to their membership in a group much smaller than a nation, such as a tribe, a province, or town, or a religious group. This situation is the result of the fact, mentioned earlier, that a strong sense of national identity was lacking in most of Asia, where political boundaries were very fluid and the mass movements of peoples frequent.

The Nomads

Among the thousands of different groups of peoples that exist in Asia, nomads are especially interesting to anthropologists because they preserve a very ancient way of life, possibly more ancient than that of the farmers, who are settled on the land. Nomadic life can be described as vertical or horizontal. In vertical nomadism, common in Iran and Afghanistan, a tribe moves in the autumn to low-lying plains, where there will be grazing during the winter. In spring the tribe moves up into the highest mountain valleys, where grass survives throughout the heat of summer. In Saudi Arabia and Mongolia nomadism is horizontal: the people wander throughout the year across fairly level semidesert lands within well-defined tribal territories.

Nomads everywhere lead a largely self-sufficient way of life. Their flocks of sheep, goats, horses, camels, and yaks provide milk, butter, cheese, and meat in addition to wool to be woven into clothing, tents, and carpets. The carpets, woven by the women, the wool, and the animals themselves are sold in the bazaars of towns for grain and other necessities. Possessions and furnishings are few, for their total cannot exceed the load that can be carried by the animals belonging to a family. The men are renowned as hunters and warriors, who have a low regard for farmers. For many centuries the nomadic tribes furnished the backbone of Asian armies, and many dynasties were founded by a tribal leader who was able to establish his authority over a large area.

Class Distinction and Caste

In addition to groups distinguished by ethnic divisions or particular ways of life, Asia has strongly defined social classes. In every historical period there was a privileged elite and a depressed majority. Nobles and lords received titles and rewards in return for contributing armed men to the armies of rulers, and such titles were passed on from generation to generation. Each such provincial lord had the obligation to keep public order in his region. For example, when silk caravans crossed Asia they traversed the territories of these local lords, who had to protect the caravans against bandits or, if they failed to do so, to pay for stolen merchandise.

A special form of class distinction, that of castes, developed very early in India, after the Indo-Aryan invaders arrived about 1500 B.C. In time four castes evolved: the Brahmans, or priests (the highest caste); the Kshatriyas, or lords and warriors; the Vaishyas, or landowners and merchants; and the Sudras, or peasants. Below the Sudras were the untouchables, who were forced to do the most menial tasks and to have no social relations of any kind with the castes. Modern India has made great efforts to end discrimination against the millions of untouchables. In fact, the clause in the Indian Constitution of 1950 outlawing such discrimination sounds very similar to guarantees of the civil rights of minorities and races in other countries, including the United States. In Japan there is also an "untouchable" group, the Etas, who carry out degrading tasks. If we regard a caste as a hereditary social class from which one cannot escape by one's own efforts, by marriage, or by other means, we can say that castes are common to few societies. Castes do have their counterparts, however, in rigid class distinctions, but in nearly all such cases members of a depressed class may escape from it.

An Arab boy from Jordan.

Nomads in Afghanistan.

An Indian Brahman, or priest.

Japanese strolling on the Ginza in Tokyo.

A place sacred to Hindus—the Ganges River at Banares in India.

A classical dance being performed at a Shinto shrine in Kyoto, Japan.

Jerusalem, a city holy to Jews, Christians, and Muslims.

The Shwe Dagon Pagoda in Burma, one of many great monuments to Buddhism.

Languages of Asia

As people are classified by race or group, languages are classified by major families and each such family may have an extensive family tree. The names of the language families found in Asia are similar to, but not identical with, those of the ethnic divisions listed previously. They include Indo-European, Dravidian, Semitic, Sino-Tibetan, Malayo-Polynesian, and Altaic. Languages within the same family may sound so dissimilar that it is hard to believe that they are related. Within the Indo-European family are Persian, spoken in Iran and Afghanistan; Pushtu, Pashai, and Waigeli, spoken in Afghanistan; Hindi, Gujarati, Punjabi, Sindhi, Marathi, Bengali, and others, spoken in India; and English, French, German, Russian, and most of the other languages of Europe and America. Dravidian includes Telugu, Tamil, and Malayalam, spoken mainly in south India. Semitic languages, which include Hebrew and Arabic, are spoken in Southwest Asia. The Sino-Tibetan family includes Chinese, Tibetan, Thai, and languages of Burma and Laos. Malayo-Polynesian languages are spoken in Malaysia, Indonesia, and the Philippines. The Altaic family includes Turkish, Mongolian, and languages spoken by various peoples of Inner, or Central, Asia and North Asia. Japanese and Korean are sometimes classified separately and sometimes together.

The languages spoken by the men of victorious armies who settled in conquered regions of Asia sometimes replaced those long in use. Also, expanding religious movements sometimes brought in new languages. Where the same language is spoken over a very wide region, very different dialects evolve and people in different areas are not able to understand each other. Thus, in China the people of the north are unable to understand the speech of the south, although the written language remains identical. An Arab, too, may find the dialects of other Arabs unintelligible, although Arabic has only a single classical, written language. The more extensive the area in which a language is spoken, the more dialects come into being. In India there are 14 major languages and hundreds of dialects, some of them spoken by fewer than 1,000 people.

Religions of Asia

The major religions of the world have hosts of adherents in Asia. Figures available are only estimates, but there are about 400,000,000 Hindus, nearly all in India; approximately 240,000,000 Buddhists, with the heaviest concentrations in Tibet, Sri Lanka, the peninsula of Indochina, and Japan; and over 350,000,000 Muslims in Southwest Asia, Pakistan, Bangladesh, India, Malaysia, and Indonesia. China did have vast numbers of Confucians and Taoists, but during the course of many years in which religious affiliation and practices have been discouraged, no figures have been released. Japan has Shinto as its oldest active faith. Long ago it was formalized from its origins in nature and ancestor-worship, and at one time it was the state religion. The oldest faith of Korea is Sinkyo, which has much in common with Shinto, but Confucianism has many more followers. There are, in addition, millions of animists in Asia, who worship the elements and natural phenomena.

Christian missionaries have been least successful in converting the followers of monotheistic religions—those who believe in a single god. Thus, for example, they have had very little success among Muslim or Jewish communities. Catholic missionaries penetrated Asia by the 16th century, but not until early in the 19th century did any significant numbers of Protestant missionaries appear. In that century preachers were followed by teachers, doctors, and nurses, who introduced modern medicine and methods of teaching to Asia. There may be 60,000,000 Christians throughout Asia, with the local churches increasingly headed by their own people rather than by foreign Christians.

REGIONS OF ASIA

In order to relate the peoples of Asia to the land, the continent has been divided into six regions—Southwest, South, Southeast, East, Inner, and North.

Southwest Asia

Southwest Asia includes the countries of Jordan, Lebanon, Israel, Syria, Turkey, Iraq, Iran, Kuwait, Saudi Arabia, Yemen (Sana), Yemen (Aden), Cyprus, and the tiny oil-rich sheikhdoms that rim the eastern shores of the Arabian Peninsula—Bahrain, Qatar, the United Arab Emirates, and the Sultanate of Oman. Most of the inhabitants of this area are Muslims, followers of the religion of Islam, which arose in Arabia. Semitic, Turkic, and Aryan ethnic groups are represented. The Arabs, who are Semites, as are the Jews, occupy most of the area, with the exception of Turkey, Iran, Israel, and Cyprus. The region has been the birthplace of Judaism, Christianity, Islam, and Zoroastrianism.

Within Southwest Asia are concentrated three of the world's surviving monarchies—Jordan, Saudi Arabia, and Iran. The ruling families of

Hagia Sophia in Istanbul, Turkey, overlooking the strait of Bosporus.

A shepherd and his flock in the island nation of Cyprus.

The Negev desert of Israel.

For many of the states of Southwest Asia oil is the major source of income.

Jordan and Saudi Arabia trace their lineage back to the tribe of Mohammed, the prophet of Islam. In Iran (which together with Afghanistan forms the mountain-studded Iranian plateau) the institution of the monarchy goes back to the 6th century B.C. The Persian Gulf and all the states that border it—Saudi Arabia, Kuwait, the various sheikhdoms, Iraq, and Iran—produce enormous quantities of petroleum and natural gas. Tiny Kuwait has such a large income from its oil that its per capita income ranks among the highest in the world. Much of the money derived from oil goes to speed the modernizing of industry, transportation, and agriculture in these lands. Rainfall is scanty over most of the area, and farmland is largely irrigated.

South Asia

South Asia consists mainly of the triangular mass of India. To the northwest and northeast of India are Pakistan and Bangladesh. In the north Nepal, Sikkim (now part of India), and Bhutan nestle within the high Himalayas, while the fertile island of Sri Lanka and the Maldives chain lie off the southern tip of India. Afghanistan, lying between Pakistan and Iran, is primarily a mountainous land. The region has two great river systems, whose tributaries rise within a few miles of each other in the Himalayas—the Indus, emptying into the Arabian Sea, and the Ganges, flowing into the Bay of Bengal.

In general, South Asia enjoys a continental climate with the typical marked seasons and fairly limited rainfall. Over India the winter rains

Bombay, one of India's largest and most cosmopolitan cities.

A mountainous region of Pakistan.

A Hindu temple. Hinduism is the principal religion of India.

are scanty, but in the summer they become abundant as the monsoon winds blow rain clouds to the north. The area is not overly rich in the kinds of natural resources required for industrialization, although India has deposits of iron ore and coal.

Long before the Christian Era, Bharat—the Indian name for India—developed a complex way of life, a pattern of social classes, and a system of beliefs with a host of gods, which together became known as Hinduism. As a religion Hinduism is elaborate and difficult to define. It can be said, however, that most Hindus believe in a cycle of rebirth, accept their place in the caste system, and respect the Vedas, a collection of literature associated with the most ancient history of the faith.

Islam was firmly established in India by conquering armies by the early 13th century, and it was followed by millions of converts from other faiths. However, these numbers never came close to rivaling the Hindu population, and religious strife was often present. As a result, when India gained independence from the British Empire in 1947, it split into two states—predominantly Hindu India and the new nation of Pakistan, inhabited mainly by Muslims. Bangladesh, formerly East Pakistan, came into existence in 1971 following a civil war in Pakistan.

In the 6th century B.C. an Indian prince, Siddhartha Gautama, gave up worldly pleasures for a life of meditation and teaching. Gautama attained the status of Buddha, "the enlightened one," and the faith known as Buddhism was born. Today there are relatively few Buddhists in India, but nearby Sri Lanka is a stronghold of the faith.

Southeast Asia

Southeast Asia today includes the independent countries of Burma, Thailand, Laos, Cambodia, Vietnam, Malaysia, Singapore, Indonesia, and the Philippines. In the 19th century the British acquired Burma, Malaya, and Singapore. The French took over all the rest of mainland Southeast Asia, with the exception of Thailand, and it became known as French Indochina. These colonies of Great Britain and France gradually won independence in the years following World War II.

Independence, however, did not bring peace to Indochina. Vietnam, which had a Communist government in the North and a non-Communist government in the South, was the scene of especially bitter fighting. Eventually, American troops were drawn into the war, which finally ended with the victory of Communist forces throughout Vietnam, Laos, and Cambodia. Vietnam was officially unified in 1976.

The thousands of islands of Indonesia, formerly a Dutch colony, stretch about 3,000 miles (4,800 km.) from west to east in a pattern shaped like the blade of a sickle. Most of the people of Indonesia, which ranks among the most populous nations of the world, live on the large islands of Java, Sumatra, and Sulawesi, and in Kalimantan, the Indonesian part of Borneo. The Philippines number over 7,000 islands, most of them small. They have a total population much smaller than Indonesia's, concentrated in four of the islands. Rainfall is heavy throughout the two countries and simple cultivation is rewarding; many kinds of trees yield abundant fruit without any cultivation at all.

Terraced rice fields in the Philippines.

The Mekong, one of the major rivers of Southeast Asia.

A street in Singapore. The great majority of its people are of Chinese descent.

The Mekong, Chao Phraya (Menam), Salween, and Irrawaddy are the major rivers that flow down through the Indochinese Peninsula. Potentially the most important of these is the Mekong, which separates Thailand and Laos and crosses Cambodia and Vietnam. Studies have been completed for an ambitious project to develop the resources and water supply of the entire river valley.

Southeast Asia is inhabited by a great many ethnic groups. The original inhabitants were largely displaced by successive waves of immigrants moving down from the north. So strong and persistent were these movements that some of the earlier migrants were forced to move on to the islands of Indonesia. These migrations and their consequences were marked by savage wars, and today there is a legacy of distrust among the several countries of the peninsula. Two foreign ethnic minorities are found throughout the region—Indians and Chinese. The Chinese settled in urban areas and are the most active businessmen.

While Hinduism made many early inroads into Southeast Asia, as is witnessed by the great temples, such as Angkor Wat, Buddhism became a much more powerful force and today the vast majority of the inhabitants of the peninsula are Buddhists. Islam also penetrated the area, carried by the ships of Arab traders, and became strong in such areas as Malaysia and Indonesia.

East Asia

East Asia is dominated by China, one of the largest countries in the world and, with about 850,000,000 people, the most populous country in the world. To the east is the peninsula of Korea. The island states of East Asia include Japan and Taiwan (Formosa). The latter, geologically, is apparently the northern extremity of the Philippines chain.

For many years China was something of a mystery to outsiders. To a certain extent it still remains a mystery. However, the visit of the president of the United States and the entry of China into the United Nations have done much to reopen this ancient land to the world.

To themselves the Chinese people are the Han and their country is the Middle Kingdom. (Not all the inhabitants of China are Han, but the total number of other ethnic groups is very small in comparison with the entire population.) Chinese civilization goes back to before 2000 B.C., and it has been marked by great cultural achievements. When we think of China we often think first of tea, porcelains, silks, paintings, lacquerware, fireworks, carvings in ivory and jade, and other crafts. But literature and philosophy were also highly developed and were brought within the reach of many Chinese through the invention of paper and printing.

Successive Chinese dynasties, secure in their strength and proud of their culture, regarded their country as the center of the world and its people as superior to all others—the others were called barbarians. The highly developed language of the Chinese had its special influence on the country. The language contained some 40,000 separate characters, each of which had to be memorized for use in reading and writing. Learning as many as 5,000 characters was an achievement, and those able to do so and to pass difficult examinations became part of the bureaucracy that controlled the country for centuries. The thread that ran through government and family life was Confucianism, the teaching of K'ung Fu-tzu. Known to the West as Confucius, he lived from 551 to 479 B.C. Confucius stressed such virtues as love and compassion and justice with mercy, and placed great emphasis on the importance of the family. Together with Confucianism, Buddhism and Taoism have had the most adherents in China, with Muslims and Christians in fewer numbers.

The present Chinese Communist government, however, which came into power in 1949, is relentlessly opposed to all religions. In an effort to impose unity and uniformity throughout China and to cope with the country's problems, it has tried to destroy many of the old values of the Chinese, particularly the traditional tight-knit family unit. But this drive for uniformity is hampered by a number of serious obstacles. One is the difficulty of imposing authority over hundreds of millions of people spread across such a vast land. Another is the problem of feeding China's masses, since only a small percentage of the land is arable. Industrialization has been a major priority of Chinese leaders, and today China is going through a period of forced industrialization, aimed at making the country independent of imports of vitally needed manufactured goods and capable of producing items that can be sold abroad for foreign currency.

Japan consists of four large islands and about 3,000 smaller ones, forming an archipelago (or chain of islands) some 1,200 miles (1,900 km.) in length. A temperate climate and abundant rainfall favor the intensive cultivation of the mountainous terrain. In writing about Japan, superlatives are in order: Tokyo, its capital, is the largest city in the world, with a population of about 11,000,000, including its suburbs; Japan's more than 100,000,000 people live in an area smaller than California; the Japanese coastline is 1,700 miles (2,730 km.) long; and an average of about 1,500 earthquakes are recorded in the country each year.

Long influenced by China in such fields as calligraphy (beautiful writing), literature, philosophy, and art, Japan was reluctantly exposed to the Western world after centuries of self-imposed isolation. Pressure was exerted in the form of a visit, in 1853, by an American naval squadron and by American insistence that trade relations be established between the two countries. By the beginning of the 20th century Japan was on the way toward industrialization. Its leaders had realized that their country would remain a third-rate power if it continued to rely on agriculture alone. As a result, Japan today is one of the most highly industrialized nations in the world.

Taiwan, about 100 miles (160 km.) off the mainland, is the island refuge of those Chinese who did not choose to live under Communism after the defeat of the Nationalist forces in 1949. Hard work, together with financial aid from the United States, wrought something of an economic miracle there. The standard of living in Taiwan has become one of the highest in all Asia.

The people of the peninsula of Korea never achieved the world prominence of their neighbors, the Chinese and the Japanese. The Koreans did, however, develop a remarkable and distinctive culture of their own. Exposed to invasions from the mainland, they took shelter in isolation in the 17th century. Later the country was taken over by Japan. Prior to the end of World War II, it was agreed that the Soviet Union would accept the surrender of Japanese troops north of the 38th parallel of north latitude, and the United States those to the south. After the war ended in 1945, efforts to establish a united Korea failed, and two separate

Residents of Peking relaxing in view of the old Summer Palace.

Japan is a nation of many small farms.

Members of a commune near Shanghai threshing rice.

The bustling Ginza district of Tokyo on a rainy day.

Two old South Korean gentlemen in the traditional robes of the country.

states came into being—the Republic of Korea to the south of the 38th parallel, and the Democratic People's Republic of Korea, a Communist state, to the north. War broke out in 1950, when North Korean troops moved south across the parallel. A United Nations force, composed primarily of American troops, entered the fighting on the side of the South Koreans. Later, Communist Chinese forces came to the aid of North Korea. In 1953 a truce was concluded, but permanent peace has not come about and Korea remains divided.

Inner Asia

This region is sometimes referred to as Inner Asia and sometimes as Central Asia. Inner Asia may be more appropriate, since its major land-mass is that of Tibet, an area of lofty heights and vast distances, difficult of access from any direction. Its people are a very cohesive ethnic group who have lived in this region for at least 1,000 years. Tibet was long isolated from the outer world by the deliberate choice of its lamas, the teachers of the Tibetan form of Buddhism. Its towering monasteries housed thousands of monks and nuns. The Buddhist belief in rebirth was interpreted to mean that in every generation Bodhisattvas, noble beings who are eligible to forgo the cycle of birth and rebirth but who choose instead to serve mankind, are reborn in Tibet. The Dalai Lama and the Panchen Lama are identified in early childhood as being such Bodhisatt-vas, and at age 18 the Dalai Lama is formally installed as the supreme ruler of the country. In 1950 the armies of Communist China invaded and conquered Tibet, and the Dalai Lama fled to India. Although the Chinese employed the most savage measures in an attempt to eradicate Buddhism, the deeply religious Tibetans still maintain their beliefs.

The other part of Inner Asia is a region formerly known as Chinese Turkestan and Sinkiang and now as the Sinkiang-Uighur Autonomous Region, governed by the Chinese. As depressed in altitude as Tibet is elevated, Sinkiang is a region of sparse rainfall, great deserts, and interior drainage. While the Tibetans suffer from extreme cold, raging winds, and an oxygen-thin atmosphere that serves to slow down physical activity, the dwellers of Sinkiang are plagued by heat and drought.

The population of Sinkiang is largely Turkic, including such peoples as the Uighurs, Kazakhs, Kirghiz, and Taranchi. Many of them are still nomadic, following their grazing horses and sheep and living in tents. They speak a language related to Turkish and are Muslims. In race, language, and religion they have strong ties with groups living to the west, in Afghanistan, and in the Soviet Union. Sinkiang remains less well-known than Tibet, itself a land of mystery. What little is known of Sinkiang may be due to the fact that the great "silk road" that joined China with the Middle East and India in ancient times passed that way, or to its highly prized grapes and melons. As in other lands of limited rainfall, the major food crops of Sinkiang are wheat and corn. In Tibet, however, the main crop is barley, because this grain will grow and mature at great altitudes —the average altitude of Tibet is about 15,000 feet (4,500 m.).

North Asia

North Asia is largely that vast area of the Soviet Union known as Siberia, although it also includes Soviet Central Asia, lying to the north of Iran and Afghanistan. It is also reasonable to include the Mongolian People's Republic, a Communist state, which is wedged in between China and the Soviet Union, but which has much closer relations with the latter nation. Russian penetration and exploration of Siberia became very active in the 16th century, and by the 19th century the government of the czars had acquired most of the region. Even as early as the 18th century the practice of exiling criminals and political prisoners to these wilds had begun.

Siberia is often thought of as a dreary land, uniformly frozen solid throughout most of the year. Actually, its topographical uniformity lies only in the fact that all its major rivers, except the Amur, flow north into the Arctic Ocean. Western Siberia is a lowland, only a few hundred feet above sea level; from south to north it displays zones of black-earth soils, forests, and tundra. Central Siberia, bounded on the west by the Yenisei River and on the east by the Lena River, consists of much higher ground, and it is much colder. For at least 6 months of the year the average temperature is below freezing. Eastern Siberia has a monsoon climate, with the characteristic seasonal rains. A belt of tundra crosses its northern extremity, with forests to the south. Siberia offers great obstacles to human activity, but also great rewards, and the Russians have made enormous efforts to exploit its natural resources. Unlimited waterpower is available from its rivers; its forests of evergreen and hardwoods are endless; and beneath the soil are vast deposits of essential minerals.

The Mongolian People's Republic makes no news and is visited very infrequently by foreigners other than Russians. Yet it is not small. Its approximately 1,200,000 people live in a land more than three times the size of Spain. Pasture covers most of the land and the livestock population is about 25 times that of the human one. This grazing ground is the homeland of the Mongols, whose ancient capital was at Karakorum. It was from here that the horsemen of Genghis Khan and his sons and grandsons flooded across Asia in the 13th century. One group of Mongols came within a few miles of the Mediterranean and another penetrated into Central Europe.

Soviet Central Asia includes the Soviet Socialist Republics of Kazakhstan, Kirghizia, Tadzhikistan, Uzbekistan, and Turkmenistan (Turkmenia). Related peoples are also to be found in Iran, Afghanistan, and China. Uzbeks, Kazakhs, and Kirghiz are Turko-Mongols—Turkic in speech and with Mongoloid features. The Turkmen are Turkic and speak a Turkish-related language, and the Tadzhiks are Iranians who speak a Persian dialect. The five republics cover about 1,500,000 square miles (3,900,000 sq. km.) and include at least 27,000,000 people. Most are followers of the Muslim faith.

Timur—known to the Western world as Tamerlane—was born into a Turkish tribe in the 14th century and by early middle age had become the lord of Samarkand. It was from that city, now within Uzbekistan, that he set out on campaigns that resulted in the capture of such far-flung cities as Tabriz, Baghdad, Aleppo, Damascus, and Delhi. His magnificent tomb, crowned by a very high dome, still stands at Samarkand. This city and neighboring Bukhara contain many architectural monuments, including tombs, mosques, palaces, and religious schools.

Mongolian herdsmen. The animals are yaks.

ASIAN HISTORY

Intelligent man, in the persons of the hunters of the Stone Age, lived in caves in the mountain ranges of Asia. From the finds of human and animal bones and of crude stone tools and weapons, the way of life of these people, who lived between 20,000 and 10,000 years ago, has been partially reconstructed.

It is generally believed that about 7000 B.C. groups of these hunters began to move out of their caves onto open ground. To simplify greatly what then took place, it may be said that some groups became herders of newly domesticated animals, while others turned to farming. The herders were nomads—hardy people whose harsh way of life hampered their cultural development. The farmers had the better life, and those lucky enough to settle in fertile river valleys were the most fortunate. Although important early settlements far from great rivers have been uncovered, the most brilliant and enduring early cultures sprang up in valleys such as those of the Tigris and Euphrates rivers of Southwest Asia, the Indus River in what is now Pakistan, and the Yellow River of China.

Why did the people prosper and culture flourish along these river valleys? One reason was the presence of ample water for the irrigation of crops in regions where natural rainfall was scanty. Another was the ease of transporting men and goods by boat. Also, the concentration of a large population in a fairly limited area meant that capabilities for defense were strengthened and the stimulus for progress through close contacts intensified.

The attempts of these early peoples to understand the nature of the physical universe gave rise to mythologies, stories and legends relating to the origins of the earth and the heavens and to gods and heroes. The Chinese, for example, have a tradition that reaches back into the most distant past to a series of mythical heroes who made nature serve man.

The Tigris-Euphrates Civilizations.   Iraq, or Mesopotamia ("land between the rivers") as it was called by the Greeks, is watered by the Tigris and Euphrates rivers. Its southern part was the site of the very early civilization (about 3500 B.C.) of the Sumerians. The Sumerians developed a form of writing called cuneiform ("wedge-shaped"), which was inscribed on clay tablets. Their literature included epic narratives and prayers to Anu, the god of heaven, and to Enlil, god of storms. The major city, Ur, is believed by its excavators to have had a population of several hundred thousand. Fabulous treasures have been recovered from the tombs of its kings—burial furniture and musical instruments of rare wood and ivory, and quantities of delicate gold jewelry.

After the decline of the Sumerians (about 1900 B.C.), power in Mesopotamia passed to the Babylonians. The most renowned of the rulers of Babylon was Hammurabi. Ruling from about 1792 to 1750 B.C., he issued a code of laws to which modern scholars have attached his name. At the same time, dictionaries of the languages of the kingdom were compiled and astronomers and astrologers plotted the movements of the heavens. About the middle of the 16th century B.C., Babylon was destroyed by invaders from the north and east. Not for several centuries did a strong kingdom arise, when northern Mesopotamia came under the rule of the Assyrians. The Assyrian Empire, perhaps the world's first empire, survived a period of decline to become dominant again near the end of the 10th century B.C. Excavations of Assyrian palaces have uncovered hundreds of scenes carved in bas-relief on great stone walls, showing the Assyrian kings hunting lions and engaged in warfare.

Meanwhile, to the west, in what is now Lebanon, the Phoenicians were building an empire based on trade. The enduring contribution of this seafaring people, who sailed across much of the known world, was the invention of the alphabet.

A century of strife with a revived Babylon ended in the destruction of Assyrian power and the ascendancy of Babylon under Nebuchadnezzar (605–562 B.C.). But this prosperity did not endure, and Babylon fell to Cyrus, king of the Medes and Persians, in the 6th century B.C. The Medes and Persians came from the Iranian plateau to the east. But before turning there, let us look again to the west and record the emergence on the stage of history of a relatively small group of people.

The Hebrews. The Hebrews appear about 2000 B.C. when, according to the Old Testament, Terah, father of Abraham, led them from the city of Ur into the hills of northwestern Mesopotamia. Abraham made a covenant, or agreement, with Yahweh (Jehovah) that his people would become a great nation. Abraham led the early Hebrews to the frontiers of Palestine, where they probably lived as tent-dwelling nomads who herded sheep and goats. A long period of exile in Egypt was followed, in about the 13th century B.C., by the Exodus under Moses. After years of wandering the Hebrews conquered Palestine, then called Canaan. The Hebrew kingdom reached its height with the reign of Solomon (about the 10th century B.C.), which was marked by the presence of a powerful army, by activity in mining and commerce, and, most important, by the building of the Temple. After Solomon, the land of the Hebrews was divided into two kingdoms—Israel and Judah. They were overcome first by the Assyrians and then by the Babylonians, and large segments of the population were taken into captivity by the victors. After Babylon was conquered by Cyrus the Great, he issued an edict that provided money for the rebuilding of the Temple at Jerusalem and encouraged the Hebrews to return to Palestine. Some did, but it is not known how many.

The Medes and Persians. To the east of the Tigris and Euphrates rivers rises the lofty Iranian plateau, studded with mountain ranges. In approximately 1500 B.C. Aryan peoples from the heart of Asia moved into the Iranian plateau. Some of these nomadic herders turned south into the Indian subcontinent. Other tribes moved across the Iranian plateau to its western limits. These newcomers arrived in successive waves over a long period of time, and they included such tribes as the Parsa and Mada. This Aryan migration is the first of which any record exists of what was to become a pattern of movement out of remote Asia. In later centuries many other nomadic groups went west, as well as invading armies.

Before the end of the 7th century B.C. the Mada (known to us as Medes) became strong enough, allied with the Babylonians, to conquer Assyria and to dominate the related tribe of the Parsa (Persians). By the middle of the 6th century B.C. a Persian subject of the Medes, Cyrus II, renowned in history as Cyrus the Great, absorbed the Median kingdom. He established a friendly and enduring Persian-Median alliance and built an empire that stretched from India to the Mediterranean Sea. The religion of the Persians was Zoroastrianism, which emphasizes the eternal struggle between good and evil, but Cyrus showed tolerance to the many religions of the conquered lands. Darius I came to the Achaemenid

This figure of a bull found in Iran dates from about 1000 B.C.

throne (so named after Achaemenes, the ancestor of the ruling Persian family) in 521 B.C. Leading his armies across the Bosporus, he took much of northern Greece; but later campaigns were less successful, with the Persians defeated by the Greeks. Darius I also began the construction of Persepolis. Xerxes I, who succeeded his father, Darius, in 486 B.C., went on with the building at Persepolis and also campaigned against Greece.

The great empire of the Persians was finally overthrown by Alexander the Great of Macedonia, who marched across Asia to take and burn Persepolis in 330 B.C. After Alexander's early death, the empire was divided among his generals, who established ruling dynasties. These same lands were later ruled by other dynasties of Iranian origin—the Parthian (2nd century B.C. to 3rd century A.D.) and the Sassanian (3rd to 7th century A.D.). Both dynasties engaged in warfare with the Roman Empire and later with the Byzantines. The Byzantine Empire, its capital in Constantinople (Istanbul), began as the eastern half of the Roman Empire. It lasted until the capture of Constantinople by the Ottoman Turks in 1453.

The Arabs. For centuries the vast Arabian Peninsula was outside the mainstream of the civilizations mentioned. But at the beginning of the 7th century A.D., Arab armies poured out of the peninsula to overthrow the earlier empires. These Arabs brought the religion of Islam, a faith given to the prophet Mohammed in lengthy revelations from Allah (God), which were collected in a sacred book, the Koran. Hundreds of thousands were converted to Islam in a few years. Fewer were converted by the sword, as is sometimes said, than because the religion offered a social equality that had never existed anywhere in earlier times. Gradually the faith spread across western Asia and into Europe and Africa, while to the east it penetrated Central Asia, the Indian subcontinent, and Southeast Asia.

A portion of sculpture from a Hindu temple in India. It is about 2,000 years old.

The Indus Valley and Indian Civilization. Negritos may have been the original inhabitants of much of the Indian subcontinent. Then they were pushed toward the islands beyond India by Dravidians coming down from the north. Probably these Dravidians were the builders of the splendid cities of Harappa and Mohenjo-Daro, situated along the course of the Indus River and erected not too long after 2500 B.C.

The incoming Aryans drove the Dravidians to the southern part of the subcontinent and developed a civilization based on the religious ritual of the Vedas (sacred writings), on epics, on the Sanskrit language, and on the caste system. From these beginnings Hinduism was to evolve. After the 6th century B.C. records of the many Aryan states exist. A great dynasty, the Mauryan, appeared in the 4th century B.C., with Asoka (who reigned during the 3rd century B.C.) as its greatest figure. At first a ruthless soldier, after his conversion to Buddhism Asoka forsook warfare to care for the prosperity of his subjects. More than 30 of his edicts, engraved on stone pillars, have survived. Another resurgence came under the Guptas of the 4th century A.D., a time known as India's classical period, when literature, science, and art flourished.

Arab forces first landed in India early in the 8th century A.D. By 1200, after strong resistance, much of India was controlled by Muslim dynasties. Under the Mogul Empire (16th to 18th century), Muslim power and culture reached brilliant heights.

Bowl from the palace of an 18th-century Manchu emperor of China.

The Yellow River Valley and Chinese Civilization. Remains of very ancient settlements have been uncovered in the valley of the Yellow River. The finds indicate that the people worshiped a variety of gods and their own ancestors. About 2000 B.C. the first records of a stable government, that of the Hsia period, begin to appear. The Chou period (11th to 3rd century B.C.) has been called the classical age of China. It was marked in literature by excellent prose and poetry, in art by bronze vessels that are the pride of museums all over the world, and in religion and ethics by the teachings of Confucius (K'ung Fu-tzu) and Lao-tzu. Other major dynasties were to follow over the centuries, but space permits only a few words about some of them.

The Ch'in period (3rd century B.C.) consisted mainly of the reign of its first figure, Shih Huang-ti, who began the Great Wall to keep out northern invaders. The Han dynasty, which followed the Ch'in, lasted for over 400 years and was the first long-lived imperial dynasty. The period between the 1st and 7th centuries A.D. was not especially noteworthy, but there was a renewal of political power and a revival of culture in the T'ang dynasty (7th to 10th century A.D.). Early in the 13th century, during the Sung period, a Mongol warrior named Temujin, who took the title of Genghis ("the mighty one") Khan, swept down into China at the head of united Mongol tribes. A half century later all of China fell to the Mongols, with Kublai Khan, grandson of Genghis, ruling at Cambaluc

(Peking). It was to the court of Kublai Khan that Marco Polo journeyed and of which he brought back such fascinating accounts.

The Mongols were finally driven out by the founders of the Ming dynasty (14th to 17th century). During that period great Chinese fleets explored the seas, traveling as far as the east coast of Africa. The Ch'ing (or Manchu) dynasty that followed ended with the establishment of the Chinese Republic in 1912.

Japan.   The islands of Japan were probably first populated by the Ainu, now only a few thousand in number, who live in Hokkaido, to the far north. The Ainu differ in appearance from the Japanese in several ways, particularly in their greater hairiness. The people who were to evolve into the Japanese probably came from the Asian mainland and from nearby islands, displacing the Ainu. The oldest written Japanese myths, legends, and traditions appear to be associated with the Nara period (8th century A.D.), when the country's first capital was established at Nara. One of these myths accounts for the appearance of the first Japanese couple, with the sun goddess born from the eye of the male. Another legend identifies this goddess as the divine ancestor of the Japanese ruling family, which itself was considered divine until recent years.

The Great Buddha at Kamakura, Japan, dates from the 13th century.

A 16th-century painting depicting an unsuccessful Japanese invasion of Korea.

Strongly influenced by Chinese thought and culture in its early periods, Japan eventually developed a distinctive culture of its own. Politically, the country was ruled for centuries by shoguns, or warlords, while the emperor had little power and usually engaged in cultural pursuits. The last of the shoguns held power until the middle of the 19th century, when, under the Meiji Restoration, power was regained by the emperor. This same period saw the growth of Western influence in Japan, following the arrival of the American naval squadron under Commodore Matthew Perry. It also marked the beginning of modern Japan.

The European Period of Asian History

In 1498 the Portuguese navigator Vasco da Gama reached India by sailing around the southern tip of Africa. Da Gama was followed by Portuguese traders, who established themselves along the coasts of India, Ceylon (now Sri Lanka), and Malaya (now Malaysia) in the first decades of the 16th century. In the second half of the 16th century a Spanish settlement was established in the Philippines, and in the 17th century the Dutch began their colonization of the islands of Indonesia, which became known as the Dutch East Indies. Along with the traders and colonists came Christian missionaries.

Rivalry between the European powers became intense, particularly in India, where British and French interests collided. In the 17th century the English East India Company set up trading posts at Madras, Bombay, and Calcutta. But as the British moved inland, their advance was blocked by their French rivals and by local Indian rulers, and a series of wars resulted. By the end of the 18th century Great Britain had emerged as the dominant colonial power in India. From India the British expanded south and east. Ceylon was taken from the Dutch (who earlier had taken it from the Portuguese), and Burma was conquered and annexed to India in 1886. The British also moved into China. As a result of the Opium War of 1839–42, they acquired Hong Kong and special trading rights in Chinese port cities.

While the British were active in India and China, France was carving out an empire in Southeast Asia. In 1862 the French gained control of the three eastern provinces of Cochin China (now part of Vietnam), and then gradually took over the eastern part of the Southeast Asian peninsula. Eventually the French Indochinese empire included the present states of Vietnam, Cambodia, and Laos. In 1898 the United States joined the list of colonial powers in Asia by acquiring the Philippines after the defeat of Spain in the Spanish-American War.

Russian expansion into Asia, which began in the 16th century with the colonization of Siberia, reached its peak in the 19th century when the Russians advanced to the Sea of Japan. By 1860 the eastern region of modern-day Soviet Russia had been acquired and a settlement founded at Vladivostok, which became Russia's leading Pacific port. During the next 20 years the Russians moved into Central Asia.

The continuing weakness of China in the late 19th century encouraged several powers, including Japan, to seize pieces of the crumbling Manchu empire. Part of Shantung was occupied by Germany in 1897, while the Russians took over part of Manchuria. Other nations were also active, claiming special rights and spheres of influence in China. Chinese nationalists reacted strongly to this foreign intervention. In 1900 a nationalist group known as the Boxers organized an armed uprising and attacked the foreign legations in Peking. A combined military force from the various nations was rushed to Peking, and the Boxers were defeated.

After the Boxer Rebellion, Russia attempted to move from Manchuria into Korea. But the Russians were opposed by the Japanese, and the Russo-Japanese War (1904–5) resulted. The Japanese, who had adopted Western methods and military techniques beginning in the 1870's, defeated the Russians. This was the first time that an Asian nation had defeated a European colonial power in a major war, and the Japanese victory encouraged the growing nationalist movements in those

Asian lands under European domination. Japan also acquired Korea as its own colony in 1910.

The first major success of the nationalists was in China, where Sun Yat-sen and his Kuomintang Party established Asia's first republic in 1912. The breakup of the Ottoman (Turkish) Empire in 1918, following World War I, led to increased nationalist activity among the Arab countries of Southwest Asia. Governed briefly by Great Britain and France as mandates from the League of Nations, the Arab states achieved independence at about the time of World War II.

The period after World War II saw both an intensification in nationalist feeling and independence for many Asian nations. The Philippines, India, Pakistan, Burma, Korea, and Indonesia all became sovereign states in the years following the war. In 1954, after long drawn-out fighting with France, Cambodia, Laos, and Vietnam won their independence, although Vietnam remained divided for many years. Today, except for a few tiny areas, European control of Asian territory has disappeared and the nations of Asia are independent.

ASIA TODAY

Probably the most far-reaching impact of the West on Asia has been the introduction of its political theories and institutions. All the Asian nations now have constitutions, and each has proclaimed that it is building democracy or socialism. However, such terms sometimes seem loosely used. A so-called Democratic People's Republic such as that of North Korea is not at all democratic. Socialism is generally taken to mean a system of government in which the state owns all the means of production, transportation, and communications. But because of economic and technical reasons some nations in which socialism has been advocated have moved toward a "mixed economy"; that is, a combination of state-owned and privately owned businesses and industries.

Communist governments have been firmly established in several Asian countries. Not long ago, Communism was regarded as a set of unchanging principles and practices, but today there are said to be several ways to achieve Communism. China, for example, has followed a very different road from that of the Soviet Union.

Political instability has troubled many of the nations of Asia. In some countries divisiveness and antagonisms among numerous political parties have often prevented the passage of needed legislation. In others, an authoritarian leader has acquired more and more power, governing through a single-party system. In both cases, economic, political, or other crises have led to the governments being overthrown by military officers. After such a coup d'état, the military leaders are apt to create a new single-party system, with promises of eventual return to civilian rule.

Nationalist and Minority Groups

Other forces in Asia today tend to hinder national unity. Various groups within nations have agitated for autonomy, or self-government, based on language or ethnic relationships, including the Nagas of India, the Bengalis of East Pakistan, and the Kurds of Iraq. The Bengalis won independence after a civil war in 1971. Their nation is called Bangladesh.

Another problem facing a number of countries is what to do with unwelcome minorities. In the 19th century thousands of Tamils—people from southeastern India—were brought to Sri Lanka to work on plantations. Their number rapidly multiplied. Most were considered citizens neither of India nor of Sri Lanka. After years of negotiations between the two governments, India agreed to take a large number of Tamils back and Sri Lanka has agreed to grant citizenship to others. In Southeast Asia there are large Chinese and Indian minorities. Chinese are often resented by the majority populations because they may hold near monopolies in such fields as merchandising and moneylending. Both minorities tend to remain separate from the local populations.

Territorial Disputes

Aside from internal disputes that threaten national stability, there are disputes between neighboring states that often threaten to erupt into open warfare. Afghanistan and Pakistan are at odds over so-called Pushtunistan, a slice of territory in northern Pakistan. While Pakistan claims that there is no real dispute, the Afghans insist that the Pushtu people should be allowed to vote on the question of self-government for the area. Pakistan and India have gone to war over Kashmir, a region famed for its lovely lakes and snowcapped peaks. Kashmir was incorporated into India after the partition of 1947, although a great majority of its inhabitants are Muslims rather than Hindus. A border dispute between India and China brought Chinese troops into northern India in 1962. The fighting ended, but the dispute remains unresolved.

China also claims an extensive area of Soviet Asian territory, and frontier clashes have broken out between the two countries. While Indonesia has discarded its territorial claims against Malaysia, the Government of the Philippines has argued its rights to Sabah in Malaysia. Cambodia and Thailand have been at odds over borders, and until recently Iraq and Iran disputed navigational rights at the head of the Persian Gulf.

International Relations

In international affairs many of the countries of Asia describe themselves as nonaligned, or as neutrals. Under the leadership of the late Prime Minister Jawaharlal Nehru of India, some of these nations and some non-Asian ones banded together as a "third force." Their ideal was to remain aloof from the Cold War between the countries of the West and those of the Communist bloc.

Asian countries belong to numerous organizations concerned with the advancement of health, education, and science. Turkey, Iran, and Pakistan are also members of the Central Treaty Organization (CENTO), sponsored by Great Britain and the United States. The Philippines and Thailand were members of the Southeast Asia Treaty Organization (SEATO), along with Australia, New Zealand, Great Britain, France, and the United States. Cambodia, Laos, and South Vietnam were placed under the protection of SEATO, and it was to defend South Vietnam against invasion from the north that the SEATO countries sent troops to Vietnam. SEATO has since suspended its operations.

The Problem of Overpopulation

Most of Asia suffers from the joint problem of too many people and too little food. With such a relatively small percentage of their total land under cultivation, many Asian nations have long been hard pressed to feed their people. As far as is known, the population of Asia remained fairly constant for centuries, due in large part to the effects of wars, famine, floods, and plagues. Modern methods of preventive medicine have resulted in a decline in the death rate and a corresponding sharp increase in population. Most Asian countries now have an increase in population of over 2 percent a year. At this rate the population of Asia will double by the end of the century, and thus by the year 2000 the continent will have about 4,000,000,000 (billion) people.

The Asian Economy

Virtually all the countries of Asia have been trying to establish social-welfare systems in which each person has a job, adequate food and lodging, free medical care, and various social security benefits. The limited financial resources of many of the countries, however, make this goal a very difficult one to achieve. In order to industrialize themselves on the Western model, most of the Asian countries must import machinery, iron and steel, metal products, electrical equipment, chemicals, cars and trucks, and hundreds of other manufactured goods. To pay for these goods, Asian products must be sold abroad. With few exceptions, the Asian products are raw materials, such as jute, rice, tea, rubber, oil, and tin. These raw materials must compete for markets with the same items produced in other parts of the world, and prices for the raw materials have fluctuated over the years. As a result many Asian countries have an unfavorable balance of trade; that is, their income from goods sold abroad is less than the cost of items imported.

Hopes for the Future

Although this would seem to forecast a gloomy future for Asia, there is cause for optimism. The problem of overpopulation is being tackled through family planning and birth control, with some countries very active in the field. Japan was the first country to take the necessary steps and as a result its population increase is only about 1 percent a year. India is also making great efforts to establish family planning and health clinics.

Nearly every country in Asia is now working at long-term plans for economic development, aided by grants and loans from the industrialized countries of Europe and from the United States. Huge dams and irrigation canals will provide water for new farmland, while improved varieties of seeds and the use of chemical fertilizers are expected to result in much larger yields of crops. Continuing financial and technical aid, however, is needed, and this must come from the oil-rich countries of Asia and from such industrialized Asian nations as Japan, as well as from the non-Asian capitalist and Communist countries. Peace is also essential if Asia is to have a bright future.

Asia today is a continent of very young people who are benefiting from vastly expanded educational facilities. Ever-increasing numbers of doctors, engineers, teachers, scientists, and technicians are being trained, and with their skills they will help lead the Asian societies of tomorrow. National unity and national pride, as well as the willingness to work for the good of all rather than for family interests or village or tribal concerns, are emerging as the constructive forces within Asia.

DONALD N. WILBER, Editor, *The Nations of Asia*

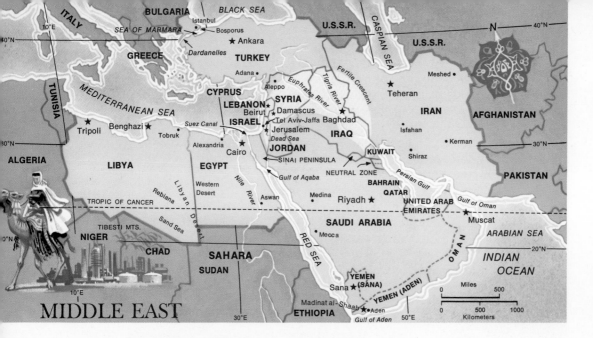

MIDDLE EAST

MIDDLE EAST

The Middle East is not a strictly defined geographical unit, but a term of convenience used mainly since World War II. The area is usually described as being bounded on the north by the Caspian and Black seas; on the east by Afghanistan and Pakistan; and on the south by the Arabian Sea, the Gulf of Aden, and the Sahara.

Civilization began in the Middle East more than 5,000 years ago—many centuries before the Golden Age of Greece and the founding of Rome. In fact, Western culture traces its origins to ancient Middle Eastern civilizations. Three of the great religions of the world—Judaism, Christianity, and Islam—began in this region. It was here also that man learned to till the soil, to build cities, and to record events in written languages.

Generally, the Middle East is considered to include the countries of Iran, Turkey, Iraq, Syria, Lebanon, Israel, Jordan, Saudi Arabia, Yemen (Sana), Yemen (Aden), Cyprus, the Arab Republic of Egypt, and Libya. Also included are the smaller states of Kuwait, Bahrain, the United Arab Emirates, Qatar, and the Sultanate of Oman. Some geographic authorities place Morocco, Algeria, and Tunisia in this region. Most of the region's people are culturally united by the religion of Islam and the Arabic language.

THE LAND

The Middle East covers an area of about 3,442,000 square miles (8,915,000 square kilometers). By and large, the terrain and climate are hostile to man and other living things. Vast stretches of the Middle East are dry, barren deserts. In some places the deserts are great oceans of sand. In others, they are bleak scrublands of sunbaked gravel and rock. Mountains and hills ring Turkey and Iran and line the coastal areas along the surrounding seas and gulfs. Supplying water to the land has always been a problem in the Middle East. Three major rivers, the Tigris, the Euphrates, the Nile, and their tributaries have been an important water

A valley in Lebanon, one of the fertile areas of the Middle East.

A camel caravan in the Nafud desert of Saudi Arabia.

source throughout history. By means of dams and canals, the people of the area have been able to irrigate and farm the land along these rivers.

For the most part, the Middle East has a hot, dry climate, with temperatures rising to well over 100 degrees Fahrenheit (38 degrees Celsius) during the summer months. The desert area, stretching from the Sahara through the Arabian Peninsula into Iran, receives little or no rain during the year. In the mountains the weather tends to be cooler, and there is a fair amount of rainfall in the winter and spring. A particular exception to the generally arid terrain is the Fertile Crescent, a strip of land that runs in a curve through Jordan, Israel, Lebanon, Syria, southern Turkey, Iraq, and Iran. Bounded on the west by the Mediterranean Sea and on the southeast by the Persian Gulf, the Fertile Crescent is a highly productive agricultural belt. It was here that man first learned to cultivate the basic food crops of wheat and barley, beginning as early as 10,000 years ago. A similar development took place a bit later in the Nile Valley, another exception to the dryness of the Middle East. Thousands of years later, the early civilizations along the Nile, Tigris, and Euphrates rivers developed methods of using the rivers to irrigate the surrounding land. These important discoveries are considered the basis for the higher civilizations that sprang up later in the Middle East, Europe, and elsewhere.

THE PEOPLE

Approximately 135,000,000 people live in the Middle East. Because much of the region is desert, a large proportion of the population is crowded into fertile river valleys, such as those of the Nile, the Tigris, and the Euphrates, and in the coastal lands of the Mediterranean Sea. Approximately half the population of the Middle East consists of Arabs or Arabic-speaking people. The two other largest groups are the Turks and the Iranians. There are also substantial numbers of Kurds, Jews, and other minority groups. Although Jews and Arabs have different cultural traditions, they both speak languages of the Semitic family. A very high birthrate and declining death rate have resulted in a rapid population increase in recent years.

Arabic is the dominant language of the region except in Turkey, Iran (formerly Persia), and Israel, which have their own distinct languages. Greek and Turkish are used on the island of Cyprus. Both the Turkish and Persian (Farsi) languages, however, have been greatly influenced by Arabic. Until 1928 the Turks used Arabic script, but then changed to the Roman alphabet. The Persian language is still written in Arabic script. Hebrew is the chief language of Israel. The Kurdish people, who live mainly in Iran, Turkey, and Iraq, have a language related to Persian. Nearly all of the people in the region are followers of Islam, the religion taught by the Arab prophet Mohammed. Lebanon is about equally divided between Christians and Muslims, while Israel is predominantly Jewish. Smaller groups of Christians and Jews also live in almost all the other countries of the Middle East.

Economy

Although oil is the most important source of income, farming is the main occupation in the Middle East, as it has been for thousands of years. However, because of the shortage of water only about 10 percent of the land is cultivated. Wheat, barley, rice, and other cereal grains are

Oil refinery in Saudi Arabia. Oil is the region's most important resource.

among the principal crops, and cotton is also important. Figs, dates, grapes, citrus fruits, and olives are also grown. Those not involved in farming work mainly as craftsmen, in light industries, or for the large oil companies. Since earliest times, Middle Eastern craftsmen have been noted for their skill in producing fine textiles, leather goods, woodwork and metalwork, jewelry, and ivory carvings. Industrialization began in the Middle East only about a century ago and is limited because of a lack of capital, skilled labor, and trained technicians. But many governments in the region are encouraging industrialization in order to absorb unemployed or underemployed farm workers.

Except for oil, the Middle East has few mineral resources. Some coal, chromium, and manganese have been discovered. Potash is mined in Israel, and phosphates in Jordan. The area produces over one third of the world's crude oil. There are large oil wells and refineries in Saudi Arabia, Iran, Iraq, and Kuwait, as well as in the Persian Gulf states and Libya. Large quantities are exported to Europe, the United States, and elsewhere. Money from foreign sales has enabled the Middle Eastern governments to raise the standard of living of their people through public works, development of new industries, and improved education.

Way of Life

Due to modern communications and the influence of European countries and the United States, the way of life in the Middle East is becoming more like that of the West. As a result, the area has lost much of its former picturesque quality, as well as some of the outward signs of poverty and misery. Many people in the cities wear Western clothes and use Western furniture and electrical appliances in their homes. After World War I, the Turkish president Kemal Atatürk led a movement to modernize Turkey. This included doing away with traditional costume, such as the Turkish fez (a brimless hat). More recently, Egypt's former president, Gamal Abdel Nasser, tried to discourage Egyptian men from wearing the *gallabiya* (or *gallabiyea*), an ankle-length cotton robe, and to adopt Western-style clothes. The old Muslim tradition of having women cover their faces with veils is also dying out. Many women are being admitted to universities and into the work force, but there are still greater restrictions placed on women than on men.

The majority of Middle Eastern people still live in small villages, where farming is the main occupation. Here life has not changed so rapidly, and many of the old ways have been preserved. Homes are usually made of sun-dried brick (adobe) or stone, and the family—including grandparents and grandchildren—all live together in generally cramped quarters. Simple, ancient devices for lifting water from rivers to the fields are still in use, though modern pipelines have been installed in many areas. Camels and donkeys are used as means of transportation, but the automobile is gradually replacing them. This has caused hardship for those Bedouin tribesmen who earn their living as camel breeders. For the most part, these desert nomads still live as they have for centuries, constantly moving about to find water and grass for their flocks.

CITIES

Since ancient times, the cities of the Middle East have been important centers of commerce and culture.

Damascus, the capital of Syria, and **Aleppo** are trading cities that go back to 2000 B.C. or earlier. Today they are Syria's leading industrial and business communities.

Beirut, the capital of Lebanon, has been an important Mediterranean port for more than 3,000 years. The city has continued to be a leading financial and trading center of the Middle East. In the mid-1970's, however, Beirut suffered very heavy damage as it became the chief battleground of the Lebanese civil war.

Jerusalem, the capital of Israel, was a mountain stronghold and a sacred shrine some 4 centuries before King David made it the capital of the ancient kingdom of Israel (around 1000 B.C.). Jews, Christians, and Muslims all regard the Old City of Jerusalem as a holy place because of its many sacred shrines. (A separate article on JERUSALEM appears in this volume.)

Mecca, the most sacred city of Islam, is located in the western coastal foothills of Saudi Arabia. Islam calls on all Muslims to make at least one pilgrimage to Mecca—the birthplace of the prophet Mohammed—if they are able to. Pilgrims to Mecca visit the Kaaba, Islam's holiest shrine, which contains the sacred Black Stone. When saying their daily prayers, all Muslims must face in the direction of the Kaaba. **Medina**, also in Saudi Arabia, is revered by Muslims as the place where Islam took root after Mohammed's flight from Mecca in A.D. 622. The city subsequently became the center of the Islamic state. It is also said to be the burial place of Mohammed.

Istanbul, once the capital of Turkey, has been historically important because it commands the entrance to the Black Sea and because it is the country's major port. It was originally the Greek city of Byzantium before the Roman emperor Constantine rebuilt it (A.D. 324–30) as Constantinople, the eastern capital of the Roman Empire.

Zina Dizengoff Circle in the center of Tel Aviv, Israel's largest city.

Throngs of Egyptians crowd the streets in the old part of the city of Luxor.

A view of Istanbul, Turkey. The city was formerly known as Constantinople.

Baghdad, the capital of Iraq, was founded in A.D. 762 as the capital of the Abbasid caliphate (dynasty) and became the center of Islamic art and commerce. It was the historic setting for the golden age of Islamic civilization in the Middle East. The city is located on the Tigris River, at a point where it is closest to the Euphrates.

Cairo, the capital of the Arab Republic of Egypt, is the largest of the Middle Eastern cities, with a population of some 4,000,000. It is located on the Nile River, just before the river divides into two branches and forms the fertile Delta. Near Cairo are the three pyramids of Giza and the great statue known as the Sphinx. Cairo is also the site of Al-Azhar University, which is considered the greatest Islamic institution of higher learning. **Alexandria** is the most important port city of Egypt. It was built by the Greek general Alexander the Great in the 4th century B.C. Alexandria is now also a sea resort and serves as the summer capital of Egypt.

Teheran, the capital of Iran, is one of the youngest cities in the Middle East. It became the capital of Iran in the late 18th century and has developed into the country's most important industrial, communications, and cultural center.

Amman, the capital of Jordan, is located on the site of the ancient Greco-Roman city of Philadelphia. It became the capital of the then Emirate of Transjordan after World War I, and has since become an important urban center.

Tel Aviv-Jaffa, Israel's largest city and its commercial and cultural capital, began in the early 1900's as a suburb of the ancient port of Jaffa. A growing metropolis, it is today one of the most modern cities in the Middle East. The old city of Jaffa was incorporated into Tel Aviv in 1950.

Tripoli and **Benghazi** are the two capitals of Libya and the country's largest cities. Both are major ports on the Mediterranean Sea. The two cities are being greatly modernized as a result of income from the overseas sale of Libyan oil.

HISTORY

The earliest recorded civilization known to man was founded by the Sumerian people of Mesopotamia (now Iraq), in an area bounded by the Tigris and Euphrates rivers. The Sumerians invented written language (about 3500 B.C.), probably the wheel, and also a calendar based on the farming seasons. The first Middle Eastern civilizations were made possible by farming methods that enabled people to settle in permanent towns and villages instead of having to search constantly for food as wandering nomads. Armies were formed, codes of law and government were established, and cities were built. Usually the cities were built around temples, where priests sought the favor of the gods by making offerings of goods collected from the people. Each person had to contribute a portion of his crop, some silver, or another valuable product to the priests, and so the idea of taxation came into being. In the cities, people lived in houses made of sun-dried bricks or stone. They worked as weavers, potters, and silversmiths, and at other crafts, and they traded their products for food and whatever else they needed.

Because it is a bridge between the continents of Africa, Asia, and Europe, the Middle East became important as a trade route. Caravans using donkeys—and much later, horses and the Arabian camel—as pack animals carried goods from the Middle East northwest into Europe and northeast into Central Asia. The great desert of the Sahara and the dense forests of Central Africa were a barrier to southward movement and trade, but in the last centuries B.C. commercial contact with East Africa was made by sea and with West Africa by caravan across the Sahara. Along with their native products, the Middle Eastern traders brought their technology and culture to the peoples of Europe and Asia. These traders used measured weights of silver as currency about 1,500 years before the invention of coined money.

The rich cities of the Middle East and the trade routes they controlled were a valuable prize. Many wars were fought and empires built by rival kingdoms, which sought to dominate the entire region. Beginning around 3200 B.C. the Egyptians developed a great civilization along the banks of the Nile River. Egyptian architects and artists created magnificent temples and the large tombs known as pyramids. The greatest of these pyramids, built for the pharaoh (king) Cheops in the 27th century B.C., took 20 years to construct and required over 2,000,000 blocks of stone, each weighing many tons. Egyptian armies (after 1500 B.C.) conquered and plundered neighboring lands, which provided their pharaohs with slaves and wealth. Other peoples, including the Babylonians, the Hittites, the Assyrians, and the Persians also built great kingdoms.

Around 1200 B.C. the Hebrew people, who had fled from forced labor in Egypt, moved into Palestine. A Hebrew kingdom was established, which then split into two parts—Israel and Judah—after the death of King Solomon (about 925 B.C.). The Hebrews had developed a religion based on the belief in one God (known as Yahweh or Jehovah) instead of many gods. Israel was conquered by the Assyrians in the 8th century B.C., and in the 6th century the Babylonians overran Judah. But although most of the Hebrews were driven into exile and scattered among many lands, they preserved their religion and culture in the centuries that followed.

The collapse of the Assyrian Empire during the 7th century B.C. was followed by the rise of the Persian Empire, the greatest of its time. By the 6th century B.C. the Persians controlled a huge region that extended from the Aegean Sea to what is now Pakistan. A vast network of roads and way stations were built to connect the cities of this large empire. Messengers could travel hundreds of miles in a matter of days by changing horses frequently at the station houses. The Persians followed the religion of the prophet Zarathustra (Zoroaster), who preached that the world was a battleground between two opposing spirits, one representing "truth and light" and the other "darkness and lies." This idea later influenced Jewish and Christian religious thinking.

In the 4th century B.C., Greek armies led by Alexander the Great conquered the Persian Empire. The Greeks ruled the area for about 2 centuries, until the Romans established their authority. When the Roman Empire split in two in A.D. 395, the Eastern, or Byzantine, empire continued to rule over most of the Middle East from its capital at Constantinople (now Istanbul). By this time many of the region's inhabitants were Christian, but in the 7th century A.D. a new religion known as Islam took root. The Arab prophet Mohammed, the founder of Islam, set up a religious community at Medina, in what is now Saudi Arabia. There Mohammed preached total submission to the will of God (Allah). After Mohammed's death in 632, his followers (known as Muslims) spread the teachings of Islam through conquest, example, and persuasion. In the process, Muslim armies swept across North Africa and into Spain, and east to the borders of China and India.

Towards the end of the 11th century, the Seljuk Turks conquered the eastern region of the Islamic world, including part of the Middle East. During the next centuries the Crusaders from Europe, the Tatar Mongols,

Roman ruins in Jordan mark the time when the Roman Empire ruled the area.

and finally the Ottoman Turks invaded the area. By the middle of the 16th century the Ottoman Turks had built an empire that included nearly all of the Middle East, much of North Africa, and southeastern Europe as far as Austria. The Ottoman Empire lasted from the 14th century to the end of World War I (1918). For several centuries Constantinople, the empire's capital, was a thriving center of trade and commerce. But though the Turkish sultans and the ruling class of military, political, and religious leaders enjoyed great riches, most of the peoples of the empire remained poor and backward.

While Europe underwent a social, political, and technological revolution in the 18th and 19th centuries, the Ottoman Empire was slow to change. Corruption became widespread, and in the 19th century the empire was referred to as "the sick man of Europe." By then Europeans were moving into the Middle East for commercial purposes and as technical advisers and financiers. They introduced Western ideas into the region—a process that began in earnest with Napoleon's invasion of Egypt in 1798. The opening of the Suez Canal in 1869 gave Europeans a direct sea route to India and the Far East. The Middle East became economically important to European imperial powers such as France and Great Britain and later to American oil companies.

The final collapse of the Ottoman Empire came at the end of World War I. By that time, the empire had been reduced to Turkey and the Arab countries of the Middle East. After 1918 these Arab countries became protectorates of Great Britain and France, under the mandate (authority) of the League of Nations. The discovery and development of rich oil fields in the Middle East after 1900 increased the value of the region to the European powers and the United States. Between the end of World War I and the 1960's the Arab nations gained their independence from France and Great Britain.

Israel was created by a 1947 United Nations resolution that partitioned (divided) the mandated land of Palestine into separate Jewish and Arab states. Both Jews and Arabs had historical claims to Palestine, and the Arabs regarded the establishment of Israel as a theft of Arab land. After the expiration of the British mandate over Palestine, on May 15, 1948, Arab armies moved into the region in support of Palestinian Arabs. The Arab forces fought a 7-month war with the Israelis and were defeated. An armistice was arranged through the United Nations in 1949. In 1956, 1967, and 1973 warfare again broke out between Israel and its Arab neighbors. Although ceasefires were arranged after each outbreak, the truce following the 1973 war brought only an uneasy peace. In 1975 and 1976 a civil war in Lebanon posed yet another threat to the Middle East's delicate balance.

Today the Middle East is once again an area of conflict between powerful rivals. Both the United States and the Western powers and the Soviet Union and the Communist bloc are interested in the area because of its important location and its valuable oil deposits. In recent years some Arab states have made great efforts to improve living conditions through what they call Arab Socialism. But although progress has been made, many economic and social problems remain.

GEORGE KIRK, University of Massachusetts; author, *A Short History of the Middle East*

Reviewed by YASSIN EL-AYOUTY, St. John's University
Fellow, The Middle East Association for North America

The Blue Mosque in Istanbul dates from the 17th century.

TURKEY

"Whoever is the master of Constantinople is the master of the world." These words were spoken by Napoleon Bonaparte. In Napoleon's day the master of Constantinople was the Ottoman Empire, the predecessor of modern Turkey. But the Turkey of today, in whose domain Constantinople (now known by its Turkish name—Istanbul) lies, is far from being the master of the world. The once great empire of the Ottomans is gone. Nevertheless, Turkey remains one of the most strategic sites on the globe, for it is a gateway to the Middle East, and it still controls the vital straits that link the Black Sea with the Mediterranean Sea.

The land that is now Turkey has been important almost since the beginning of recorded history. It was the birthplace of one of the earliest civilizations, that of the Hittites, who were among the first people to work iron. The partly legendary city of Troy was located on the Aegean coast of Turkey, and along this same coast the genius of ancient Greece flourished. The region has been a battleground of empires. The Persians sought to master the area. Alexander the Great made it the pivot of his own empire. The Romans, already masters of most of Europe, established themselves at Constantinople, becoming rulers of much of the then known world.

The first Turks, among them the Seljuks, migrated to Turkey from beyond the Ural Mountains. The empire of the Seljuks was besieged from the west by the Crusaders and from the east by the Mongols of Genghis Khan. Finally, the region fell to another branch of the Turks, the Osmanlis, or Ottomans, ancestors of the present-day Turks. The Ottomans captured

Constantinople and built an empire that became one of the great powers of the world. But, like all empires, that of the Ottomans declined too. And with the decline of the empire, European powers, especially the British, French, and Russians, sought to control Constantinople, or at least to prevent each other from controlling it. Eventually, the Ottoman Empire was dismembered and modern Turkey emerged.

THE LAND

Although only a part of the once vast empire, Turkey is still a sizable country. Its total area is a little over 300,000 square miles (780,000 square kilometers). Nearly all of Turkey—about 97 percent—lies in Asia. Asian Turkey, a huge peninsula, is called Anatolia (in Turkish, Anadolu), from the Greek word meaning "rising sun." Another name for it is Asia Minor. The European part of Turkey is called Thrace. The two parts are separated by a series of three interlinked waterways—the strait of Bosporus, the Sea of Marmara, and the strait of the Dardanelles—that connect the Black Sea with the Aegean Sea, an arm of the Mediterranean. The straits are Turkey's most historic region, a place where history and legend are interwoven. The Dardanelles is the most famous part of the straits. Not far from its western end lie the ruins of ancient Troy.

Most of Turkey consists of a highland, the Anatolian plateau, framed by mountain ranges and crisscrossed by swiftly moving streams. The plateau is a harsh country, with hot summers and cold winters and generally limited rainfall. Much of the soil is barren. Pasture and prairie cover most of the land. About 17 percent of the plateau is cultivated, and another 12 percent is forest. The chief crops grown here are wheat, barley, maize (corn), potatoes, and sugar beets. Some of the land is devoted to vineyards and olive groves. Sheep and goats are raised both for food and for their wool and hair. The hair of Angora goats is known as mohair.

The plateau is rich in mineral resources. Turkey is one of the world's

This Anatolian farmer is working part of his wheat crop.

TURKEY

The Fortress of Rumelihisar—on the Bosporus

FACTS AND FIGURES

REPUBLIC OF TURKEY—Türkiye Cumhuriyeti—is the official name of the country.

CAPITAL: Ankara.

LOCATION: Southeastern Europe and Southwestern Asia. **Latitude**—35° 49′ N to 42° 06′ N. **Longitude**—26° 03′ E to 44° 49′ E.

AREA: 301,381 sq. mi. (780,576 sq. km.).

PHYSICAL FEATURES: Highest point—Mt. Ararat (16,945 ft.; 5,164 m.). **Lowest point**—sea level. **Chief rivers**—Kizil Irmak, Sakarya, Menderes, Seyhan, Ceyhan, Gediz, Orontes. **Major lakes**—Van, Tuz.

POPULATION: 35,700,000 (estimate).

LANGUAGE: Turkish.

RELIGION: Muslim (predominant), Christian, Jewish.

GOVERNMENT: Republic. **Head of state**—president. **Head of government**—prime minister. **Legislature**—Grand National Assembly. **International co-operation**—United Nations, North Atlantic Treaty Organization (NATO), Central Treaty Organization (CENTO).

CHIEF CITIES: Istanbul, Ankara, Izmir.

ECONOMY: Chief minerals—chromite, copper, coal, petroleum, manganese. **Chief agricultural products**—cotton, tobacco, wheat, barley, maize (corn), potatoes, sugar beets, nuts, olives, fruits, mohair, wool. **Industries and products**—textiles, iron and steel, cement, paper, mining, food processing, handicrafts. **Chief exports**—tobacco, cotton, fruits, nuts, minerals. **Chief imports**—machinery, motor vehicles, iron and steel products, petroleum, chemicals.

MONETARY UNIT: Turkish lira.

NATIONAL HOLIDAY: October 29, Republic Day.

NATIONAL ANTHEM: *Istiklal Marsi* ("March of Independence").

largest sources of chromite, and there are large copper deposits. The iron content of Turkish ore is high, and the country has rich seams of coal and some petroleum. One of the Turkish mineral specialties is meerschaum, which is used to make smoking pipes. Much of the country's mineral wealth, however, has not been adequately explored and utilized.

In contrast to the hard Anatolian plateau are the coastal areas, narrow bands of "softer," more fertile land, with a milder climate and heavier rainfall. Along the northern coast, on the Black Sea, the world-famous Turkish tobacco is cultivated, as well as another Turkish specialty, hazelnuts, and, more recently, tea. Some of Turkey's industry is also located here. The Aegean coast in the west is probably the most beautiful part of Turkey, the bright blue sea reflecting the pleasantly contoured hills. This is one of Turkey's most prosperous regions: it has some of the best harbors; its cities are centers of trade and finance; and the sun-drenched soil provides rich crops of cotton, tobacco, grapes, and figs. The southern, or Mediterranean, coast might be called Turkey's Florida. It produces cotton and citrus fruits, and the Turkish Government hopes to develop it into a resort area.

THE PEOPLE

In this land of great diversity live some 35,700,000 people. Most are Turks. In the southeastern mountains there are some 2,000,000 Kurds. There are also several hundred thousand Arabs and smaller numbers of Circassians, Greeks, Armenians, Georgians, and Jews. Nearly all the Turks, Kurds, and Arabs are Muslims, followers of the religion of Islam.

How does one recognize a Turk, say in a group of Europeans? The founder of the Turkish Republic, Kemal Atatürk, resembled a Swede, with blue eyes and a fair complexion. At the other extreme, there are Turks, though not too many, who have distinct Mongoloid characteristics. The majority of Turks could not easily be picked out of the group of Europeans. Many look like people of southern Europe.

Language. The Turkish language, however, is different from most of the languages of Europe, which belong to the Indo-European family. Turkish is classified under a language group known as Ural-Altaic, so named after two mountain ranges, the Urals on the dividing line between Europe and Asia, and the Altai Mountains in the heart of Asia. Some scholars, though, believe that Turkish more properly should be included in the separate Altaic family. An interesting trait of Turkish is that it is an agglutinative tongue; that is, it "glues together" several parts of speech. For example, a Turk would use one word, *yazmadim*, for the four English words "I did not write."

The Turks seem to have come originally from the heart of Asia. They were close neighbors of the Mongols, whose language also belongs to the Altaic family. From their original homeland the Turks migrated in many directions. Today there are Turkic-speaking peoples not only in Turkey, but scattered over a vast area, from the Arctic to the Mediterranean, from the Danube Valley to eastern Asia. Millions of Turkic people live in Soviet Central Asia.

Way of Life

Village Life. Most Turks are farmers. Some are herdsmen who raise sheep and goats. The traditional Turkish farm village consists of houses built of sun-dried mud brick, without any adornment. Because for centuries Anatolia was a convenient invasion route, villagers preferred to build their homes so that they could not be seen easily—between hills or in depressions in the land, and blending in with the surrounding landscape. Sometimes houses are built on spurs of rocky ground, so as not to encroach upon arable land. The houses are usually as austere and functional inside as out, containing a few cooking utensils, the traditional *mangal*, or charcoal grill, and the necessary furniture and bedding; but they are enlivened by the bright colors of Turkish carpets.

Nearly every village has more ambitious houses, the homes of the locally eminent families. These houses are likely to be built of stone. There usually is a separate village house used for social occasions, and even the small villages often have a coffeehouse where the men gather in the twilight hours after the day's work. Larger villages—usually those with access to roads—have a *sinema*, or movie theater. Visits of dancing bears are a popular folk entertainment. Wrestling is a favorite sport with men, while *futbol* (soccer) is very popular with boys as well as men. The monotony of everyday life is also broken by feasts of circumcision and weddings.

A farmer and his family. In the background women are at work in the fields.

A village on the bleak Anatolian plateau.

A wrestling match at Edirne in Thrace. By tradition the wrestlers are oiled.

Life in the Cities. Life in the large towns and in the cities is of course quite different. Such major cities as Istanbul, Ankara, and Izmir have a variety of social, educational, and cultural institutions—theaters, concert orchestras, museums, universities, art exhibits—as do large cities almost everywhere. The difference between town and country is also noticeable in the dress of the people. In the traditional rural areas, women wear pajama-like bloomers, usually while working in the fields. Although the wearing of veils by women has been prohibited, a country woman will still cover her face with a loose-hanging scarf at the approach of an unfamiliar male face. A veiled woman in the city, however, would be as out of place as a veiled woman on an American street or a Paris boulevard. Western clothes are usually worn by city people.

Food and Drink. Feminine beauty in the East is still associated with plumpness. It is also a status symbol. People who can afford to eat well like to show it, particularly in the traditional rural areas. The Turks like to eat meat except for pork, which is forbidden by Islamic law. The American turkey is not a favorite food in Turkey. (The Turks call it *hindi,* because they claim it originated in India.) Instead, mutton, lamb, and beef are preferred. Shish kebab—pieces of lamb broiled on skewers, with onion and green pepper—is very popular. *Döner-kebab*—slices from the side of a lamb cooked on a spit—is also a favorite. Another popular food is pilaw, which is a rice dish made with meat or fish. Some of the Turkish sweets are famous in many parts of the world, particularly halva (made with crushed sesame seeds) and baklava (a pastry usually filled with nuts and honey). The favorite alcoholic beverage is raki, which is distilled from either grain or grapes, looks like water, tastes like licorice, and is quite strong. Turkish coffee also has acquired well-earned fame, and tea is a popular drink.

This smiling waiter is serving tea to his customers in the city of Adana.

Major Cities

Istanbul.   At the southern extremity of the Bosporus, where the narrow strait curves into the wider Sea of Marmara, lies ancient Constantinople—and modern Istanbul. Constantinople was founded in the 4th century A.D. on the site of the old Greek city of Byzantium by the Roman emperor Constantine I (the Great). The name means "city of Constantine." Constantinople became the capital of the Roman Empire and then of the Byzantine Empire, successor to the Roman Empire in the East. One of the wealthiest and most magnificent cities of the Middle Ages, Constantinople was also virtually impregnable, its strong walls withstanding many sieges. Only three times in its long history was the city captured, the last time, in 1453, by the Ottoman Turks.

While Constantinople stood on the European side of the Bosporus, modern Istanbul stretches across both sides of the strait. The Golden Horn, an inlet of the Bosporus and the harbor of Istanbul, separates the old and new sections of the European part of the city. Bridges span the Golden Horn, and a bridge across the Bosporus was completed in 1973. The new bridge connects the Asian and European parts of the city for the first time. Istanbul is Turkey's largest city (with a population of over 2,200,000), its chief port, and the commercial, financial, and cultural center of the country. It was also the capital until 1923, when the seat of government was transferred to Ankara.

One of the two most famous structures in Istanbul is Hagia Sophia, which was to the Eastern Church what St. Peter's in Rome is to the Roman Catholic Church. Completed in the 6th century, its great dimensions, and especially its boldly vaulting dome, have made it admired for centuries as the greatest example of Byzantine architecture. After the capture of Constantinople by the Turks, it served as a mosque. It is now a museum. The city's second most renowned structure is the magnificent Blue Mosque. Istanbul's hundreds of mosques remind the visitor that he is in Muslim land. And the seraglios (palaces and harems) of the former sultans recall the days of the Ottoman Empire.

Istanbul. The Galata Bridge crosses the Golden Horn and connects the old and new parts of the city.

Hagia Sophia, one of Istanbul's most famous sights.

Ankara, the capital of Turkey. The statue is of Kemal Atatürk.

Ankara. Turkey's capital and its second largest city, with over 1,200,000 people, Ankara is situated in the central part of the Anatolian plateau. It is a striking illustration of the old and new Turkey. Although the city goes back to ancient times, modern Ankara dates from its establishment as the capital in 1923. The old part of the city has meandering lanes and crowded bazaars, the new has modern buildings and wide boulevards. As well as being the seat of government, Ankara is an important industrial city and the heart of the region producing the famed Angora goats. (The city was formerly known as Angora.) The city's cultural and educational life centers around the several colleges of the University of Ankara, the state conservatory of music, ballet, and drama, and the national library. The tomb of the founder of the republic, Kemal Atatürk, is in Ankara.

Izmir. A seaport on the Aegean coast, Izmir (formerly known as Smyrna) is one of Turkey's most important commercial cities, with a population of over 500,000. As much as any city in Turkey, it is rich in history. It was one of the ancient Greek cities along the eastern Aegean and is said to be the birthplace of the poet Homer. Izmir is particularly important in Turkey's import-export trade. It is also popular with tourists. Other important Turkish cities include Adana, Bursa, and Eskisehir.

Not far from Izmir lie the ruins of a number of famous historical cities: **Pergamum** (modern Bergama), which in Roman times had a theater capable of seating over 4,000 people; **Sardis**, capital of the kings of Lydia, the name of one of whom, Croesus, has become a synonym for great wealth; **Ephesus**, which contained one of the ancient Seven Wonders of the World, the Temple of Artemis (Diana). Along the Mediterranean coast lies **Tarsus**, the birthplace of Saint Paul.

Ruins at Ephesus. An ancient Greek city, it later was a Roman provincial capital.

HISTORY

The Ottoman Empire.   The Ottomans were so called after the legendary founder of their dynasty, Othman, or Osman. In the centuries following their arrival in Turkey, they expanded rapidly, crossed the Bosporus into Europe, and conquered most of the territory of the decaying Byzantine Empire. Only Constantinople, behind its strong fortifications, held out. In 1453, after a short siege, during which the Turks moved part of their fleet overland to the Golden Horn, the city fell. Napoleon's statement about the master of Constantinople being the master of the world almost became a reality. From their strategic location on the straits, the Turks pushed westward into Europe, southward into Africa, and eastward toward the Indian Ocean. In a remarkably short time their empire extended across three continents. In Europe they reached the walls of Vienna before being turned back. To the kingdoms of eastern Europe they were the gravest threat since the earlier invasions of the Mongols.

The Ottomans were so successful because they were a military people, rigidly regimented and accustomed to hardship. Above all, they were united. Their foes often were not. In their forays into Europe the Turks impressed the most promising Christian boys into their armed forces. They were given special training, converted to the Muslim faith, and became the nucleus of the *yeniçeri* ("new troops"), or Janissaries.

The Ottomans reached the height of their power in the 16th century under Sultan Suleiman I (the Magnificent), when the empire stretched from Hungary to the Caspian Sea, and from southern Russia to North Africa. But from this high point began a slow but steady decline.

The decline of the Ottoman Empire was caused by the growing power of the surrounding European states, by the growth of nationalism

among the peoples of the empire, which led to disunity, and by corruption and inefficiency in both the army and government. The once dreaded Janissaries became unfit for fighting, although they were able to depose sultans whom they did not favor. The Sublime Porte, the name for the government of the Ottoman sultans, fell into debt and faced bankruptcy. The empire was known as the "sick man." The greatest threat to the Ottoman Empire came from its northern neighbor, Russia. It was an age-old ambition of the Russians to possess the straits and thus have access for their fleet into the Mediterranean. Only intervention by Great Britain and France, who feared the power that Russia would acquire, thwarted this ambition. Nevertheless, during the 19th and early in the 20th century the Ottomans lost most of their European territory, as Greece, Serbia (later to become part of Yugoslavia), Rumania, and Bulgaria won independence. A reform movement, that of the Young Turks, developed during this period and had some success, but could not stem the decline.

The Creation of the Republic. The final breakup of the Ottoman Empire came after World War I. In the war the Turks had been allied with the defeated Central Powers, chiefly Germany and Austria. As a result of the peace treaty (the Treaty of Sèvres), the empire lost all its remaining non-Turkish territory, including the Arab lands of the Middle East. The treaty also called for the occupation of Izmir and its surrounding region by Greece and for the neutralization of the straits. In effect, the land of the Turks was reduced to the hard central core of Anatolian highland. Constantinople was occupied by British and French troops, and Italian forces invaded southern Anatolia.

Opposition to the treaty was led by the Nationalists under Mustafa Kemal, an army general. Kemal rallied the Turkish Army, which eventually drove out the invading armies. The last Ottoman sultan was deposed and, in 1923, the Republic of Turkey was established. Kemal became its first president. The Treaty of Sèvres was replaced by the Treaty of Lausanne (1923), which established the present boundaries of Turkey.

The Reforms of Kemal Atatürk. The Turkey that Kemal took over was a backward state, afflicted with disease, ignorance (90 percent of the people were illiterate), and poverty. He immediately began a series of reforms that covered the entire spectrum of Turkish life. A crash program in sanitation and education was begun. Bright children barely out of primary school were installed as teachers. Land was taken from religious foundations and from large estates and distributed among the peasants. An ambitious program of industrialization was started.

Under the Ottoman system, the chief of state, the sultan, was also the religious head, or caliph. Kemal effected the separation of Church and State. Since the wearing of Oriental headgear, the turban and the fez, was associated with the old regime, Kemal had them replaced with Western-style hats. Women previously had been treated almost as slaves, and one man could have as many as four wives. Kemal outlawed this practice of polygyny, and women were given the same rights as men. Under Ottoman rule many of the laws of the land were based on the Koran, the centuries-old holy book of Islam. Kemal had the legal system reformed, based on the most progressive laws of various European countries.

The Ottoman Empire had used the Arabic alphabet. Kemal felt it was hard to learn. To encourage literacy and to bring his country closer to the West, he had the Roman alphabet adopted. As a political forum for the country, Kemal earlier had established the Grand National Assembly. Kemal's own political party was called the Republican People's Party. It was not until later years, however, that opposition parties came into existence.

Under the Ottoman Empire, the people had no family names. Kemal ordered all Turks to adopt family names. Kemal himself was offered the name Atatürk—"father of the Turks."

Turkey Since Atatürk. Kemal Atatürk died in 1938, a year before the start of World War II. Turkey was neutral during most of the war, but joined the Allies in 1945. After the war, when the old Russian threat reasserted itself, American president Harry S Truman offered Turkey aid under the Truman Doctrine. Later, Turkey became a member of the North Atlantic Treaty Organization (NATO). In recent years the Russian threat has subsided, and Turkey has developed friendlier relations with the Soviet Union.

Turkey went through a period of political crisis in the early 1960's. Inflation, corruption in the government, and political suppression led to the overthrow of the government of Premier Adnan Menderes by Turkish army officers. A military government was established under General Cemal Gürsel. Menderes was tried and executed. A new constitution was drawn up, and Gürsel became president, serving until his death in 1966. Since then, in spite of occasional political unrest, parliamentary elections have continued with orderly changes of government.

A different kind of crisis developed in 1974, over the island of Cyprus. (See the article CYPRUS in this volume.)

The Government. The Turkish parliament, the Grand National Assembly, consists of two houses, the National Assembly and the Senate. All members are elected by the people, except for 15 senators appointed by the president. The president, who is elected by the Grand National Assembly for a 7-year term, is the head of state. The president appoints the prime minister, who heads the Council of Ministers. The prime minister and Council of Ministers must have the support of the parliament.

EMIL LENGYEL, Fairleigh Dickinson University; author, *Turkey*

CYPRUS

Cyprus, the easternmost island in the Mediterranean Sea, is a small country with a turbulent history. Almost 80 percent of the approximately 630,000 inhabitants of Cyprus are Greek, many of whom have long hoped for *enosis,* or union with Greece. Yet the nearest part of Greece, the island of Rhodes, is about 300 miles (480 kilometers) away, and Greece has not controlled Cyprus since the 4th century B.C. Turkey lies only 40 miles (64 km.) away. About 18 percent of the people are Turkish. They have opposed *enosis* and have been engaged in a bitter conflict with the Greek Cypriotes. The situation has been aggravated by the involvement of Greece and Turkey in Cyprus. And so this beautiful island, which should be a haven of peace, has instead been torn by strife.

Metaxas Square in downtown Nicosia. The city of Nicosia is the capital of Cyprus and the cultural and economic center of the island.

Ruins of a castle high in the Kyrenia Mountains of northeastern Cyprus.

THE PEOPLE AND THEIR LAND

The Greek Cypriotes speak Greek and most of them belong to the Greek Orthodox Church, although their church administration is completely independent of the church in Greece, and they elect their own archbishop. The Turkish Cypriotes speak Turkish and are Muslims. Despite their separate traditions and loyalties, and despite recent conflicts, the Cypriotes, Greek or Turkish, lead similar lives.

The Cypriotes, a kind and friendly people, are devoted to their land and to their families. Their lives are not easy, but they take great pleasure in sitting on a café terrace and gossiping over a cup of strong, black coffee or a glass of the local sweet red wine. Everyone in Cyprus is intensely interested in news—whether it be local gossip or important political developments. The sleepy and relaxed pace on this Mediterranean island is gradually changing, but very slowly. Education is a serious matter and children spend long hours in school and often travel considerable distances, particularly to secondary schools.

Most of the people live on the Messaoria plain, a belt that lies between the two mountain ranges that cross the island from west to east. In the north is the Kyrenia range and in the southwest are the Troodos Mountains.

The climate in Cyprus is generally mild, with hot, dry summers, and some rain in the winter. In the mountains there is considerable snow, as well as rain, so that there is enough moisture to permit forests of pine, oak, and cypress to flourish.

Until recently the shortage of water in Cyprus was a serious problem. Springs were so valuable that a single spring might be owned by dozens of people—usually members of one family—and rights to water use were often sold. A household might purchase one hour from the owner of one spring, and another hour from another owner, in order to obtain enough water to live. Since 1960 an extensive dam-building program has helped solve the water problem of Cyprus.

Cyprus is mainly an agricultural country. About half of the people till the soil of their small farms for their livelihood, and a large percentage of the island's land is cultivated. Though much of the land is barren rock, there is great beauty in the cultivated areas, as well as many spots where wild flowers grow lavishly. Huge carob trees, groves of olive, lemon, orange, and grapefruit trees, and fields of barley fill the landscape. And in the spring, which comes early in this warm land, there are patches of magnificent beauty as cherry, almond, apricot, quince, and peach trees burst into bloom.

ECONOMY

In ancient times Cyprus was rich in metals. Its most important metal, copper, was as coveted a metal as uranium and plutonium are today. The Latin for copper (*cuprum*) was first used for the copper found

A fast-growing beach resort at Famagusta on the southeast coast.

FACTS AND FIGURES

REPUBLIC OF CYPRUS—Kypriaki Dimokratia (Greek), Kibris Cumhuriyeti (Turkish)—is the official name of the country.

CAPITAL: Nicosia.

LOCATION: Island in the eastern Mediterranean Sea. **Latitude**—34° 33′ N to 35° 41′ N. **Longitude**—32° 17′ E to 34° 35′ E.

AREA: 3,572 sq. mi. (9,251 sq. km.).

PHYSICAL FEATURES: Highest point—Troodos Mountain (Mount Olympus) (6,406 ft.; 1,953 m.). **Lowest point**—sea level. **Chief river**—Pedias.

POPULATION: 630,000 (estimate).

LANGUAGE: Greek and Turkish (official); English.

RELIGION: Greek Orthodox, Muslim.

GOVERNMENT: Republic. **Head of government**—president. **Legislature**—house of representatives. **International co-operation**—United Nations.

ECONOMY: Chief minerals—iron and copper pyrites, chromium, asbestos. **Chief agricultural products**—potatoes, citrus fruits, carobs, grapes, wheat, olives. **Industries and products**—minerals, wine, cement, canned fruit, lace, embroidery, pottery. **Chief exports**—wine and alcoholic beverages, minerals, potatoes, citrus fruits, grapes, carobs. **Chief imports**—transportation equipment, fuel, fabrics, agricultural equipment, meat and other food, machinery.

MONETARY UNIT: Cyprus pound.

NATIONAL HOLIDAY: October 1, Independence Day.

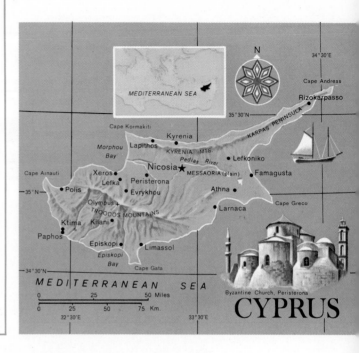

CYPRUS

Byzantine Church, Peristerona

in Cyprus. Now most of the metal deposits are found in the Troodos Mountains. Iron and copper pyrites and chrome are the leading mineral resources today. But with the exception of asbestos, they are fast dwindling and will probably be exhausted within the next decade.

While Cyprus has no heavy industry, there is some manufacturing. Cement is made and fruit is canned. Vineyards on the island supply the grapes for the local wine makers. Beer and other beverages are manufactured. There are also small home industries—lacemaking, pottery, embroidery, and buttonmaking. The Cypriotes have to import machinery, fuel, meat, fabrics, and medicines, and they find it difficult to pay for these imports by their exports of minerals, fruits and other agricultural products, and wine. They are hoping that tourists will provide the revenue to make up the trade deficit.

Cyprus has much to attract tourists. The climate is dry and healthful, with sunshine nearly all year round. There are many fine beaches all along the shoreline in the north and south of the island, and the scenery is often spectacular. Hotels have been built, beaches have been improved, and historic sites have been made more accessible. An entirely new resort area has been created on the coast near Paphos.

NICOSIA, THE CAPITAL, AND OTHER CITIES

Nicosia, the capital and largest city of Cyprus, has a population of more than 100,000 people. It is the economic, cultural, and political center of Cyprus and a modern oasis in the middle of the Messaoria plain, which has no other urban center. All the other important cities of Cyprus are on the coast.

Limassol and **Famagusta**, on the south and southeast coasts, are the second and third largest cities of Cyprus. Both have lovely natural harbors. Famagusta, a city filled with interesting historic sites, is thought to be the place chosen by Shakespeare as the setting for *Othello*. The town, whose history goes back to the 3rd century B.C., was the capital

A Barnabite monk working on an icon for a church.

of Cyprus under the Venetians, during the 15th and 16th centuries. It contains ruins of medieval fortifications and Venetian palaces. It is dominated by the Gothic Cathedral of St. Nicholas, which was begun in the 15th century and is now a mosque. During World War II Famagusta, which served as a British naval base, was heavily bombed.

HISTORY

The earliest history of Cyprus is lost, but excavations prove the existence of a Stone Age culture. The first Greeks appeared about 1500 B.C., and ever since, Cyprus has seemed a Greek land. The Phoenicians settled on Cyprus about 800 B.C., and subsequently the Cypriotes fell under the control of Assyria, Egypt, and Persia. Alexander the Great brought Greek rule back to Cyprus for 10 years, but on his death in 323 B.C. Cyprus became an Egyptian dependency. In 58 B.C. it became a Roman colony.

When the Roman Empire was divided into two parts, Cyprus became a part of the Eastern, or Byzantine, Empire. During this time the island was invaded many times by Muslims from the Middle East and North Africa. In 1191 Richard I of England (Richard the Lion-Hearted), on his way to the Holy Land as the leader of the Third Crusade, took Cyprus and granted it to Guy de Lusignan, a French nobleman. The Lusignans ruled for 300 years until the 15th century, when Venice gained control of Cyprus. The Venetians stayed for 82 years, and they were followed by the Turks, who remained 300 years. It was during this period that many Turks immigrated to Cyprus.

At the Congress of Berlin in 1878, Turkey transferred the administration of the island to Great Britain in return for British protection of Turkey. In 1914, when Turkey joined the war against Britain, the British formally annexed Cyprus as a part of the British Empire. During that war and during World War II, Cyprus was an important naval base.

MODERN CYPRUS

During much of the 20th century there has been a strong movement in Cyprus to join Greece. The majority of Cypriotes consider Greece their mother country because it gave them both their language and their religion. For years the Greek Orthodox Church in Cyprus has been close to the government and has always favored union with Greece, no matter who controlled the island. During the long occupation by the Turks the Greek language and even the local government were allowed to continue, and the Greek Orthodox archbishop of Cyprus became the unifying figure for the people. For centuries the Church thus preserved the continuity of Greek culture in Cyprus.

A Greek Cypriote underground movement, known as EOKA, started a guerrilla campaign against the British during the 1950's. It became a threat to stability in 1955, and the next 4 years were marked by increasingly serious acts of violence. The Turkish community, in opposing the Greek drive for *enosis,* also resorted to violence. Tension between Greece, Turkey, and Great Britain grew. In 1960 the British granted independence to Cyprus. Most of the people preferred to join Greece, but the treaties signed by Greece, the United Kingdom, Turkey, and leaders of the Greek and Turkish Cypriote communities forbade *enosis* and also forbade partition of Cyprus into Greek and Turkish sectors. The British retained control over their military bases.

A constitution provided for a Greek Cypriote president, elected by the Greek people, and a Turkish vice-president, elected by the Turkish citizens of Cyprus. Legislative and governmental offices were divided, with about 70 percent going to the Greeks and 30 percent to the Turks. Archbishop Makarios III was elected president.

In 1963 bitter fighting broke out between Greek and Turkish Cypriotes over proposed changes to the constitution. The United Nations brought about a cease-fire and was active in keeping the peace. However, the Turks withdrew from the government and set up their own administration. In 1967 there was another clash between the two communities, but after that relations between them became easier, until 1974.

In 1974 Greek Cypriote forces led by Greek army officers overthrew the government of Cyprus. Archbishop Makarios fled the country. Turkey, fearing that *enosis* was the reason behind the coup, sent its troops to the island. Turkish forces continued their military action until they controlled the northern part of the island. The new Greek Cypriote regime proved short-lived, and Makarios returned as president. The Turkish Cypriotes have demanded a government made up of a federation of Greek and Turkish communities, but the Greek Cypriote government has refused.

The latest crisis in Cyprus makes reconciliation between Greek and Turkish communities even more difficult. Until the problem of a government satisfactory to both sides is solved, the threat of further conflict hangs over this ancient land.

ALFRED PERLÈS, Poet and author

The Barada River flows past Martyrs' Square in Damascus.

SYRIA

Syria is an Arab land in the heart of the Middle East. Some 5,000 years ago, a great seafaring people called the Phoenicians established trading centers along the coast of present-day Syria and Lebanon. Since the days of the Phoenicians, trade has been an important part of Syrian history. As a result of its location on a traditionally important trading route—linking Asia with Europe and Africa—Syria has attracted traders and invaders from all four points of the compass. From the west came the armies of Egypt, and later the Greeks, the Romans, and the Crusaders. From the south came the Arabs, who made the most lasting impression on Syrian culture and history. In later centuries, fierce Mongol warriors and Turks invaded the area from the east and north. Conqueror followed conqueror, and Syria remained under foreign rule until 1946, when full independence was achieved and the modern nation of Syria was born.

THE LAND

The Syrian Arab Republic has an area of 71,500 square miles (185,180 square kilometers). Except for a small coastal area fronting the Mediterranean, Syria is surrounded on all sides by other lands—Turkey to the north, Iraq to the east, Jordan to the south, and Lebanon and Israel to the west. Syria is divided into four natural belts running parallel to the Mediterranean. First is the coastal plain, a series of sandy tracts and cultivated land that extends from the Mediterranean to the mountain country farther inland. Next come the western mountain ranges, which are forested and well-watered by numerous springs. The two principal ranges are the Alawite (or Nusairiyya) Mountains, which begin at the Turkish border and extend south, and the Anti-Lebanon, a mountain range that forms the western border with Lebanon. The third belt is the central depression, consisting mainly of a rugged, hilly plateau that trails off into the Syrian Desert. Lastly comes the northeastern mountain region called Al Jazirah, which includes a fertile valley area.

Most of Syria's fertile land is in the western part of the country or in the great river valleys. The Euphrates River and its tributaries make up Syria's most important river system and its longest waterway—flowing from Turkey through Syria and into Iraq. Syria's other principal river is the Orontes. Syria's climate is influenced both by the cool, rain-bearing Mediterranean Sea breezes and the hot, dry desert winds from the east. The Syrian summer is hot and long, lasting from May to September. How-

Beehive-shaped huts are typical of many Syrian villages.

ever, highly varied weather conditions occur in Syria, ranging from the extreme heat of the desert and semidesert areas to the cold winters in the higher mountain ranges. Along the coast the climate is generally mild. Rainfall averages about 30 inches (76 centimeters) annually in the west but is much lower in the east and almost nonexistent in the desert.

Economy. Syria has relatively little mineral wealth, and the principal natural resource is its fertile land. However, oil is extracted and refined in the northeastern part of the country, and there are deposits of gypsum, coal, lead, and copper. Farming is the main activity, and the chief crops are wheat, cotton, barley, sugar, and tobacco. Syria is noted for its fine fruits, particularly figs, grapes, melons, and apples. To provide more land for farming, a number of irrigation projects are under way. One of the most important is the dam on the Euphrates River, which is being financed partly by the Soviet Union. There are only a few industries in Syria, the most important being the weaving and embroidering of textiles. Soap, cement, and pottery are manufactured, and Syria's skilled handicraftsmen produce a wide range of goods made from metal, wood, and leather.

THE PEOPLE

Syria's population of about 6,000,000 is predominantly Arab. Among the minority groups, the Armenians, the Druzes, and the Kurds are the largest. As a result of the country's geographical situation and historical connections with other lands, Syria has received people of many racial and ethnic groups. Semitic peoples, particularly the Aramaeans and the Canaanites, drifted into Syria beginning around 3000 B.C. Later, the Arabs (both before and after the rise of Islam in the 7th century A.D.) moved into the area, spreading their language and customs. Other peoples who mixed with the population of Syria over the centuries included the Hittites, the Persians, the Greeks, and the Romans. Turkic people later came in several waves. Since the time of the Crusaders (during the 11th and 12th centuries A.D.), there has been a sprinkling of Europeans along the coastal areas.

Religion. Most Syrians are Muslims, the great majority of whom are members of the Sunni branch of Islam. There are also several smaller religious sects, including the Alawites and the Druzes. Christianity found its earliest followers in this part of the world, and today Syrian Christians representing 12 denominations number over 600,000. Syria's Jewish population has steadily declined in recent years and now is believed to be only a few thousand.

Language. For the most part, the people of Syria speak Arabic. Armenians generally speak only their own language, although many of them have a basic knowledge of Arabic. In some communities, especially along the northern border, Turkish or Kurdish is used along with Arabic. To the north of Damascus, there is a small group of villages whose people speak Syriac as well as Arabic. Syriac is still used in the rites of some Christian churches. French and English are the most widely spoken European languages.

Education. Free elementary and secondary education is provided by the state, and all children are required to attend school for a 5-year period. However, many children living in outlying villages or the desert encampments of the Bedouin nomads do not receive schooling. Syria's

FACTS AND FIGURES

SYRIAN ARAB REPUBLIC is the official name of the country.

CAPITAL: Damascus.

LOCATION: Southwest Asia. **Latitude**—32° 19′ N to 37° 20′ N. **Longitude**—35° 37′ E to 42° 22′ E.

AREA: 71,498 sq. mi. (185,180 sq. km.).

PHYSICAL FEATURES: Highest point—Mount Hermon (9,232 ft.; 2,814 m.). **Lowest point**—sea level. **Chief rivers**—Euphrates, Orontes, Yarmuk, Jordan.

POPULATION: 5,900,000 (estimate).

LANGUAGE: Arabic (official), Armenian, Kurdish, Turkish, Syriac (Aramaic), French, English.

RELIGION: Muslim, Christian, Druzes.

GOVERNMENT: Republic. **Head of state**—president. **Head of government**—prime minister. **International co-operation**—United Nations, Arab League.

CHIEF CITIES: Damascus, Aleppo, Homs, Hama, Latakia.

ECONOMY: Chief minerals—petroleum, gypsum, phosphates, lead, copper, coal. **Chief agricultural products**—wheat, barley, millet, cotton, grapes, olives, sugar beets, figs, apples, pears. **Industries and products**—textiles, cement, leather, sugar refining, handicrafts. **Chief exports**—raw cotton, textile goods, sugar, livestock, vegetables and fruit. **Chief imports**—base metals and manufactures, mineral fuels and oils, machinery and electrical materials, precious metals, chemicals and pharmaceuticals, vehicles.

MONETARY UNIT: Syrian pound.

NATIONAL HOLIDAY: April 17, Evacuation Day.

NATIONAL ANTHEM: *Al Nashid al-Suri* ("Protectors of the Nation").

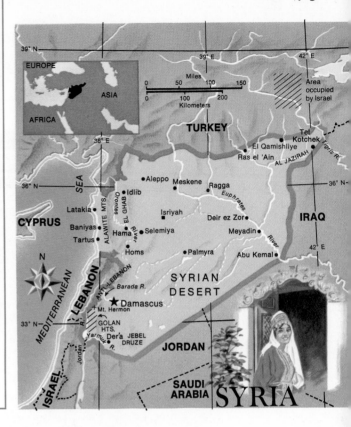

principal university is Damascus University, founded in 1924. Other institutions of higher learning are the Arab Academy in Damascus and the University of Aleppo. There are also a number of technical schools, teacher training colleges, and other specialized institutes.

Way of Life. Syria is mainly a nation of farmers and shepherds, and more than two thirds of its people work the land or breed livestock. Although Syria's cities are the country's centers of culture and commerce, most people live in small villages. A large number of Syrians are members of Bedouin tribes, which wander from one grazing spot to another with their flocks of sheep and goats and their horse and camel herds. The Syrian Government is attempting to settle the nomadic Bedouins in permanent villages.

Living conditions vary according to whether people are in towns, villages, or the Bedouin encampments. Bedouin nomads generally live under harsh conditions. When he is wandering, the Bedouin's home is a tent made of camel and goat hair, and therefore known as *bayt shar* ("house of hair"). His furniture consists of a few cushions and blankets. Mats made of reeds or handwoven rugs cover the tent floor, and a small kerosene lamp may be used to provide light. The traditional dress of the Bedouin consists of the *abah* (or *abayyah*), an ankle-length sleeveless cloak, and a white shirt reaching to the feet, which is worn under the cloak. The *kafiyyah*, a rectangular cloth headdress held in place by a ropelike band, is used to cover the head. Bedouins eat a simple fare of milk, dates, and boiled cereal.

In the villages, people live mainly in one-room houses made of stone or sunbaked bricks. Wealthier families have two-room houses with tile

Many Syrians wear a mixture of traditional Arab dress and Western clothing.

roofs and glass windows. Poorer homes have roofs made with mud and thatch and only wooden shutters on the windows. Water for drinking and cleaning usually is obtained from the village well. Only a few villages close to major towns have electricity. The average villager wears a mixture of Arab and Western-style dress.

In the larger towns, the life style is more westernized. Homes tend to be better furnished, Western dress is more common, and the diet is somewhat fuller. Traditional Arab foods include heavily spiced lamb and chicken dishes mixed with rice. All of the larger towns and cities have permanent marketplaces, or souks, which are important social as well as commercial areas. Here the village farmer and artisan come to sell their goods from open stalls. In smaller villages where there are no permanent bazaars, market days take place once or twice a week in the maidan, or village common. Colorful village festivals are held after the spring planting and before the harvest. The national holiday is Evacuation Day (April 17), which celebrates the departure of the French Army in 1946.

Cities. **Damascus**, one of the world's oldest cities (population about 600,000), is Syria's capital and chief urban center. Situated in an oasis watered by the Barada river, the city contains many ancient monuments, including the Great Mosque built during the Omayyad Arab dynasty and the tomb of Saladin, the Muslim leader who defeated the Crusaders. The city's world-famous craftsmen are especially noted for their fine woodwork and Damascus silk and satin goods.

Aleppo, the country's second most important city (population about 550,000), is the commercial and industrial heart of Syria. Other cities include Latakia, a major port on the Mediterranean Sea, and Homs and Hama in Syria's wheat and cotton belt. Another historically important town is Palmyra, which is a leading fig producer.

The ruins of Palmyra, said to have been founded by King Solomon.

HISTORY

Throughout history Syria has been a battleground, and its fertile valleys have supplied many invading armies. When the ancient Egyptians expanded their empire, they occupied the southern part of Syria (around 1600 B.C.). Their rivals and opponents, the Hittites, controlled the northern sector. In succeeding centuries, the Babylonians, Assyrians, and the Aramaeans fought over the area. By the end of the 11th century B.C., the Aramaeans ruled much of Syria. They established their capital at Damascus and built a large fortress at Aleppo. Syria derives its name from Syriac, the Aramaean dialect spoken around Damascus.

In the 6th century B.C., the Persians absorbed Syria into their vast empire. Syria next came under Greek rule, following the defeat of the Persians in the 4th century B.C. by Alexander the Great. By this time, Damascus was already a major city, noted both for its commerce and culture. Syria's scientists pioneered in the field of astronomy, and its engineers developed new irrigation methods. Under Roman rule, which began in the 1st century B.C., Syria prospered. But during the Byzantine period, after A.D. 330, the area fell into a period of decline.

Early in the 7th century A.D., Arab armies inspired by the prophet Mohammed (the founder of Islam) conquered Syria. The Arabs gave Syria its unifying national characteristics—the Arabic language and the Islamic religion. For the next few centuries Syria flourished. During the period of the Omayyad Arab dynasty (A.D. 661–750), Damascus became the capital of the Arab Islamic empire and a crossroads between Europe and Asia. In its marketplaces jewels, silk, spices, and perfumes from the East were offered for sale along with textiles, wines, and glassware from the cities of Southern Europe. The artisans of Damascus and Aleppo were master craftsmen who fashioned beautiful ornaments out of silver, gold, brass, and copper. Aleppo, with its imposing citadel, controlled both the military and trade routes connecting northern Iraq with the Mediterranean coast and the central plateau of Anatolia (now Turkey). Syrian merchants ranged far and wide with their caravans.

After the 10th century, the Islamic empire began to fall apart as a result of quarreling among the rival dynasties. By then Syria had lost its position of importance. Once again the area became a battleground for invading Seljuk Turks, Crusaders, and Mongols. In 1401, while Syria was under the Mameluke sultans of Egypt, the Mongols under Tamerlane laid waste to the land and sacked Damascus and Aleppo.

In 1516 the Ottoman Turks captured Aleppo. For the next 4 centuries Syria was part of the Ottoman Empire. Under the harsh rule of the Turks, Syria became a poor and backward nation. Villages were abandoned, most of the cities decayed, and the population of the country dwindled. During World War I (1914–18), the Arabs rebelled against Turkish rule, and at the end of the war the Ottoman Empire was broken up. For a short period an independent kingdom was established in Syria. But in 1920, under a League of Nations mandate, France was given the right to occupy and govern Syria until the country was considered ready for independence.

Throughout the centuries of foreign domination, the Syrian people had continuously struggled for independence. During the period of French rule Syrian nationalists redoubled their efforts, and there were several uprisings against the French. Meanwhile, general conditions in the country improved as the French built roads and schools, modernized the cities, and established industries. During World War II (1939–45), the Free French Government of General Charles de Gaulle granted a measure of independence to Syria. However, the French continued to maintain considerable influence over the government through the presence of their troops. Finally, in 1946, the last French troops left, and Syria became completely sovereign. Except for a period between 1958 and 1961, when Syria joined with Egypt to form the United Arab Republic, the country has remained independent. Syria, along with other Arab nations, has fought in three wars against Israel—in 1948–49, in 1967, and in 1973.

GOVERNMENT

Syria was proclaimed a republic while under French mandate, and this form of government was continued after independence. Parliamentary government was more or less in effect until the early 1960's. But since 1949, Syria has been beset by internal political troubles. Military takeovers have led to a series of dictatorships. In 1963, the Arab Socialist Resurrection Party (Baath Party) seized power and subsequently established the National Council of Revolution to rule the country. Syria thus became a socialist state ruled by a single party. Economic dislocation brought about by land reform and nationalization of industry, plus a continuing struggle for power by rival factions, kept Syria economically and politically unstable. Recently, however, efforts have been made to stabilize the country and to make it more democratic. At present the most important political office is that of the president. He is assisted by a prime minister and cabinet. In 1971 Syria joined Egypt and Libya in a loose union called the Federation of Arab Republics.

NICOLA A. ZIADEH, The American University of Beirut

LEBANON

Although it is one of the smallest nations in the world, Lebanon has been the scene of important events since ancient times. Carved on a cliff at the mouth of Dog River north of Beirut are the names of 18 great men of history who passed through the country to conquer or be conquered. Among the names are those of Ramses II of Egypt, Nebuchadnezzar of Babylon, Alexander the Great, and the Roman emperor Caracalla. The 19th and final inscription, carved in 1946, commemorates the independence of Lebanon.

Under the Phoenicians, Lebanon's early inhabitants, the country held a unique position as a trading and seafaring nation. Modern Lebanon also gained a unique position because of its commerce, its generally high standards of living and education, and its role in world affairs. But a civil war, which erupted in 1975, divided the country, caused great destruction and loss of life, and endangered the peace of the region.

THE LAND

Lebanon stretches north and south along the eastern shore of the Mediterranean Sea. Unlike most of its neighbors in the Middle East, Lebanon has no desert. Parallel mountain ranges run the length of the country. The western range, the Lebanon mountains, rises close to the sea, leaving only a narrow coastal plain. The Lebanon mountains reach a height of over 10,000 feet (3,000 meters). The eastern range, the Anti-Lebanon, is almost as high. Between them is the Bekaa, a fertile plateau.

Modern buildings overlook ancient Beirut's rocky Mediterranean coast.

Lebanon enjoys great diversity of landscape. Bays break the coast-line, and at some points the mountains reach the sea. In the north, on Mount al-Mukammal, a single grove remains of the majestic ancient cedars of Lebanon, from which the Phoenicians built their galleys and which King Solomon used in building his Temple in the 10th century B.C. But the cedar remains the symbol of Lebanon, appearing on the flag and on coins and stamps.

At Baalbek the enormous columns of Roman ruins rise against the sky. Baalbek today is the scene of an annual festival of music, dance, and theater. Along the coast mighty Crusader castles recall the knights who once made their homes there.

The two largest cities, Beirut, the capital (population about 700,000), and Tripoli (about 128,000), are both seaports. Zahle, the third largest city, lies in the Bekaa.

Economy.   Lebanon has few natural resources. But by the time of the civil war, the energy and industry of its people had given it a generally high standard of living. Trade and tourism, along with insur-ance, real estate, and banking firms, brought in most of the national income. Agriculture was important, too, especially in the fertile Bekaa. From the air it looks like a vast carpet of varying shades of green, each shade representing a different crop—wheat, barley, corn, alfalfa, and potatoes. In recent years fruit orchards and chicken and dairy farming have been introduced to the area. Because of the scarcity of flat farming land, farmers have patiently terraced the steep hillsides and lower foot-hills, and planted them with fig trees and grapevines. The lower hills, valleys, and coastal plain are dotted with olive trees, which supply the

Farming and livestock raising provide a fifth of Lebanon's national income.

FACTS AND FIGURES

REPUBLIC OF LEBANON—Al-Jumhuriyah Al-Lubnan-iyah—is the official name of the country.

CAPITAL: Beirut.

LOCATION: Southwest Asia. **Latitude**—33° 05′ N to 34° 45′ N. **Longitude**—34° 58′ E to 36° 55′ E.

AREA: 4,015 sq. mi. (10,400 sq. km.).

PHYSICAL FEATURES: Highest point—Qurnat al-Sawda (10,131 ft.; 3,088 m.). **Lowest point**—sea level. **Chief rivers**—Litani (al-Litani), Orontes (al-'Asi).

POPULATION: 2,500,000 (estimate).

LANGUAGE: Arabic (official), French, English.

RELIGION: Christian, Muslim.

GOVERNMENT: Republic. **Head of state**—president. **Head of government**—prime minister. **Legislature**—Chamber of Deputies. **International co-operation**—United Nations, Arab League.

CHIEF CITIES: Beirut, Tripoli, Zahle, Saida (Sidon).

ECONOMY: Chief agricultural products—citrus fruits, apples, barley, peas, potatoes, grapes, olives, bananas, wheat. **Industries and products**—cement, olive oil, wine, oil refining, tourism. **Chief exports**—vegetables, fruits. **Chief imports**—fuel, cereals, metals, machinery, vehicles, iron and steel products.

MONETARY UNIT: Lebanese pound.

NATIONAL HOLIDAY: November 22, Independence Day.

NATIONAL ANTHEM: *Kullu na lil watan lil 'ula lil 'alam.* ("All of us for the country, glory, flag.")

farmers with olives and olive oil, staples of their diet. Also on the coastal plain, orange, lemon, and banana trees and date palms abound.

Industry, a growing part of the economy, includes the making of olive oil, wine, textiles, and cement. At Tripoli and Saida (Sidon) there are refineries for oil brought by pipeline from Iraq and Saudi Arabia.

THE PEOPLE

Peoples from every Middle Eastern country and group are found in Lebanon. Recent immigrants include Palestinian refugees and Syrians. The national language is Arabic, but French and English are widely used. Within the two main religious groups, Muslims and Christians, there are many recognized communities. Most Muslims belong to either the Sunni or the Shi'a (Shi'ite) branches of Islam. There are also many Druzes, whose religion is a mixture of Islam and other faiths. Most Christians belong to the Maronite Church, one of the Eastern Catholic churches. Most of the others belong to the Orthodox Eastern Church. Today somewhat less than half the people are Christians.

Everyday foods in Lebanon include lamb, chicken, fish, rice, cracked wheat, olives, a great variety of fruits, and round, flat loaves of bread. *Kibbeh,* which is made of cracked wheat and meat or fish pounded together and served cooked or raw, is considered the national dish.

Beirut. Before Lebanon's capital was devastated by civil war, it was a bustling city. People from the Lebanese mountains and plains crowded its sidewalks, along with business tycoons, Beirut intellectuals, oil-rich sheikhs from other Arab countries, and tourists.

Before the civil war Beirut was one of the busiest cities in the Middle East.

From Beirut's airport, the main stopover between Europe and Asia, a wide coastal boulevard winds between beaches and ultramodern luxury hotels and apartment houses to downtown Beirut, where the narrow ancient streets were once crammed with cars from every country.

Beirut had become a financial center of the Middle East. Many American and European business firms had offices there. To help transact business, a large number of local banks operated alongside foreign ones. The cultural side of the country is represented by art academies, museums, the National Library, and universities. The Lebanese University and the Arab University are the newest. The University of St. Joseph was founded in 1875 by French Jesuits, and the American University of Beirut was founded by missionaries in 1866.

Life in the Villages.   The mountains are a different world. Depending on the location of springs, villages perch on mountain ridges or nestle in narrow valleys. Each village has its own identity, based on religion, size, and distance from Beirut. Constructed of local limestone, the traditional village houses are small and flat-roofed. Some newer houses are larger and have pitched red tile roofs and arched casement windows.

Before the war, people from Baghdad to Cairo spent their summer vacations in the cool mountain villages. Many communities had modern villas, hotels, nightclubs, swimming pools, and television. On higher slopes there were ski resorts. Western dress could be seen in villages as well as in the capital. The more isolated Druze and Muslim settlements, however, kept many of the traditions and customs of the past.

In spite of the individuality of the villages, the rugged, ruddy-faced mountaineers have traits in common. They lead simple lives and work hard. They also share a deep love for the land their forefathers patiently cultivated and terraced before them. Owning their own land, instead of working as tenant farmers, adds to the villagers' pride and fierce loyalty to land and village. Along with independence of spirit, the people also keep alive the traditions of friendliness and hospitality.

Education. A high literacy rate indicates a deep respect for education. All children must have at least 5 years of primary schooling, in either a government school or a private school. There are both government and private schools at the secondary as well as the university level.

HISTORY AND GOVERNMENT

The earliest civilization in Lebanon was that of the Phoenicians, who developed the first alphabetical system of writing. In the golden days of the Phoenicians (from the 12th to the 9th century B.C.), great city-states grew up—Arwad, Byblos (from which the word "Bible" comes), Sidon, and Tyre. From these city-states the Phoenicians sailed westward in their double-deck ships, carrying products to trade—fruits, glass, jewelry, and purple-dyed wool and linen. The purple dye was extracted from a shellfish, the murex, found along the coast.

The centuries-old port of Byblos was one of the great Phoenician city-states.

The ruins of a temple built by the Romans still stand at Baalbek. The Roman occupation of Lebanon began in 64 B.C. and lasted several centuries.

According to legend, disappointed lovers leaped from Beirut's Pigeon Rock.

Later a succession of powers, the Assyrians, the Persians, and the Greeks under Alexander the Great, dominated the Phoenician city-states. Contact with the West entered a new phase with the Roman occupation, which began in 64 B.C. Under the Roman Empire the country enjoyed several centuries of peace and prosperity. By the 4th century Christianity had emerged as the prevailing religion. In the 7th century Muslims brought the new religion of Islam from the Arabian Peninsula.

Lebanon became part of the Ottoman (Turkish) Empire in 1516 and remained under Muslim rule for almost 400 years. But with its own capable princes, Lebanon managed its political and economic affairs independently. Trouble between the Christians and Druzes led to a civil war in 1860. But European powers intervened and guaranteed that Lebanon would be self-governing. A Christian governor-general was appointed by the Ottoman Government.

Independence.   When strife continued, Lebanese Christians began to emigrate to Brazil, the United States, and other countries. After World War I, Ottoman rule over Lebanon ended, and it became a French mandate under the League of Nations. In 1925 Lebanon became a constitutional republic, although France kept some powers. Those powers were ended by constitutional amendments in 1943, and Lebanon became completely independent on November 22 of that year. By unwritten agreement, the president of the country was to be a Maronite Christian, the prime minister a Sunni Muslim, and the speaker of the legislature a Shi'a Muslim.

Independent Lebanon faced a problem. The Arabic language and the Muslim population drew the country toward the larger Muslim world. The Christians—influenced partly by Christians who had emigrated— tended to look westward. The Muslims feared being controlled by the West. The Christians feared control by neighboring Arab countries. When a revolution seemed near in 1958, Lebanon's president asked the United States for help, and marines were sent. But peace was restored without military force.

Civil War.   Lebanon tried to maintain a moderate Arab position in the Arab-Israeli wars in 1956, 1967, and 1973. This policy was opposed by Lebanese radicals, mainly Muslims, and by Palestinian refugees in Lebanon. From their bases in Lebanon, both of these groups attacked villages in Israel. When Israeli forces counterattacked into Lebanon, the Lebanese Government refrained from sending its small army against the counterattackers. As a result, tension grew between the Palestinian radical groups and the Lebanese Government. Tensions also were increasing between Muslims, including the Palestinian guerrillas, and Christians. Because they were once the majority, the Christians had more representatives in the legislature. The Muslims, now in the majority, demanded a greater share of political power. Full-scale civil war broke out in 1975, with terrible destruction and loss of life. In 1976 Syria sent troops in an effort to separate the warring parties, and an uneasy peace was restored. After a Palestinian attack on Israel from a Lebanese base, Israeli troops invaded southern Lebanon in 1978. Thousands were made homeless as Israeli troops advanced to the Litani River. The troops withdrew after the United Nations sent in a peacekeeping force, and the Lebanese again began rebuilding their war-torn country.

VIOLA H. WINDER, Author, *The Land and People of Lebanon*

The Shrine of the Book in Jerusalem houses the Dead Sea Scrolls.

ISRAEL

The State of Israel, which was established in 1948, is one of the smaller nations of the world. But despite its size, Israel's first 25 years were filled with great accomplishment in the face of many difficulties. During this period the country more than tripled its population, increased its agricultural and industrial production by several times, and absorbed hundreds of thousands of new immigrants—all while fighting four wars against hostile Arab neighbors.

Although modern Israel has had a short life, the history of its people goes back to Biblical times. The ancient Hebrews were the forefathers of the Jews who established the new nation of Israel in 1948. When the Hebrews were driven from Palestine by the Romans in the 1st and 2nd centuries A.D., they were scattered to the far corners of Europe and North Africa. Nevertheless, these exiled Jews (the Diaspora) kept their close identity with the Holy Land. Throughout the centuries, Jewish customs, traditions, religious beliefs, practices, and prayers reflected the ties of Judaism with Palestine. During the Middle Ages and in modern times when Jews were persecuted in Europe, often it was hope of returning to Palestine that kept alive the fires of their faith.

During the 19th century, Jewish emotional ties with Palestine were channeled into a political movement, Zionism. The Zionist movement gained many supporters due to Jewish persecution by the Russian czars at the end of the century. After the rise of Hitler in Germany during the 1930's and the spread of Jewish persecution throughout Europe, Zionist demands for a Jewish state became even stronger.

Between World War I and World War II, Palestine was governed by Great Britain under a mandate from the League of Nations. After World War II, when clashes between Jews, Arabs, and the British threatened to plunge Palestine into civil war, Great Britain requested the United Nations to intervene. In November, 1947, the United Nations recommended division of Palestine into a Jewish state, an Arab state, and an international zone, which was to include Jerusalem. While the Jewish population accepted the partition recommendation, the Arabs fought against its imposition. The result was the first Arab-Israeli War. During this struggle, on May 14, 1948, Israel proclaimed its independence.

THE LAND

The area of Israel, within the frontiers established by the armistice agreements signed with the Arab states in 1949, is almost 8,000 square miles (20,700 square kilometers). In the south lies the Negev. The Negev is largely desert, uninhabited except for the towns of Beersheba (sometimes called the capital of the Negev) at its northern base and the Red Sea port of Eilat (Elath) at the southern tip and a few small mining towns and collective settlements. However, the Negev provides Israel with its greatest mineral potential as well as its trade routes to East Africa and the Far East. The regions north of the Negev include the heavily populated and agriculturally rich Mediterranean coastal plain and the more sparsely settled northern hill region of Galilee.

In addition to the Mediterranean Sea, which forms its western coastline, Israel borders on two large bodies of water. They are the Sea of Galilee (also known as Lake Tiberias and Lake Kinneret), a lake of great historic interest, and the Dead Sea. The Dead Sea is a salt lake noted for its mineral wealth and lack of animal life and for being the lowest point on the earth's surface—about 1,300 feet (400 meters) below sea level. The principal rivers of Israel are the Jordan, the Kishon, and the Yarkon. There are no high mountain peaks in Israel. However, a chain of mountainous hills, with an average height of 2,000 feet (600 m.), stretches from Lebanon southward to Sinai.

Areas under Israel's control were greatly expanded after the Six Day War in June, 1967. As a result of its victory, Israel occupied the Syrian Golan Heights, the Jordanian West Bank, and the Egyptian-held Gaza Strip and Sinai Peninsula. Occupation of these areas increased by over four times the total land under Israel's control.

Climate. Israel's climate is typically Mediterranean. Summers are hot and sunny; winters are chilly and damp. Temperatures range from a low of 40 degrees Fahrenheit (4 degrees Celsius) in the mountainous regions during January, to over 100 degrees F. (38 degrees C.) in the southern Negev during the hot month of August. Snow is not uncommon in the hills of Galilee and there have also been winter snows in Jerusalem, the coastal region, and northern Negev.

Annual rainfall, concentrated in the winter and early spring, varies from year to year, at times bringing floods. During other years there have been severe droughts. At Eilat, the driest measuring point, there was only

The arid region around the Dead Sea.

Sabra, a cactus native to Israel, is also the name for native-born Israelis.

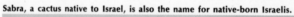

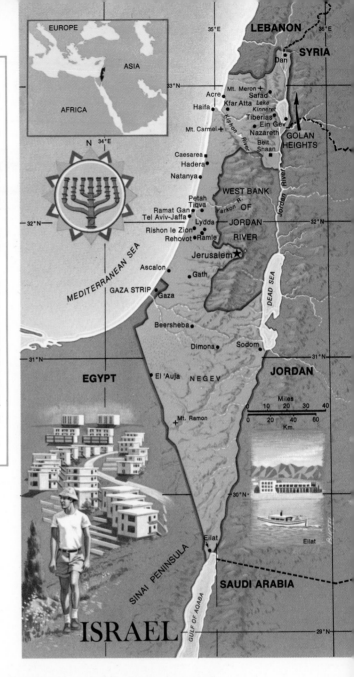

FACTS AND FIGURES

STATE OF ISRAEL—Medinat Yisrael—is the official name of the country.

CAPITAL: Jerusalem.

LOCATION: Southwest Asia. **Latitude**—29° 29' N to 33° 17' N. **Longitude**—34° 11' E to 35° 26' E.

AREA: 7,992 sq. mi. (20,700 sq. km.).

PHYSICAL FEATURES: Highest point—Mount Meron (3,963 ft.; 1,208 m.). **Lowest point**—Dead Sea, about 1,300 ft. (400 m.) below sea level. **Chief rivers**—Jordan, Kishon, Yarkon. **Major lakes**—Sea of Galilee (Lake Kinneret), Dead Sea.

POPULATION: 3,000,000 (estimate).

LANGUAGE: Hebrew and Arabic (both official); English, French, and various other languages are also spoken.

RELIGION: Jewish, Christian, Muslim.

GOVERNMENT: Republic. **Head of state**—president. **Head of government**—prime minister. **Legislature**—Knesset. **International co-operation**—United Nations.

CHIEF CITIES: Tel Aviv-Jaffa, Jerusalem, Haifa.

ECONOMY: Chief minerals—potash, phosphates, petroleum, bromine, salt, copper. **Chief agricultural products**—citrus fruits, sugar beets, cotton, dairy products. **Industries and products**—food processing, diamond polishing, chemicals, textiles, cement, ceramics, machinery, gasoline, paper, plastics, precision instruments, tobacco. **Chief exports**—citrus fruits, polished diamonds, textiles. **Chief imports**—machinery, diamonds, grains, iron and steel products.

MONETARY UNIT: Israeli pound.

NATIONAL HOLIDAY: April/May, Independence Day.

NATIONAL ANTHEM: *Hatikvah* ("The Hope").

½ inch (1.27 centimeters) of rain in a recent year, while at the point of heaviest rainfall, in Galilee, a little over 48 inches (122 cm.) was recorded.

What the Land Produces. Although much of Israel is sandy or covered with only a thin layer of soil, when irrigated the land can be quite productive. Crops produced are typical of the Mediterranean. "A land of wheat and barley and vines and fig trees and pomegranates, a land of olive oil and honey," says the Bible. While Israel does produce these items, Israeli farmers using modern methods of irrigation have greatly increased the range of crops. Citrus fruits are the country's most valuable agricultural commodity and one of its largest exports. In recent years a fairly large European market has been developed for cut flowers. Cotton has also been introduced as an industrial crop to supply the country's textile industry. Dairy products are plentiful, although there are shortages

A kibbutz, or collective settlement.

of animal fodder (food) and wheat grains, which must be imported. Thus, while providing most of Israel's food, agricultural production falls short of making Israel completely self-sufficient.

THE PEOPLE

The growth of Israel as a nation has come about through a melting-pot process. Its approximately 3,000,000 people have come from some 70 countries, representing all the inhabited continents of the world. Of various backgrounds, cultures, and customs, and speaking over 20 different languages, these people have all played a part in the development of modern Israel. A little over 40 percent of the country's Jews were born in Israel; almost one third immigrated from Europe and America; the rest came from Asian and North African countries. There are about 400,000 Israeli Arabs, plus another 1,000,000 Arabs in the areas occupied during June, 1967.

Over 80 percent of the people live in urbanized areas. About 3 percent live in collective settlements (kibbutzim) and another 14 percent live in various other types of rural settlements such as co-operatives and farm villages. One percent of the population are Arab Bedouin—nomads and seminomads who raise sheep, goats, and camels. A third of the Jewish population of Israel live in the three large cities of Tel Aviv-Jaffa, Haifa, and Jerusalem. The majority of the rest of the people live or work in or near these three main urban centers.

Tel Aviv-Jaffa.   The largest of Israel's cities, with a population of about 400,000, Tel Aviv-Jaffa lies on the Mediterranean coast. Jaffa is a seaport more than 4,000 years old. According to the Bible, Jonah was

One of the many outdoor cafés in Tel Aviv, Israel's largest city.

The city of Haifa. The Mediterranean Sea can be seen in the distance.

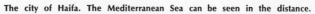

sailing from Jaffa (Joppa) when he was swallowed by a great fish. Tel Aviv is a relatively new city, founded in 1909 as a suburb of Jaffa. As large numbers of immigrants settled there, a sprawling city grew up on the sand dunes. Today Tel Aviv is the center of Israel's commerce, light industry, and entertainment. It has beaches, modern hotels, shops, cafés, theaters, concert halls, and museums.

Haifa. One of the most beautiful cities in Israel, Haifa climbs from its harbor on the Mediterranean up the green slopes of Mount Carmel. Although Haifa's history goes back at least to the 3rd century, the modern city developed after Jewish settlers started to arrive from Europe in the late 19th century. Today Haifa has over 200,000 people and is Israel's busiest port and major industrial center. The campus of the Israel Institute of Technology (Technion) and the impressive Shrine of the Báb—world center of the Bahai faith—are on Mount Carmel.

Jerusalem. An ancient city and a modern capital, Jerusalem lies high in the Judean hills. In the old sections of the city are holy places of Judaism, Christianity, and Islam. On nearby hillsides a modern city, including government buildings and the Hebrew University, has grown up. (A separate article on JERUSALEM appears in this volume.)

The Economy

Although living in a continual state of war, Israel's people have developed a relatively high standard of living. Just over 12 percent of the people are employed directly on the land, in farming or other agricultural pursuits. Many of the rest work in such activities as industry and crafts, commerce and banking, and education and social services.

Israeli industry includes food processing, diamond polishing, and chemical, textile, cement, ceramic, and machinery production. It is a mixed economy: privately owned plants employ about 76 percent of the labor force; state-owned industries and those owned by Histadrut (the General Federation of Labor) employ the balance. During the past 20 years, Israel's development has been greatly assisted by foreign aid, mostly from the United States.

Israel has over 200 collective settlements called kibbutzim (singular, kibbutz), which comprise only 3 percent of the population but have been in the vanguard of agricultural and social development. Members of the kibbutzim, which range in size from 60 to 2,000 people, share equally the profits of their labor. In most of these collectives, members eat in a common dining room, share their living facilities, and have little private property. Children are cared for and raised in collective nurseries and schools, but they have close and warm ties with their parents, whom they see frequently.

Education

Primary education is free and compulsory for children between 5 and 16. The number of students in Israeli schools has increased nearly six times since 1948–49. The number of colleges and universities has more than doubled, and every large urban area has one or more institutions of higher education. The largest are the Hebrew University in Jerusalem and Tel Aviv University.

Academic subjects resemble closely those studied in American or European schools and colleges. The chief language of instruction is He-

Part of the campus of Tel Aviv University.

brew, although English, French, Arabic, and other languages are also taught. The country's more than 70,000 Arab pupils, taught mostly in government schools, use Arabic as their language of instruction. Because of language and curriculum differences, most Arab children attend separate schools, although there are some Arab students in the Hebrew University.

Culture and the Arts

Since Israel was first established as a fulfillment of the centuries-old dream of a Jewish return to Palestine and rebirth of a Jewish state, great emphasis has been placed upon the revival and development of Hebrew language and culture. The overwhelming majority of Jews use Hebrew, one of the two official languages (the other is Arabic), as their chief means of communication, although many speak various other languages. A rich Hebrew literature and drama has developed, and there are a number of motion-picture companies producing Hebrew-language films.

The Government

Israel has a parliamentary government. Legislative authority is vested in the Knesset, or parliament, whose 120 members are elected for a period of 4 years. The president, who is elected by the Knesset, is the head of state. The prime minister heads the Cabinet, or executive branch of the government, which is responsible to the Knesset.

The Israeli Way of Life

Israeli society is the most democratic in the Middle East. Its press is free except for censorship of military operations; there are strong opposition political parties and freely expressed public criticism of the government. There is a strong feeling for social and political equality, and the gap between rich and poor is less than in most countries.

Israelis lead a varied life. They dress very much like Europeans or Americans, but with emphasis on informal styles. At one time open-necked khaki shirts and shorts were nearly a uniform for men. Now the shorts have been replaced by full-length trousers, although the open-necked shirt is still common, even among men attending formal functions. On the other hand, some Israeli women have become very style conscious. Nearly every newspaper runs a fashion page with pictures of clothes manufactured in Israel that can compete with the most fashionable New York and Paris styles. The average woman, however, dresses quite practically, as required by her work at home, on the farm, in the office, or in a factory.

Israel is not noted for its cuisine. The early socialist philosophy of its leaders de-emphasized luxuries, and until recently it was difficult to find an outstanding restaurant in the country. An average family has some kind of boiled or roasted meat perhaps daily, although poorer Israelis eat meat less frequently. On holidays chicken is a common dish. Generally there is an ample supply of fresh fruits and vegetables, and salads are eaten even at breakfast. One of the most popular dishes is eggplant, which is cooked in dozens of different ways.

On holidays thousands of people take to the road, either individually or in organized tours, to visit the many beautiful natural settings and ancient archeological sites that abound in the country. It is said that every Israeli is an amateur archeologist, as evidenced by the deep attachment to the past among people of all ages. Many, young and old, make a habit of collecting shards (bits of ancient pottery) and other relics found scattered throughout the land.

Although Israel is a Jewish state, probably less than half the Jewish population is religious in an orthodox, or strict, way. Nevertheless, the Jewish character of the country is expressed in the observance of Saturday as the weekly day of rest and worship, and in the numerous holidays that commemorate the main events in the 4,000-year-long history of the Jewish people. The two most important religious holidays are the New Year (Rosh Hashanah) and the Day of Atonement (Yom Kippur).

Approximately three quarters of the Arab population live in rural areas, and they lead a life much the same as farmers everywhere. They are tied closely to the land and their life is regulated by the seasons of the year. City life for Arabs is much like the city life of the Jewish population.

HISTORY

Although Israel was established in 1948, the Zionist movement from which it developed was founded at Basel, Switzerland, in 1897, by Theodor Herzl, a Jew born in Budapest, Hungary. The return to Palestine had been a central point in Jewish religion and thinking ever since the dispersal of the Jews by the Romans and the destruction of the Temple in Jerusalem, and throughout the centuries some Jews returned to live in the Holy Land. Herzl and his followers believed that a national home, or state, was necessary to save the Jews from further persecution such as they had suffered in the Christian countries of Europe for nearly 2,000 years. The Zionist movement, named because of its followers' aspiration to establish a Jewish home in Zion (Palestine), grew in numbers until, by the outbreak of World War I in 1914, it had great influence in many Western countries.

At this time there were about 85,000 Jews (a little less than 20 percent of the population) in Palestine, which was then part of the Ottoman

The Western Wall in Jerusalem is all that remains of the Second Temple, which was destroyed by the Romans in A.D. 70. It is especially holy to Jews.

(Turkish) Empire. The majority of the population were Muslim and Christian Arabs.

Palestine and Great Britain. During World War I, Great Britain, which was fighting the Turks, attempted to rally worldwide Jewish support for its cause. It organized special Jewish units to fight in Palestine and issued the Balfour Declaration. This was a letter from Arthur James Balfour (later Lord Balfour), the British Foreign Secretary, to Lord Rothschild. The letter stated that the British Government would assist in the establishment "of a national home for the Jewish people" in Palestine, provided it would not prejudice "civil and religious rights of existing non-Jewish communities ... or the ... status enjoyed by Jews" in other countries.

After Allied victory and the collapse of the Ottoman Empire at the end of World War I, Palestine became a British mandate. Terms of British control over Palestine included implementing the Balfour Declaration. The British governors of the country, however, soon became involved in the growing conflict between Zionist aspirations and demands of the Arabs for independence. Attempts to find a compromise or to bridge the gap between the two groups failed. During the interval between the wars, the Arabs waged a civil war against the British and against Zionist settlements in the country.

Despite opposition by Arab nationalists the Jewish community in Palestine grew rapidly, especially after the arrival of over 200,000 refugees from Hitler's persecution in the 1930's. Their number increased until by 1946 Jews were one third (over 600,000) of the total population. Schools, farms, hospitals, factories, and all the necessary institutions and establishments for an independent government were developed.

During World War II violence in Palestine had come to a halt, but it flared up again at the end of the war. When it became known that some 6,000,000 Jews (one third of the world's total) had been murdered in concentration camps by the Nazis, Zionist demands for independence became stronger. After several attacks against the British by Jewish underground groups, the British Government presented the Palestine problem to the United Nations, which recommended partition.

Arab nationalists, insisting on self-determination, refused to accept partition, and again war raged through the country. On May 14, 1948, as the last British troops were about to leave, the Jewish leaders of Palestine declared their independence and established the State of Israel. Within hours, armies from the neighboring Arab nations joined the Palestinian Arabs in a war against the new Jewish state. In the fighting, Jordan (then known as Transjordan) annexed the west bank of the Jordan River and the Old City of Jerusalem.

Israel Since Independence. After the Arab armies were defeated, Egypt, Lebanon, Syria, and Jordan signed armistice agreements with Israel, bringing hostilities to a halt in 1949. But the Arab states have declared their continuing hostility, refusing to negotiate a peace settlement with Israel. One of the obstacles to peace has been the situation of the Arabs who became refugees during the fighting. The Arab countries have insisted that they be allowed to return to Israel. The Israelis contend that the refugee problem must be part of an over-all peace plan.

In 1956 a second war broke out after Egypt announced nationalization of the Suez Canal, refused Israel passage through the canal, and blockaded the Straits of Tiran. In the 1956 war Israel was joined by Great Britain and France. Although Egypt was defeated, the Israeli, French, and British troops were ordered by the United Nations to withdraw from the occupied territory. Egypt continued to refuse Israel passage through the Suez Canal, although passage through the Straits of Tiran continued, with United Nations troops stationed at the straits. A United Nations peace-keeping force also was stationed in the Sinai.

In 1967 war broke out once more. At Egyptian insistence the United Nations forces had been withdrawn; the blockade of the Straits of Tiran was re-imposed; and Egyptian soldiers and tanks moved into the Sinai following the departure of the United Nations troops. Reacting to this and to continued threats, Israel attacked Egypt, again invading the Gaza Strip and Sinai. Egypt was immediately joined by Jordan and Syria. Within 6 days all three countries were defeated and large parts of their territory were occupied. The fourth war between Israel and the Arabs came in 1973 after attacks by Egyptian and Syrian forces. It saw heavy fighting at the Suez Canal and in the Golan Heights of Syria.

The United Nations Security Council called upon Israel to withdraw from the occupied areas and on both sides to end "all claims or states of belligerency" and to respect and acknowledge the "sovereignty, territorial integrity and political independence of every State in the area and their right to live in peace within secure and recognized boundaries free from threats or acts of force." An Israeli withdrawal from some territories was negotiated after the 1973 war, but Israel still insists on over-all peace talks with the Arabs before withdrawing from others.

DON PERETZ, Director, Southwest Asia and North Africa Studies Program
State University of New York at Binghamton

The Dome of the Rock, one of Jerusalem's many ancient holy places.

JERUSALEM

Jerusalem—a city 17 times destroyed and 17 times reborn, according to one historian—has been a holy site to Jews, Christians, and Muslims throughout the history of their faiths. Today Jerusalem is the capital of the state of Israel. The city lies about 2,500 feet (750 meters) above sea level in the Judean hills. During the summer the days are warm and the skies are cloudless. In January, the coldest month, the temperature sometimes drops to freezing. The city receives most of its rainfall during the winter, and occasionally there is snow.

The Ancient Holy Places. Jerusalem consists of an ancient section (the Old City), whose history goes back some 4,000 years, and of a modern section that has grown up on nearby hillsides. The old fortress city lies within crenelated brownstone walls built in the 16th century by Suleiman

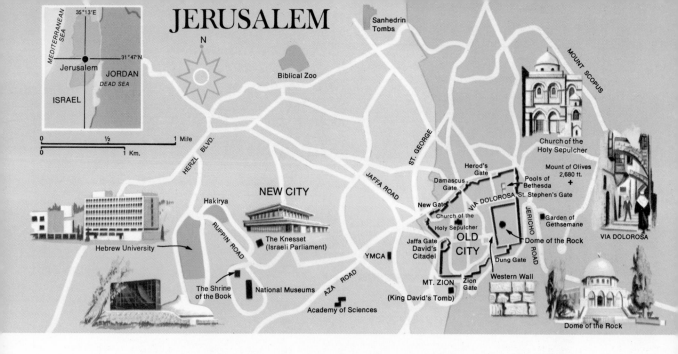

the Magnificent. A number of gates placed at intervals in the wall lead into old Jerusalem. Its streets, most of them too narrow for cars, are thronged with pilgrims who come from all over the world to visit their ancient holy places. The most revered site of Judaism is the Western Wall, or Wailing Wall, the last remnant of the ancient Second Temple. For Christians the holiest site is the Church of the Holy Sepulcher, built about 335 by the Emperor Constantine, to mark the place of Jesus' crucifixion, burial, and resurrection. Leading to the church is the Via Dolorosa, or Way of Sorrows, the path along which Jesus is said to have carried the cross. Muslims consider Jerusalem the third holiest site in Islam, after Mecca and Medina. It is believed that the Prophet Mohammed ascended on a visit to heaven from a rock close to the Western Wall. The rock is believed to be the same one on which Abraham offered his son Isaac as a sacrifice. Today a gold-domed mosque, the Dome of the Rock, shelters the spot. On Mount Zion, just south of the Old City walls, are the reputed Tomb of King David and the Upper Room, where according to tradition the Last Supper was held.

The Modern City.   Since the end of the 19th century, and more recently, after the declaration of Israel's independence in 1948 and the selection of Jerusalem as the capital in 1950, thousands of people have poured into the city to live and work. It has modern housing, shopping centers, schools, hotels, and government buildings. Three impressive groups of buildings occupy neighboring hilltops in the western part of the new city. On one hill are the Knesset (parliament), completed in 1966, and a government center called Hakirya ("the city"). To the north is the campus of Hebrew University, with the large National Library in the center. To the west is the Israel Museum, which includes an archeological museum; a sculpture garden; the Bezalel Museum, an art museum; and the Shrine of the Book, where the Dead Sea Scrolls are housed.

There is also a modern Arab city containing government buildings, schools, and a commercial center. This city developed after World War I and has expanded rapidly since 1948.

History.   The city of Jerusalem is mentioned scores of times in the

Old Testament. King David captured the city of Jerusalem, then called Jebus, from the Jebusites about 1000 B.C. and made it the capital of the Kingdom of Israel. It was there that his son King Solomon built the First Temple. The Babylonians overran Jerusalem in 586 B.C., destroying the Temple and driving the Jews into exile. The Jews returned when the city came under the rule of the Persians, and in 516 B.C. they completed the Second Temple on the site of Solomon's Temple. In succeeding centuries Greeks, Egyptians, and Seleucids conquered the city. In 167 B.C. the Jews, led by Judah Maccabee, started a revolt that finally ended in the liberation of Jerusalem. It remained Jewish until the Roman conquest in 63 B.C. It was during the period of Roman domination that Jesus of Nazareth began to preach. His message, found subversive by the Roman rulers, led to his crucifixion in a period of revolt against the Roman garrisons. In A.D. 70 Roman forces poured into the country, put down the revolt, destroyed the Second Temple, and killed or took captive thousands of Jews.

From the early part of the 4th century, when the Roman emperor Constantine gave his support to Christianity, until early in the 7th century, Jerusalem was a Christian city. In 638 Arab armies captured the region, and Jerusalem came under Muslim rule. The city became the principal goal of the Crusaders, who captured it in 1099. Battles raged between Muslims and Christians until 1244, when Jerusalem once again fell to the Muslims. The city remained under their authority through the Ottoman period (1517–1917). At the end of World War I, Jerusalem became the capital of the British Mandate of Palestine.

Left: A view of Jerusalem. Right: Jews praying at the Western Wall (the Wailing Wall), the holiest site of Judaism.

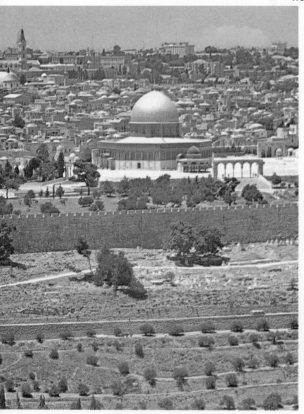

The Via Dolorosa during the Easter holiday. The participants are retracing Christ's steps on the way to Calvary.

The period of the mandate was marked by clashes between Arab and Jewish nationalist groups. Each group wanted Jerusalem as the capital in an Arab or a Jewish state. At the end of World War II the British called upon the United Nations to deal with the problem. In November, 1947, the United Nations General Assembly recommended that Palestine be partitioned into an Arab and a Jewish state, and that Jerusalem and its surrounding areas, including Bethlehem, be established as a separate international city controlled by the United Nations. The recommendations were never carried out because of the war that broke out in 1948 between Israel and the Arab states after Israel proclaimed its independence.

The war ended with an armistice agreement and the city divided between Israel and Jordan. The city remained divided by barbed wire, armed outposts, and no-man's-lands until the Six Day War of June, 1967, when the part of Jerusalem held by Jordan was captured by the Army of Israel. Although the United Nations voted against the incorporation into Israel of Arab sectors of Jerusalem, the Israeli Government declared the whole city unified within the Israeli capital. Today its 200,000 Jews and 70,000 Arab residents are no longer separated by physical barriers, but many Arab residents of the city oppose absorption into Israel.

DON PERETZ, Director of Southwest Asia-North Africa Program
State University of New York at Binghamton; author, *The Middle East Today*

JORDAN

Jordan is both an ancient land and a young country. It became an independent kingdom in 1946. Before that, the area was known as Transjordan—the land across the Jordan River. Over the centuries various nomadic peoples have wandered across the rolling semidesert east of the Jordan. In this way the region became populated by Semitic people from the Arabian Peninsula. It was in Transjordan that the Hebrew people first settled on their way to Canaan (later Palestine), after Moses led them out of Egypt. Today the Hashemite Kingdom of Jordan is one of the Arab states of the Fertile Crescent, a narrow stretch of land that runs in an arc from the Mediterranean Sea in the west to the Persian Gulf in the southeast.

THE LAND

The total land area of Jordan, including the West Bank of the Jordan River now occupied by Israel, is about 38,000 square miles (98,000 square kilometers). The country is divided into two main regions separated by the Jordan River, which flows into the Dead Sea. West of the Jordan is a fertile zone of hills and valleys. Here are located the Biblical cities of Jerusalem, Bethlehem, and Jericho. To the east of the river is a fertile plateau, which gradually drops off into rocky wasteland. More than four fifths of the eastern region is desert.

Jordan's climate is similar to that of other eastern Mediterranean lands. Winters are rainy and relatively cold, particularly in the highlands, while the other seasons are extremely hot and dry. In the Jordan Valley,

Bedouin Arab women draw water from oasis well. The Bedouins roam the desert searching for pastureland to feed their sheep, goats, and camels.

summer temperatures rise to 120 degrees Fahrenheit (49 degrees Celsius). Rainfall in the eastern plains is approximately 8 inches (20 centimeters) a year. This makes the land fertile enough for desert nomads to graze their flocks. In the western portion of the country, annual rainfall is about 15 to 25 inches (38–64 cm.), and farming is possible.

Economy.   Jordan's resources are limited, and the economy is based on farming, cattle raising, light industry, and tourism. The country's farmers grow wheat, barley, lentils, and a variety of fruits and vegetables, including figs, citrus fruits, grapes, tomatoes, cucumbers, and potatoes. Cattle, goats, sheep, and camels are raised mainly in the western part of the country. Industry is on a very small scale. There are some food-processing plants and a few factories manufacturing tobacco products, cement, and soap. Although deposits of several minerals are known to exist, only phosphates are mined. Since Jordan is not economically self-sufficient, foreign aid is needed. Financial and technical assistance comes from both Arab and Western sources. After the June, 1967, war with Israel, the West Bank —Jordan's most productive region—came under Israeli control. Jordan's economy, including its important tourist trade, suffered greatly.

THE PEOPLE

Nearly all of Jordan's approximately 2,000,000 people are Arabs. About half of these people, including more than 500,000 Arab refugees, are Palestinian in origin. The refugees, who left Palestine after the Arab-Israeli wars of 1948 and 1967, live in special camps built and supported by the United Nations. In addition to Arabs, there are about 14,000 Circassians, who immigrated to Jordan from the Russian Caucasus in the 19th

Arab street vendor displays his wares in ancient city of Bethlehem.

FACTS AND FIGURES

HASHEMITE KINGDOM OF JORDAN—Al-Mamlaka al-Urduniya al-Hashemiyah—is the official name of the country.

CAPITAL: Amman.

LOCATION: Southwest Asia. **Latitude**—29° 17' N to 33° 23' N. **Longitude**—34° 53' E to 39° 18' E.

AREA: 37,738 sq. mi. (97,740 sq. km.).

PHYSICAL FEATURES: Highest point—5,755 ft. (1,754 m.). **Lowest point**—about 1,300 ft. (400 m.) below sea level. **Chief river**—Jordan.

POPULATION: 2,145,000 (estimate).

LANGUAGE: Arabic.

RELIGION: Islam.

GOVERNMENT: Constitutional monarchy. **Head of state**—king. **Head of government**—prime minister. **Legislature**—National Assembly. **International co-operation**—United Nations, League of Arab States.

CHIEF CITIES: Amman, El Zerqa, Jerusalem, Nablus, Irbid, Hebron, Jericho, Bethlehem.

ECONOMY: Chief minerals—phosphates. **Chief agricultural products**—wheat, barley, citrus fruits, lentils, tomatoes, cucumbers, grapes, livestock (sheep, goats, cattle, camels). **Industries and products**—tourism, phosphates, fishing, textiles, soap, cement, tobacco and cigarettes, handicrafts. **Chief exports**—phosphates, vegetables and plants, fruits, tobacco, cement, stone, fertilizers, olive oil, cigarettes, hides and skins, wool and animal hair. **Chief imports**—motor vehicles, bicycles, boiler and mechanical apparatus, electrical machinery, sugar and confectionery, mineral fuel and oil.

MONETARY UNIT: Jordanian dinar.

NATIONAL HOLIDAY: May 25, Independence Day.

NATIONAL ANTHEM: Al-Nashied al Malaki ("The Royal Anthem").

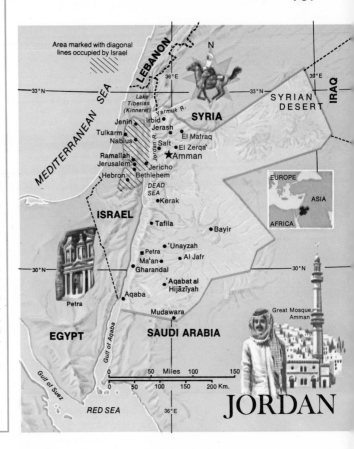

century. More than 90 percent of the people are Muslims. There are also about 150,000 Christians. Arabic is the official language of the country, and the people speak with an accent similar to that of Syrians and Lebanese. Because of past ties with Great Britain, many educated people, including businessmen and government officials, also speak English.

Way of Life. Despite differences among city-dwellers, villagers, and nomadic Bedouin tribesmen, Jordanians are united by their religion, language, and historic traditions. Certain traditional values and customs, including hospitality, personal honor, and loyalty to kin, are still important. In rural areas family ties are stronger than feelings of nationalism, and it is common for several generations of one family—along with cousins, aunts, and uncles—to live together. Under this arrangement, each immediate family of parents and children occupies a separate room in a house, or one of several small houses grouped together in a particular section of a village.

About half of Jordan's people live in small farming villages. Their homes are flat-roofed buildings made of stone or baked-earth bricks. One room in the house is used for livestock and farm tools. Most villages have one or more small squares (sahah), where open markets and social events are held. Large apartment houses have been built in the bigger towns, but the majority of city people still live in small family dwellings. Bedouin tribesmen, who move from one place to another, live in tents, which they carry with them on their camels and donkeys.

Jordanian dress varies from Western-style clothes in the cities to traditional Arab costume in the villages and the desert encampments of the

Bedouin nomads. Arab male dress consists of a black or brown cape called an *abayyah,* and the *kafiyyah,* a folded cloth headdress. Loose, flowing white and black robes are worn by the desert-roaming Bedouins. Most Jordanian women wear colorful shawls, gaily embroidered jackets, and long skirts. Many men in the cities wear the traditional *kafiyyah* along with Western-style suits.

The food eaten by Jordanians is similar to that of other Arab lands. Appetizers are made from mashed chick-peas or eggplant mixed with sesame oil, lemon juice, and spices. Main courses often consist of lamb or chicken combined with squash, eggplant, or okra and served with rice and pine nuts. A flat, round bread is eaten with these dishes. The favorite desserts are baklava and *katayif,* pastries made with nuts and syrup.

Education. Jordanian children are required to attend school for a minimum of 9 years. There are presently over 2,000 primary, intermediate, and secondary schools, including some 400 Muslim private schools. Total enrollment in these schools is approximately 450,000, and there are over 13,000 teachers. An additional 70,000 refugee children are being taught in schools supervised by the United Nations Relief and Works Agency (UNRWA) for Palestinian Refugees in the Near East. Jordan has about 30 institutions of higher learning, including teachers colleges, technical schools, and the University of Jordan, founded in 1962.

Cities. Amman, the capital of Jordan, is the country's largest city (population about 350,000). It is situated in a valley and extends onto the surrounding hills. Amman is built on the site of the ancient Greco-Roman city of Philadelphia. After World War I, Amman became the capital of the emirate of Transjordan and developed into a major urban center. Jordan's only seaport is Aqaba, which is located at the southern tip of the country, along the 10-mile (16 kilometers) strip of coast fronting the Gulf of Aqaba.

Re Faisal, a main thoroughfare of Amman, Jordan's capital and largest city.

Other cities include Nablus (population about 63,000), El Zerqa (population about 120,000), and Irbid (population over 60,000).

Several ancient cities, famous for their historic and religious importance, are located in the western region of Jordan now occupied by Israel. They include the Old City of Jerusalem, and Jericho, Hebron, and Bethlehem. The Old City of Jerusalem, still enclosed by walls built by the Turks in the 16th century, contains many holy places. The most renowned are the Church of the Holy Sepulcher, on the traditional site of the crucifixion of Jesus Christ; the Western (or Wailing) Wall, containing part of the original wall of the Second Temple of the ancient Hebrews; and the Dome of the Rock, where Muslims believe the prophet Mohammed ascended to heaven on a white horse. (A separate article on JERUSALEM appears in this volume.) In nearby Bethlehem is the Church of the Nativity, which according to tradition is built on the spot where Christ was born.

Government. Jordan is a constitutional monarchy. The king is the head of state, and he exercises his authority through a prime minister and a cabinet called the Council of Ministers. The National Assembly is the lawmaking body. It consists of a senate, whose members are appointed by the king, and a house of representatives elected by popular vote. Only males over the age of 18 can vote in Jordan. Local government is conducted by eight provincial governors appointed by the king.

HISTORY

The history of the area now known as Jordan began about 9,000 years ago, when a tribe of hunters established a permanent farming settlement —the world's oldest known community—on the site of what later became the city of Jericho. In Biblical times the area of modern Jordan contained the kingdoms of Moab, Edom, Ammon, and Gilead. After the 10th century B.C., the western part of the region was included in the Hebrew kingdom of Judah. Later an important trade route passed through the area, connecting western Arabia with the eastern Mediterranean ports. Along this route stood the famous city of Petra, whose buildings were carved into the surrounding red sandstone cliffs. The ruins of this once-flourishing caravan town can still be seen.

For many centuries, the Jordan region passed from one kingdom and empire to another. Both the Greeks (in the 4th century B.C.) and the Romans (from the 1st to the 4th century A.D.) ruled over this land. In the 7th century A.D., the area was conquered by the Arab followers of the prophet Mohammed and became a part of the Muslim empire. The Arab invaders introduced Islam to the people of the region, and although Jordan was later dominated by the Ottoman Turks for 4 centuries (1516–1918), it remained an Arab land.

The Ottoman Empire was broken up at the end of World War I (1918), and the League of Nations made Transjordan—the eastern part of present-day Jordan—a mandate (protected territory) of Great Britain. In 1923 Great Britain established the semi-independent emirate (principality) of Transjordan under the rule of Emir Abdullah ibn Hussein—a member of the Hashemite family, which claims descent from the prophet Mohammed. Jordan thus became a separate country for the first time in its history.

Transjordan achieved full independence in 1946, and the name of the country was changed to the Hashemite Kingdom of Jordan. In 1948, shortly after the United Nations partitioned Palestine into an Arab state

El Khaznah temple, one of the ruins of Petra, an ancient caravan town. Rose-colored buildings were carved out of surrounding sandstone cliffs.

and a Jewish state, Jordan's Army joined with other Arab armies in the first Arab-Israeli War (1948–49). When a truce between Israel and the Arab states was signed in 1949, Jordanian troops controlled the West Bank of the Jordan River (formerly part of eastern Palestine). This area was annexed by Jordan in 1950.

King Abdullah was assassinated in 1951. After a brief reign by Abdullah's son Talal, King Hussein I (Talal's son)—the present monarch—came to the throne. The young King, who assumed full power at the age of 18, has had to face many problems. Israel's occupation of the West Bank of the Jordan River—after the 1967 Arab-Israeli War—resulted in severe economic losses. The additional burdens imposed by thousands of new refugees, continuing trouble with Israel, and the emergence of strong Palestinian nationalism have all weighed heavily on the monarchy. In 1970 Jordan's Army put down a Palestinian guerrilla uprising. Under pressure from other Arab countries, however, Hussein has since agreed to recognize the Palestine Liberation Organization, the main guerrilla group, as sole spokesman for Palestinian rights in the West Bank.

ALAN R. TAYLOR, The American University, Washington, D.C.
Reviewed by YASSIN EL-AYOUTY, St. John's University
Fellow, the Middle East Association for North America

SAUDI ARABIA

The hot sun blazes down out of a clear blue sky. Sand dunes stretch as far as the eye can see. On the horizon a mirage looms deceptively. A group of black tents can be seen in the distance with a few palm trees swaying nearby. And a lonely caravan of Bedouins and their camels appears moving along a desert trail.

This image of Saudi Arabia, reflected in countless books and motion pictures, is the one most familiar to Westerners. In a sense it is true. But it is so narrow a picture that it becomes misleading. Saudi Arabia is a vast country—in area it is almost four times the size of France—and although much of the land is desert, there are also coastal plains, high mountains, and large cities. It is a rich country, its wealth derived from the great reservoirs of oil lying beneath the surface. Historically it is unique in having been the birthplace of Mohammed, the prophet of Islam, a faith that spread beyond Saudi Arabia to much of Asia and Africa.

THE LAND

Saudi Arabia's area is about 830,000 square miles (2,150,000 square kilometers). It occupies most of the Arabian Peninsula, which it shares with the small Arab states along the Persian (or Arabian) Gulf in the east and with Yemen (Sana), Yemen (Aden), and Oman in the south. It is a diversified land. In the west, stretching for about 1,000 miles (1,600 kilometers) along the Red Sea, is a narrow coastal plain, the Tihama.

A Bedouin style coffee break.

Rising boldly from this plain are high mountains varying from 4,000 to 12,000 feet (1,200–3,650 meters) in elevation. These mountains slope gradually toward the east, forming the central plateau, which extends to the coastal region along the Persian Gulf.

The western regions of Saudi Arabia comprise the Hejaz and Asir; the central region is called Nejd (or Najd); and the eastern region is known as the Eastern Province (formerly al Hasa). In the south lies a forbidding desert known as the Rub' al Khali (the "empty quarter"), which covers about one third of the country. The Rub' al Khali is a sea of sand and ridges, which may receive no rain for years. In the north is the Nafud, the second great desert of Saudi Arabia. These two deserts are connected by a belt of sand ridges called the Dahna.

It is estimated that about 15 percent of Saudi Arabia can eventually be farmed. However, only about 1 percent of this is now under cultivation, including the chain of oases that dot the central plateau. Cultivation is by irrigation, the water drawn mainly from wells. Saudi Arabia has no permanent rivers or lakes. Its wadis, or desert valleys, are formed by temporary streams that appear after infrequent but sudden rains.

Climate.   In the summer, except at high altitudes, the days are hot and dry, with temperatures soaring to well over 100 degrees Fahrenheit (38 degrees Celsius). At night the temperatures drop sharply. Winter temperatures are less extreme. Along the coastal plains the climate is milder but more humid. Rainfall, which comes during the winter months, averages only 6 to 7 inches (15–18 centimeters) a year and sometimes takes the form of rainstorms.

Terrace farms near Abha are green islands of fertility in a harsh land.

FACTS AND FIGURES

THE KINGDOM OF SAUDI ARABIA—al-Mamlaka al-'Arabiya as-Sa'udiya—is the official name.

CAPITAL: Riyadh.

LOCATION: Arabian Peninsula in southwest Asia. **Latitude**—16° 11' N to 32° 09' N. **Longitude**—34° 34' E to 55° 41' E.

AREA: 829,997 sq. mi. (2,149,690 sq. km.).

PHYSICAL FEATURES: Highest point—Mount Razih (11,999 ft.; 3,657 m.). **Lowest point**—sea level. **Chief mountains**—Asir highlands, Hejaz highlands. **Chief deserts**—Nafud, Rub' al Khali.

POPULATION: 7,200,000 (estimate).

LANGUAGE: Arabic.

RELIGION: Muslim.

GOVERNMENT: Monarchy. **Head of state**—king. **International co-operation**—United Nations, Arab League.

CHIEF CITIES: Riyadh, Jidda, Mecca.

ECONOMY: Chief minerals—oil, gold, silver, gypsum, rock salt, iron ore. **Chief agricultural products**—dates, fruits, vegetables, grains, livestock (camels, goats, sheep). **Industries and products**—oil and oil products, building materials, fertilizers, plastics. **Chief exports**—oil and oil products, dates. **Chief imports**—foodstuffs, vehicles, machinery and appliances, textiles and clothing, building materials.

MONETARY UNIT: Riyal.

NATIONAL HOLIDAY: September 23, Unification Day.

NATIONAL ANTHEM: *Al-Nasheed al Watani* ("National Anthem").

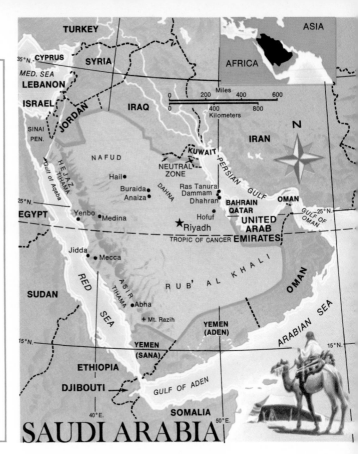

SAUDI ARABIA

THE PEOPLE

The vast majority of Saudi Arabia's more than 7,000,000 people are Arabs. Almost all are Muslims of the Sunni branch of Islam and belong to the strictly orthodox Wahhabi sect. Islam is the state religion and plays an important role in the everyday life of the people. Arabic is the official language of the country.

Way of Life. The people of Saudi Arabia may be divided according to their way of life into two main groups: those living in settled communities and those who are nomads or seminomads. The first live either in farming villages in the oases or in towns or cities. The Arabian farmer grows crops not only for his own and his family's immediate use but also to trade and sell. Dates, vegetables, and grains are the chief crops. The farmer's nomad counterpart, the Bedouin, moves constantly to find grazing land for his camels, goats, and sheep and local markets for his products. The Bedouin's chief means of transportation is the camel, and his camels, goats, and sheep provide him with milk, wool, meat, and leather. The nomad's home is his tent, which is usually made of camel or goat hair and furnished with rugs and pillows. Today more and more Saudi Arabians are becoming industrial workers, chiefly in the oil industry.

Traditional male Saudi Arabian dress consists of an undergarment reaching to the ankles called a *thobe* over which is worn an *aba,* or long gown, and a head covering called a *ghutra,* which is held in place by an *agal.* Women wear long, loosely fitted robes and are usually veiled in accordance with a widespread Muslim practice. In the cities women often wear Western-style dresses in their homes.

Education. Saudi Arabia has two modern universities, one at Riyadh and the other at Jidda. An Islamic university at Medina offers courses in Islamic studies. Scholarships to these institutions are available.

Below the university level, there is free public education. The government is establishing schools so quickly that every 3½ days a new free public school opens. Today about 80 percent of all Saudi boys of primary school age are enrolled in school, and about 70 percent of the girls.

Before the coming of modern public education in the 1930's, children were taught in the *kuttab,* a village school run by students of the Koran (the holy book of Islam). Here pupils learned to read and memorize the Koran and mastered some writing and arithmetic as well. In a few remote villages, the *kuttab* is still the only school.

Cities. The capital of Saudi Arabia and its largest city is **Riyadh**, which has a population of about 300,000. It lies in the east central part of the Nejd plateau, Saudi Arabia's heartland. Riyadh was once a fortress city of fawn-colored, windowless mud houses. Today it is growing fast and becoming more modern thanks to the discovery of large reservoirs of underground water. The walls have been pulled down and many old quarters replaced by modern boulevards, parks, and buildings.

Mecca, the holy city of Islam and the birthplace of Mohammed, lies about 40 miles (64 km.) inland from the Red Sea. Mecca's population of 250,000 swells greatly during the Hajj, the annual Muslim pilgrimage. The pilgrim's goal is the Kaaba, Islam's most holy shrine, a small granite building enclosed within the Great Mosque. Muslims believe the Kaaba was built by the prophet Abraham and his son Ishmael. The sacred Black Stone inside the Kaaba is said to bear Abraham's footprint.

Medina, Islam's second holiest city, with a population of about 60,000, lies about 200 miles (320 km.) to the northeast. It is the site of the tomb of Mohammed and the place where he lived during the last 10 years of his life.

Mecca's Red Sea gateway is **Jidda** (population 300,000), Saudi Arabia's most cosmopolitan city and chief port. Jidda is enjoying an industrial boom and is modernizing. But in many districts the old multistory stucco houses, built in a distinctive Arab style, still stand.

Across the Arabian Peninsula, on the Persian Gulf coast, are the "oil cities" of Dammam and Dhahran. Dammam is the capital of the Eastern Province and the port from which Arabian oil is shipped all over the world. Dhahran is the headquarters of ARAMCO (Arabian American Oil Company).

ECONOMY

Before oil was struck near Dammam in 1938, Saudi Arabia was a poor country. Its economy was based mainly on agriculture and animal herding. Its main source of income was a tax levied on religious pilgrims. Though taxes on pilgrims have now been abolished, the Hajj is still the second largest source of income. The pilgrims spend several million dollars in Saudi Arabia annually on lodging, transportation, and purchases.

Today the economy is dominated by the oil industry, which provides about 85 percent of the country's income. Saudi Arabia is the world's fourth largest oil producer. Exports other than oil are dates, skins and hides, and wool. Imports include foodstuffs, textiles, heavy machinery, automobiles, iron, and steel. The government considers its chief tasks to

The ancient fortress town of Riyadh has become a modern capital city.

be the modernization of the country and the diversification of the economy to lessen its dependency on oil.

The government is surveying the peninsula's mineral resources. The major ones are gold, silver, gypsum, rock salt, iron ore, copper, sulfur, and phosphates. Among new industries are fertilizers, plastics, chemicals, paints, and dyes. Cement and building materials are produced, and the building industry is the country's second in importance.

The cube-shaped Kaaba within Mecca's Great Mosque is Islam's holiest shrine.

In the agricultural sector, the arable land near the oases is being expanded with improved methods of water storage. The government has built storage dams and reservoirs to entrap rainwater in the wadis.

HISTORY

Semitic peoples have lived in the Arabian Peninsula for at least 3,000 years. Though some strong kingdoms have flourished, the peoples of the desert lived more typically in nomadic tribes organized on feudal lines. Wars were common, which made loyalty to the tribe important.

About A.D. 570 the harsh desert world produced one of the great religious leaders of all time, Mohammed. Mohammed received "calls" from God to preach a new religion. Rejected by the leaders in Mecca, his home, Mohammed fled to Medina in A.D. 622. This flight, known as the Hegira, marks the first year of the Muslim calendar. When Mohammed later returned to Mecca, it became Islam's capital and most holy city.

A few years after Mohammed's death in 632, the capital of Islam was moved out of the Arabian Peninsula, first to Damascus in Syria and then to Baghdad in Iraq. With the rise of the Ottoman Turks in the 16th century, the coastal plains of Arabia acknowledged Turkish sovereignty. But the interior remained the domain of feuding tribal sheikhs.

In the 18th century, Mohammed ibn-Abd-al-Wahhab established Wahhabism, a movement designed to purify Islam. Soon Wahhabism spread over the peninsula. The Ottoman sultan, fearing its influence, sent Egyptian armies to stamp out the movement. Though the Wahhabis were defeated in 1818, the movement, led by the Saudi family, continued. In 1902 Abdul-Aziz ibn-Saud, founder of the present royal family, recaptured the family's capital, Riyadh, from the Rashidis, his rivals. Later he drove the Turks from the eastern areas, defeated the Hejaz and Asir, and by 1926 controlled over three fourths of the peninsula. These areas were unified by 1932, and the name "Kingdom of Saudi Arabia" was given to the new country. Ibn-Saud was proclaimed king. Upon his death in 1953 he was succeeded by his eldest son, Saud. Saud abdicated in favor of his brother Faisal, who became king in 1964. Following the assassination of King Faisal in 1975, the throne passed to Crown Prince Khalid.

Government. Saudi Arabia is an absolute monarchy. The king, in his capacity as president of the Council of Ministers, exercises executive and legislative power. Financial matters and international agreements are voted upon by the Council before receiving the final approval of the king. There is a Consultative Council, the Majlis-al-Shura, which advises the monarch. Shari'a, the Law of Islam, is the law of the land.

International Relations. As an Arab country, Saudi Arabia's foreign policy is geared to the needs and defense of the Arab countries. However, Saudi Arabia is politically more conservative than such countries as Egypt, Algeria, Syria, and Iraq.

With the withdrawal of British forces from the Persian Gulf, Saudi Arabia has been playing a more important role in that area. Saudi Arabia exerts great influence in the Muslim world. The annual pilgrimage, which brings hundreds of thousands of Muslims from more than 90 lands, offers an ideal occasion for Saudi diplomacy.

YASSIN EL-AYOUTY, St. John's University
Reviewed by ABDUL HAMEED AL-GAREE, Acting Consul General
Royal Consulate General of Saudi Arabia, New York

Sections of Sana, the capital city, date from the 16th century.

YEMEN (SANA)

In all the Arabian Peninsula Yemen is the country with the greenest hills and most fertile land. Yemen lies in an area the ancient Romans called Arabia Felix ("fertile" or "fortunate" Arabia).

But how fortunate a land has it been? For centuries its people have been divided—one tribe against another, one branch of Islam against another. For nearly 1,000 years they were under the harsh rule of their kings, the imams, and more than once Yemen was conquered by another country. The last foreign power to rule was Turkey. When its rule ended after World War I, the imams had total power again. The imam made the laws, meted out punishments, and was the spiritual head of a sect comprising half the country's Muslims. But in 1962 a great change shook Yemen. An army opposed to the king attacked the royal palace. Imam Mohammed al-Badr, 67th of his line, fled for his life, and the revolutionaries declared the country a republic.

The People

Yemen is fairly thickly populated. About 5,000,000 people live in an area of 75,290 square miles (195,000 square kilometers). The Yemenis are Arabs. Their language is Arabic, and Islam has been their religion since the 7th century. For centuries there were many thousands of Jews in Yemen, but most of them have settled in Israel.

Until recently the Yemenis had almost none of the advantages of life in the 20th century. In the whole country there were only a few hospitals. There were religious schools, but no modern system of public education. Gradually, though, this picture has started to change. With aid from other countries Yemen is building hospitals and infirmaries. Several Arab coun-

A crowded market in Umran.

tries are helping to build schools and to provide teachers for them. A road linking major cities was finished in 1965.

Most of the people in Yemen make their living by farming or keeping livestock—sheep, goats, cattle, and camels. The government is trying to help by making more land available for farming, by teaching modern farming methods, by supplying better seed, and by breeding better livestock. There is fishing along the Red Sea coast.

Industry is growing, but so far there are only a few factories in the country. A modern textile factory makes cloth from cotton grown in Yemen, and a cement factory uses limestone found in the country. The Yemenis are famous for handicrafts, especially leatherwork and sword-making. Salt is mined, and explorations for oil are under way.

Families in the hinterland have always been grouped into tribes, and a Yemeni feels great loyalty to his tribe and his chief. Yemenis are also known for being independent and ready to fight if they have to. Under the imams they heard little about the rest of the world and met few outsiders. Yet the Yemenis are friendly to strangers, smiling and witty.

Daily life is harsh for most people. The poorest live in simple huts. Women in the villages work unveiled in the fields. In the cities, the women are wrapped from head to toe in bright shawls, and they are still heavily veiled when they leave their homes. The veils covering their faces are of brilliant colors and have ornamented openings for the eyes. These women have traditionally spent most of their time at home with other women. There are signs, though, that the old ways are beginning to change. Women can now go to the movies in the largest cities, and some women are even beginning to work outside their homes. Girls are trained as nurses' assistants, for example, and near Sana women have gone to work for the first time in a textile factory—wearing overalls and face veils.

Yemeni men wear tight-fitting caps embroidered with gold or silver threads. Over these they wind brightly colored turbans. Their robes,

FACTS AND FIGURES

YEMEN ARAB REPUBLIC—al Jamhuriya al Arabiya al Yamaniya—is the official name of the country.
CAPITAL: Sana.
LOCATION: Southwest Asia. **Latitude**—12° 40′ N to 17° 02′ N. **Longitude**—42° 34′ E to 46° 38′ E.
AREA: 75,290 sq. mi. (195,000 sq. km.).
PHYSICAL FEATURES: Highest point—Jebel Hadur (12,336 ft.; 3,760 m.). **Lowest point**—sea level.
POPULATION: 5,000,000 (estimate).
LANGUAGE: Arabic.
RELIGION: Muslim.
GOVERNMENT: Republic. **Head of state**—Chairman of the Presidential Council. **Legislature**—Consultative Assembly. (Military coup, 1974.) **International co-operation**—United Nations, Arab League.
CHIEF CITIES: Sana, Taiz, Hodeida.
ECONOMY: Chief minerals—salt, limestone. **Chief agricultural products**—coffee, kat, cereals (durra, wheat, barley, corn, rice, oats), cotton, vegetables, fruits. **Industries and products**—textiles, cement, handicrafts (leatherwork, swordmaking). **Chief exports**—coffee, kat, salt, hides and skins. **Chief imports**—machinery, petroleum products, yarns.
MONETARY UNIT: Riyal.
NATIONAL HOLIDAY: September 26 (1962), Proclamation of the Republic.
NATIONAL ANTHEM: *Alslam al-Watani Leljomhoryah al-Arabeah al-Yamaneah* ("Peace to the Yemen Arab Republic").

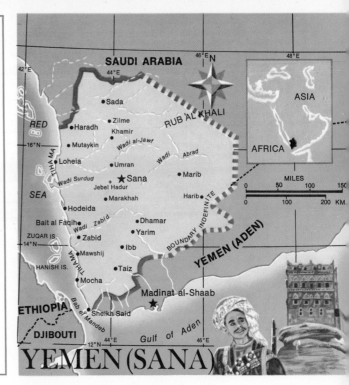

which are full and reach below the knees, are worn with embroidered leather belts. Into his belt a Yemeni proudly thrusts his curved dagger, or *jambiya*. It is common for groups of men to meet in the afternoon and chew the leaves of a woody shrub called kat. The juice of a kat leaf acts as a stimulant. Round, flat loaves of bread made of wheat, corn, or durra are a staple food. The meat most commonly eaten is mutton. Rice, vegetables, and fruit are also part of the diet.

The Land

Yemen lies in the southwestern part of the Arabian Peninsula and is bordered by the Red Sea, Saudi Arabia, and Yemen (Aden). The country's eastern boundaries have never been officially set.

Along the Red Sea are Yemen's lowlands—a narrow strip of land called the Tihama. It is a hot, sandy region with sparse vegetation and an occasional oasis. The growing of cotton, vegetables, and fruits in this region is increasing. The once famous coffee port of Mocha is there. (Even today people use the name of the city as another word for coffee.) But the crop no longer leaves the country through Mocha, and the city's streets are filling with sand. To the north is the modern port of **Hodeida**, a city of 45,000 people.

East of the coastal strip the land rises into highlands 4,000 to 12,000 feet (1,200–3,600 meters) high. The high country of Yemen is green because of the rain that falls in the summer as the monsoon winds blow from the Indian Ocean. More than 15 inches (38 centimeters) of rain falls in the summer, and the days are warm. An average temperature in July is 70 degrees Fahrenheit (21 degrees Celsius).

The farmers of Yemen learned long ago how to make good use of their hilly land by planting crops on terraces. Coffee has always been the main crop for export, but now farmers are growing more and more kat to ship to nearby countries. (Kat must be used soon after it is picked.)

Sana, Yemen's capital, and **Taiz**, which was the capital from 1948 to 1962, both lie high in the hills. Taiz has 80,000 people and·Sana 124,000. A wall of mud brick with a number of gates surrounds the ancient city of Sana. Within the walls are some of the "skyscrapers" typical of Yemen. These are tall, rectangular houses of stone and mud brick, four or five stories high, or sometimes higher. The windows are decorated with tracery in delicate designs.

In the eastern part of Yemen the land slopes off into a desert. Marib, a city still only partly excavated by archeologists, is in this eastern region. Three thousand years ago camel caravans took incense, spices, and jewels across the base of the Arabian Peninsula, north to the Mediterranean Sea. Marib was one of the resting places along the caravan route. The ruins of a great dam built about the 8th century B.C. can be seen at Marib, and part of a marble temple, built about the 5th century B.C., is still standing.

History

The land that is now Yemen was the kingdom of Saba (Biblical Sheba) from the 10th to the 2nd century B.C. A series of kingdoms followed it. The last kingdom was overcome by Christian Ethiopians in the 6th century A.D. By that time Christian and Jewish groups were already living in the country.

In the 7th century Arabs from the north introduced Islam. But before long the country was divided between followers of the two branches of Islam. Even today the people of the highlands belong to the Zaidi sect of the Shi'a branch of Islam, while the people of the lowlands belong to the Shafi'i sect of the Sunni branch. (The imams have all been Zaidis.)

For long periods between the 16th and the 20th centuries, the Turks had control over Yemen. The last period of Turkish rule ended in 1918.

In 1958 Yemen joined with the United Arab Republic (U.A.R.)—at that time a union of Egypt and Syria—in a federation called the United Arab States. But in 1961 the federation was dissolved, and in September, 1962, Yemen's government was overthrown by republicans.

The deposed Imam set up headquarters in the mountains of northern Yemen and organized an army of rugged fighters. Saudi Arabia and Jordan supported the Imam, while the U.A.R. sent troops to help the republicans. Late in 1967 troops of the U.A.R. withdrew from Yemen, and Saudi Arabia promised to stop helping the supporters of the Imam. But Yemen began to receive increased support from the U.S.S.R., Syria, and the People's Republic of China; and Saudi Arabia continued its help to the Imam. For several months early in 1968, the Imam's supporters laid siege to the capital. The siege failed, and by the middle of 1969 the republicans announced the end of the civil war. For the first time in nearly 7 years, there was the hope of peace in the ancient land.

Early in 1970 Saudi Arabia and Yemen agreed to halt all propaganda campaigns, and later in the year Saudi Arabia recognized the Yemen Arab Republic. The Imam went to live in London as an exile.

Government. The Constitution of Yemen provides for a legislature, the Consultative Assembly. Most of the members are elected directly by the people. The Assembly in turn elects the 3-man Presidential Council. In 1974 the government was overthrown in a military coup.

Reviewed by ABDALLAH AL-HAMMAMI
Permanent Mission of the Yemen Arab Republic to the United Nations

The port of Aden.

YEMEN (ADEN)

The port of Aden, rising in a haze of heat out of the green Gulf of Aden, is the heart of the People's Democratic Republic of Yemen. The trade that gives life to the country passes through this port. To the north and east, stretching in a barren strip along the base of the Arabian Peninsula, lies the body of the country, a string of former semi-independent states.

In 1967, the State of Aden, which includes the port and was previously a British colony, joined with the various Arab states to form the independent republic of Southern Yemen. The name was later changed to the People's Democratic Republic of Yemen. The country is sometimes called Yemen (Aden) after its chief city, in order to differentiate it from the Yemen Arab Republic (Sana) to the north. In 1972 the two countries signed an agreement to unite into a single state.

Aden has long been an important refueling port and a center of trade. By contrast, in the vast expanse of land to the east few people live, very little grows, and life has not changed for generations.

The Land

Yemen (Aden) occupies 111,075 square miles (287,683 square kilometers) along the Gulf of Aden, and includes several offshore islands. The country is bordered by the Yemen Arab Republic (Sana), Saudi Arabia, and Oman. Along the coast there is a hot, dry strip of sand and volcanic rock. Rising from this narrow plain are mountain ranges that level off into a high plateau. This rises to 8,000 feet (2,438 meters) in the west at Mount Djehaff and slopes down in the east. In the northwest there may be as much as 30 inches (76 centimeters) of rain a year, but in the northeastern

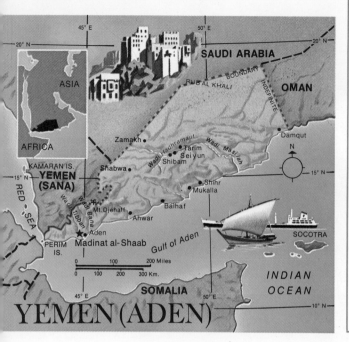

YEMEN (ADEN)

FACTS AND FIGURES

PEOPLE'S DEMOCRATIC REPUBLIC OF YEMEN is the official name of the country.

CAPITAL: Madinat al-Shaab

LOCATION: Southern coast of Arabian Peninsula. **Latitude**—12° 35′ N to 19° N. **Longitude**—43° 30′ E to 53° 05′ E.

AREA: 111,075 sq. mi. (287,683 sq. km.).

PHYSICAL FEATURES: Highest point—Mount Djehaff (8,000 ft.; 2,438 m.). **Lowest point**—sea level.

POPULATION: 1,200,000 (estimate).

LANGUAGE: Arabic.

RELIGION: Muslim.

GOVERNMENT: Republic. **Head of government**—presidential council. **Legislature**—People's Supreme Council. **International co-operation**—United Nations, Arab League.

CHIEF CITIES: Aden, Mukalla.

ECONOMY: Chief agricultural products—cotton, sorghum, millet, sesame, dates, wheat, barley. **Industries and products**—refueling of ships, transshipment of cargo, oil refining, livestock raising, fishing, cigarettes, textiles. **Chief exports**—cotton, dried fish, hides and skins, tobacco. **Chief imports**—grains, sugar, edible oils, various manufacured goods.

MONETARY UNIT: Dinar.

NATIONAL HOLIDAY: October 14.

NATIONAL ANTHEM: *Assalaam al Watani Jamhuriat al Yaman al Janoobyah al Shaabyah* ("National Anthem of the People's Republic of Southern Yemen").

part of the country the land merges into a desert—the Rub' al Khali, or Empty Quarter, of Saudi Arabia. Cutting through the mountains and plateau are wadis, dry riverbeds that fill with water after it rains. The water collects in wells along the wadis and is used to water crops.

The People

In all of the country there are only about 1,200,000 people. A quarter of them are clustered in the 75 square miles (194 sq. km.) of Aden State. The population is almost entirely Arab. There are a few Indians and Somalis. The people in the western part of the country are closely related to the Arabic-speaking people of Yemen (Sana) in the north.

Throughout history there have been feuds over the possession of villages, fertile lands, and wells. But even though the people can be awesome enemies, they are generous and hospitable friends, sharing the general Arab love of conversation and poetry. Most families live in low houses of mud brick and stay settled in one place, but there are nomadic groups in the northeast. In the interior, Arab women in dark blue robes work unveiled in the fields. The usual dress for men is a long shirt worn with a belt. A curved dagger is placed under the belt. In Aden men wear saronglike cotton skirts, which they call *futa*. Throughout the country people enjoy strong black coffee served in tiny cups. Lamb, rice, and round, flat loaves of bread are all part of the diet.

The British maintained good schools and hospitals in Aden, but there were hardly any schools or clinics in the interior until recently. Now there is a free public system of education, and the government is expanding health services.

Outside Aden, about 90 percent of the people make their living by farming. They raise sorghum, millet, dates, and sesame, as well as some vegetables and fruits for their own use. In the west, a fine-quality long-staple cotton is grown as a cash crop. The raising of livestock—sheep, goats, cattle, and camels—is also important.

Camels, here being washed in the sea, are an important form of transportation.

Business and industry are centered in Aden. Since the opening of the Suez Canal in 1869, the port has been a refueling stop for ships sailing the Red Sea. Aden, one of the world's famous ports, is the place through which goods enter and leave the country. It is also a fishing center, and it has an oil refinery and a few cigarette and textile factories.

History

Aden has been important as a port since ancient times. It has had its own local rulers, and it has been ruled by the kings of Yemen, the Turks, and the British. The British gained control of Aden in 1839. The states around Aden were ruled by their own sultans, sheikhs, or emirs. Beginning in 1839, the British made a series of treaties of protection with the hinterland states. Eventually, 20 states signed treaties with Great Britain. This group formed the Protectorate of South Arabia. Between 1959 and 1965 Aden and 16 of the protectorates joined and formed the Federation of South Arabia, which was composed of a cluster of sultanates, sheikhdoms, and emirates. The British Government promised independence in 1968. When independence came in November, 1967, the remaining four states became part of the new country.

Before independence, two Arab groups, the National Liberation Front (NLF) and the Front for the Liberation of Occupied South Yemen (FLOSY), fought for independence starting on October 14, 1963. Both groups strongly opposed the British-influenced government of the Federation of South Arabia. Qahtan al-Shaabi, an NLF leader, became the country's first president. NLF became the only party after independence. In 1969 the president was replaced by a presidential council, and in 1971 the People's Supreme Council became the legislative body. In 1972 Yemen (Aden) signed an agreement with its northern neighbor, Yemen (Sana), to unite and form a single nation.

Reviewed by PERMANENT MISSION OF
PEOPLE'S DEMOCRATIC REPUBLIC OF YEMEN TO THE UNITED NATIONS

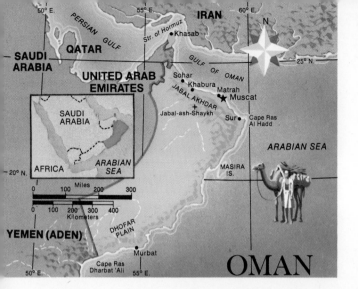

OMAN

OMAN

The Sultanate of Oman (formerly Muscat and Oman), at the southeast tip of the Arabian Peninsula, existed for many years in self-imposed isolation. But the discovery of oil in commercial quantities has helped propel the country into the modern world.

The Land. Oman has an area of 82,030 square miles (212,457 square kilometers). Its coastline extends nearly 1,000 miles (1,600 kilometers) along the Gulf of Oman and the Arabian Sea. The land itself is topographically diverse. A narrow coastal plain rises sharply to form a mountainous interior, the highest point of which is somewhat over 9,000 feet (2,700 meters) in the Green Mountain (Jabal Akhdar) region. The mountainous area slopes to a plateau fringed by desert. Oman has one of the hottest climates in the world—temperatures of 130 degrees Fahrenheit (54 degrees Celsius) are not unusual. An annual rainfall of 3 to 6 inches (8–15 centimeters) provides some water for agriculture.

The People. The population, estimated at 565,000, is largely descended from Arabian peoples. About half the people are members of the Ibadi sect of Islam. Their leader is known as the imam. The rest belong to various other Muslim sects.

Most of the people are farmers. Dates, coconuts, and cereal grains are the chief crops. In the Batina, the fertile section of the coastal plain, are the date gardens for which the country is famous. There is also some livestock raising, particularly in the province of Dhofar, known especially for its camels. The discovery of oil has initiated industrial development.

The towns along the coast serve as focal points for trade. The most important trading center includes the town of Muscat, the official capital (population about 6,000), and the adjacent town of Matrah.

History and Government. In 1650 the Sultan expelled the Portuguese, who had controlled the area for over 100 years. During the 19th century the British gained influence in Muscat and Oman through treaties opening up trade, but the sultanate maintained its independence. Great Britain continues to lend some financial support and is involved in oil ventures. The sultan is the absolute ruler, but there has been a history of tension between the sultan and the imam, who does not recognize the sultan's authority over the Ibadis. In 1970 the old Sultan was deposed by his son who announced a program of reforms for the country. Oman became a member of the United Nations in 1971.

Reviewed by GEORGE KIRK, University of Massachusetts

UNITED ARAB EMIRATES

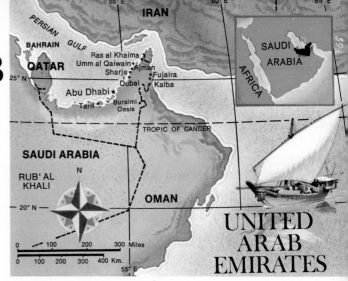

UNITED ARAB EMIRATES

The United Arab Emirates lies in the southeastern part of the vast Arabian Peninsula on a 400-mile (640 kilometers) strip of coastline bordering the Persian Gulf. This area was once called the Pirate Coast because it was a favorite haunt of pirates. More recently it has been known as the Trucial States or Trucial Oman. The United Arab Emirates is made up of seven small states. They are Abu Dhabi, Dubai (Dibai), Sharja, Ajman, Umm al Qaiwan, Ras al Khaima, and Fujaira.

The Land and Its History. The area of the country is 32,278 square miles (83,600 square kilometers). Most of this is flat desert. The highest point is in the eastern region, where the land rises to the Oman Promontory. Until the 19th century the economy was based mainly on piracy, perhaps the only means of survival in a dry and dusty climate unsuitable for agriculture. Inlets along the coast offered many hiding places for pirate ships. During the 18th and early 19th centuries British shipping was often a victim of these pirates.

In 1820 the first of a series of agreements (or truces) was concluded between Great Britain and the ruling sheikhs. It outlawed piracy and the slave trade that went hand in hand with it, and granted Great Britain control of the foreign affairs of the region. These truces made the area a British protectorate and gave it its name of Trucial States. The truces between the sheikhs and Great Britain lasted until 1971, when Great Britain withdrew from the region and the seven states declared their independence and united as a single country.

The People. The population of over 185,000 is mainly Arab. The people engage in fishing, trading, pearling, and agriculture, which has developed as a result of modern farming methods. Most of the farm land is in the Buraimi Oasis in Abu Dhabi, the largest and one of the two most populous of the states. Oil is now a major source of income, especially in Abu Dhabi. A small minority of the people still cling to the ancient nomadic ways.

Government. The government of the United Arab Emirates consists of a Supreme Council of the rulers of the states and a 34-member legislature. The capital is presently at Abu Dhabi, but a new seat of government is being built. The United Arab Emirates is a member of the United Nations.

Reviewed by GEORGE KIRK, University of Massachusetts
Author, *A Short History of the Middle East*

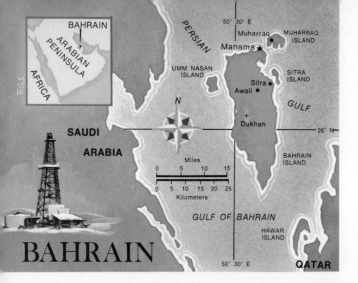

BAHRAIN

Throughout its history Bahrain has been one of the more prosperous areas on the Persian Gulf. For centuries this island country thrived as a seaport and a center for boatbuilding, fishing, and pearling. Today Bahrain's wealth comes from oil.

The Land. The sheikhdom of Bahrain is made up of several islands situated in the Persian Gulf, midway between the peninsula of Qatar and the mainland of Saudi Arabia. Most of the 230-square-mile (596 square kilometers) area is flat, sandy desert. The highest point is a 450-foot (137 meters) hill on Bahrain Island, the main island, where the capital city of **Manama** is located. Other important islands include Muharraq, site of Muharraq town, an important trade center, and Sitra. Limited agriculture, made possible largely by irrigation, produces some vegetables and fruits, but the value of the land lies in its oil.

The People. The present population of Bahrain is about 200,000. The people are descended in large measure from Arabs who migrated from Kuwait and Qatar in the 18th century. At one time the Bahrainis were known as sailors and merchants, boatbuilders and fishermen, pearl divers, and raisers of donkeys. Today most of the labor force works in the oil industry. There are immigrant workers from Saudi Arabia and other Arab states and from Iran, but native Bahrainis make up the majority. Almost all the people are Arabic-speaking Muslims.

History. After a brief period of Portuguese domination in the 16th century, Bahrain became a possession of the Iranian Safavid dynasty until its downfall in the 18th century. In 1783 Arabs from Qatar conquered Bahrain. These were the ancestors of the present ruling family. After the early 19th century Great Britain became the major influence in the area, and the piracy and slave trade that had characterized the Persian Gulf came to an end.

Government. Bahrain is a hereditary monarchy ruled by a sheikh. The constitution also provides for a National Council made up of elected members plus the members of the sheikh's cabinet, whom he appoints. Although Bahrain was self-governing, its security had been guaranteed by Great Britain. The British decided to end their commitments in the region in 1971. Negotiations to form a federation of the small gulf states—Bahrain, Qatar, and the United Arab Emirates—failed, and in 1971 Bahrain declared its independence and joined the United Nations.

Reviewed by GEORGE KIRK, University of Massachusetts

QATAR

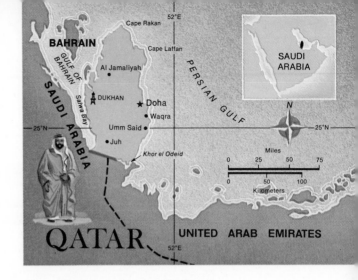

For centuries the people of the tiny sheikhdom of Qatar eked out a living by herding, fishing, and pearling. The discovery of rich oil resources, however, has brought Qatar from poverty to economic prosperity.

The Land.   Qatar is situated on the Persian Gulf coast of the Arabian Peninsula. It is bordered by Saudi Arabia and the United Arab emirate of Abu Dhabi.   Because these boundaries are disputed, the area of Qatar is usually given as between 4,000 and 8,500 square miles (10,000–22,000 square kilometers). The flat, barren land, slight rainfall, and intensely hot climate preclude agriculture to any great extent. Except in the coastal areas, the climate is quite dry. The port city of **Doha**, on the east coast, is the capital and major commercial center. About half the population of Qatar lives in Doha.

The People.   About half of the approximately 100,000 Qataris are native to the country; the rest immigrated to work in the growing oil industry. Some 45,000 people live in Doha and its surrounding areas. Almost all of the inhabitants are of Arab descent, and Arabic is the most widely spoken language. Islam is the major religion.

The working force is largely dependent on the oil industry, but the wealth that oil production has brought has also aided the development of commercial fishing and some farming through modern methods. Tomatoes, for example, have become one of Qatar's major exports.

History and Government.   The rulers of the sheikhdom of Bahrain, Qatar's neighbor to the west, governed the country for many years. They were followed by the Ottoman Turks, who exercised limited control for several centuries, and then by the British. A 1916 treaty with Great Britain recognized the sheikh as ruler of Qatar, and granted Great Britain control of Qatar's foreign affairs in return for British protection. This treaty ended when Great Britain withdrew its forces from the region and Qatar declared its independence in 1971 and joined the United Nations.

With British financial aid, Qatar has developed the harbor at Doha. A cement plant and a road connecting Qatar with Saudi Arabia have also been constructed, and the airport at Doha has been enlarged. Once a virtually ignored area of the Middle East, Qatar now ranks as an important source of the world's oil.

Reviewed by GEORGE KIRK, University of Massachusetts
Author, *A Short History of the Middle East*

KUWAIT

There are no rivers of water in Kuwait. The liquid that brings life to this desert country is a seemingly endless river of oil. Because of the tremendous revenue from its oil industry, developed since World War II, Kuwait has become one of the world's richest nations for its size. Much of the wealth of the country has been used to finance a broad range of social welfare programs for the people of Kuwait.

The Land.   The State of Kuwait occupies over 6,000 square miles (16,000 square kilometers) in the northeastern part of the Arabian Peninsula, at the head of the Persian Gulf (which Arabs call the Arabian Gulf). It is bordered on the north and west by Iraq and on the south and west by Saudi Arabia. A neutral zone of about 2,200 square miles (5,700 sq. km.) is divided between Kuwait and Saudi Arabia, and both nations have agreed to develop the resources of this area jointly.

Most of Kuwait is flat, sandy desert, only occasionally relieved by small hills. There are no rivers or lakes, and rainfall is limited. The sparse vegetation shows the effects of the intensely hot, dry climate. But underneath this barren soil lies between 15 and 20 percent of the world's estimated total oil reserves and the richest single oil deposit in the world.

Kuwait city, capital and largest city of Kuwait.

FACTS AND FIGURES

STATE OF KUWAIT—Dowlat al-Kuwait—is the official name of the country.

CAPITAL: Kuwait.

LOCATION: Middle East, northeastern corner of Arabian Peninsula. **Latitude**—28° 47′ N to 30° 04′ N. **Longitude**—46° 34′ E to 48° 18′ E.

AREA: 6,178 sq. mi. (16,000 sq. km.), not counting neutral zone.

PHYSICAL FEATURES: Largely desert, with no rivers or mountains.

POPULATION: 570,000 (estimate).

LANGUAGE: Arabic (official).

RELIGION: Muslim.

GOVERNMENT: Emirate (constitutional monarchy). **Head of state**—emir. **Head of government**—prime minister. **Legislature**—national assembly. **International co-operation**—United Nations, Arab League.

CHIEF CITY: Kuwait.

ECONOMY: Chief minerals—oil, natural gas. **Industries and products**—oil, petrochemicals, sand-and-lime brick, concrete, fishing, shipbuilding, soft drinks, tile. **Chief exports**—oil, petrochemicals. **Chief imports**—foodstuffs, building materials (iron and steel, timber, gypsum), vehicles.

MONETARY UNIT: Kuwait dinar.

NATIONAL HOLIDAY: February 25, National (Independence) Day.

NATIONAL ANTHEM: "National Anthem of Kuwait."

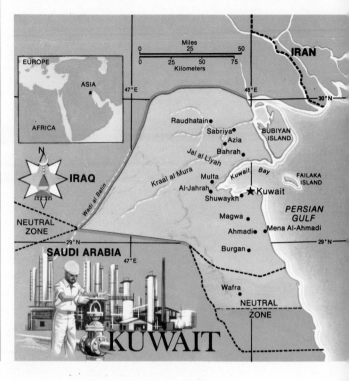

The People and the Economy.

Kuwait is a Muslim state. The official language is Arabic, although English is also widely spoken. The native Kuwaitis are descended from the Anaiza people, who migrated to the Kuwait coastline from the desert interior in the 18th century. Of the present population of about 570,000, more than half are nationals of other countries. The majority of the immigrants come from neighboring Arab states, but there are also Iranians, Indians, Pakistanis, and some Europeans. Much of the labor force is involved in the oil industry, which surpasses all other economic activities in Kuwait. Other industries include the making of chemical fertilizers, soft drinks, cement, and tile. The government has embarked on a large-scale program of water distillation and desalinization to alleviate the water shortage. But because of the lack of sufficient water for irrigation, agriculture is limited, and most of the country's food is imported. Some cereal grains and vegetables are grown in the desert oasis. Goats, camels, and sheep are raised by seminomadic herdsmen.

Before the oil industry revolutionized Kuwait's economy, most of the people were engaged in fishing, pearling, and boatbuilding. Kuwaiti boats traveled to many Asian and African ports, and the Kuwaiti sailors were well-known in many countries. Kuwait itself was largely a region of nomads—people who crossed the deserts from season to season. It is largely within the last 25 or 30 years that Kuwait has attracted an urban population, again due to oil.

All residents of Kuwait, whether native-born or not, benefit from a wide variety of social services. The government welfare program is one of the most extensive in the world and includes free medical care and hospitalization and low-rent housing. As a result of the program of free education, which includes facilities for adult education, the literacy rate in Kuwait is one of the highest in the Middle East. Much of this progress was achieved during the reign of Sheikh Abdullah al-Salim al-Sabah, who ruled from 1950 until his death in 1965, and has continued under the present ruler, Sheikh Sabah al-Salim al-Sabah.

Cities. The city of Kuwait is the capital and the largest city, with a population of over 100,000. It is also the commercial and trading center of the country. New buildings have given parts of the city a distinctly modern appearance. Near Kuwait city is the port of Shuwaykh, and to the south is the important oil town of Mena Al-Ahmadi.

History. The State of Kuwait developed from a small coastal town of the same name, settled about 1740. In 1756 the Kuwaitis selected their first emir, Sabah I. This is the origin of the present ruling dynasty. Trade was lucrative, and the community thrived. Kuwait's position at the head of the gulf made it an important port. But this prosperous life was constantly threatened by invasion from the interior of the Arabian Peninsula and by an attempt by the Turks to impose direct domination. The Kuwaitis sought the protection of the British, and in 1899 a treaty was negotiated between Kuwait and Great Britain under which the British agreed to protect Kuwait and also to take charge of the country's foreign affairs. In 1961 formal British protection was withdrawn by mutual agreement, and Kuwait became completely independent.

Government. Kuwait is an emirate (or amirate), a state ruled by an emir. The ruling power belongs to the Sabah family, which elects the emir from among its own members. Under the constitution of 1962, there is a national assembly composed of 50 members, who are elected for 4-year terms. The emir appoints the prime minister.

Reviewed by SAMI Y. SHAMMAS, Counsellor and Chargé d'Affaires
Permanent Mission of the State of Kuwait to the United Nations

Mena Al-Ahmadi is an important oil-refining town and oil-shipping port.

The city of Samarra was the capital of the Abbasid rulers of Islam in the 9th century A.D. The gold-domed mosque dates from the 17th century.

IRAQ

Iraq emerged as a new, independent nation after World War I, yet historically it is one of the oldest countries in the world. For thousands of years people have lived in its mountains and deserts and in the fertile plain between its twin rivers—the Tigris and Euphrates. According to tradition, the Garden of Eden lay somewhere in this land. What was perhaps the earliest civilization existed here. Hence the country is often called the cradle of civilization. To the ancient Greeks it was Mesopotamia, the land "between rivers."

THE LAND

Modern Iraq includes the main part of the lands watered by the Tigris and Euphrates rivers. It is bounded by Turkey on the north; Iran on the east; the Persian Gulf, Kuwait, and Saudi Arabia on the south; and Jordan and Syria on the west. The area thus encompassed is 167,925 square miles (434,924 square kilometers).

The country is divided into three main regions. The fertile plain, lying between the two rivers and stretching south and southeast, forms the bulk of the cultivated area. In the spring and summer the rivers rise, often flooding large areas of the lower plain. This region is poorly drained and much of the land cannot be put to any use. In the south, where the Tigris and Euphrates join, forming the Shatt al Arab, there is a vast marshland. The Shatt al Arab marks part of the boundary between Iraq and Iran. The question of navigation rights on the river has been a source of tension between the two countries. The northeastern part of the country is a mountainous region forming part of the Zagros range, which extends into Iran. The south and southwest are desert and semi-desert regions. (A separate article on the TIGRIS AND EUPHRATES RIVERS appears in this volume.)

The northern climate is temperate, but the plains and desert suffer extreme heat. The temperature can reach 120 degrees Fahrenheit (49 degrees Celsius). In winter the temperature may drop to freezing.

THE PEOPLE

The population of Iraq is estimated at over 9,350,000. The majority of the population is Arab. The 7th century was a time of conquests by Muslims who carried the religion of Islam from the Arabian Peninsula to other countries. Arabs entered Iraq during this period. The descendants of these 7th-century conquerors make up the present Arab population, along with people of different backgrounds who have adopted the language, religion, and customs of the Arab group. The Arabs speak Arabic and follow the Islamic religion. They are divided into the two major sects of Islam: Sunni and Shi'a.

The Kurds, who live in the northern mountains near the Turkish and Iranian borders, are also Muslims, but they are not Arabs. They came originally from Persia (modern Iran) and were an Indo-Aryan people. Their language is Kurdish. Hostility between this group and the Arab majority over the question of Kurdish rights has been a continuing source of conflict in recent years. In 1970, however, Iraq recognized

A souk, or marketplace, in Baghdad.

Baghdad, Iraq's capital, is an ancient city that once was the center of the Muslim world. Today modern houses line both sides of the Tigris River.

Kurdish nationalism and granted the Kurds the right to administer their own affairs within the Iraqi state.

Christians make up about 5 percent of the population of Iraq. They live mainly in the towns and villages of the north, especially around Mosul. Jews have lived in Iraq since Biblical times. According to tradition, Abraham came from Ur of the Chaldees in the central part of the Euphrates Valley. But there are only a few thousand Jews in Iraq today. Most of the Jewish population emigrated to Israel in 1950 and 1951. In the hills north and northeast of Mosul is a small group known as Yezidis, who may be of Persian origin. Their religion is a mixture of Islamic and earlier forms of worship.

Chief Cities.   The chief cities of Iraq are Baghdad, the capital (population, over 1,745,000), Mosul (over 243,000), the port of Basra (over 313,000), and Kirkuk (over 157,000). The country's wealth of historic sites includes the ruins of Ur, with the ziggurat (a tower surmounted by a temple), which dates from about 2500 B.C., and the remains of the Arch of Ctesiphon, an arched palace near Baghdad, dating from about A.D. 225. In the central part of the Euphrates Valley, a Shi'a region, are Najaf and Karbala, two of the most famous places of pilgrimage in the Muslim world. Al Kadhimain, a suburb of Baghdad, is another Shi'a holy place, famous for its gold-domed mosque.

Way of Life.   Nomads in the desert west of the Euphrates raise camels, sheep, and goats. The Kurds keep flocks of sheep and goats in the northern mountains, and the people of the southern marshes raise water buffalo.

In even the smallest, most remote villages, people live not in isolation, but in comfortable sociability. The houses are grouped close together, and women pay each other afternoon visits. The favorite meeting places for men are the coffee shop or the souk (marketplace).

Village houses are flat-roofed and rectangular. They are made of dried mud brick and have a door and a window. There are bright rugs or reed mats on the floor, and usually there is a chest for storing clothes and bedding. In the larger towns and cities, rows of rectangular stone or brick houses line the streets. Often on hot summer nights families sleep on the flat roofs of these houses. In modern sections of cities there are Western-style houses.

The nomadic Arabs of the desert live in tents made of goat's hair. In the southern marshes the dwellings are made of reeds. These are tied in tight bundles and bent to form supporting arches. Woven reed mats placed over the arches form a sloping roof and walls.

The traditional dress of Iraqi women is a long black robe. In the desert and villages women are unveiled, but it has always been the custom for women in towns and cities to wear a face veil. Today, however, some city women, especially the younger ones, are adopting Western clothes and giving up the use of the veil. In cities many men wear Western business suits, but the traditional loose, ankle-length robe can still be seen. In the desert men wear a camel's-hair mantle over the loose robe and a head scarf held in place with a heavy camel's-hair rope.

Dates, which are a source of sugar, fat, and protein, are a staple of the diet, along with bread baked in round, flat loaves. Lamb is the most popular meat, although few people can afford meat every day. Rice, vegetables, fruits, yogurt, and strong, sweet tea are also elements of the diet.

Social and Cultural Development. The population of Iraq, barely 2,500,000 before World War I, has risen to more than 9,350,000. The rate

The interior of a nomad's tent. Rugs cover the ground.

FACTS AND FIGURES

REPUBLIC OF IRAQ—Al Jumhuriya al 'Iraqi'a—is the official name of the country.

CAPITAL: Baghdad.

LOCATION: Southwest Asia. **Latitude**—29° 06′ N to 37° 22′ N. **Longitude**—38° 48′ E to 48° 34′ E.

AREA: 167,925 sq. mi. (434,924 sq. km.).

PHYSICAL FEATURES: Highest point—Approximately 12,000 ft. (3,600 m.), in Zagros Mts. **Lowest point**—sea level. **Chief rivers**—Euphrates, Tigris, Little Zab, Great Zab, Diyala, Shatt al Arab. **Major lakes**—Hor al Hammar, Hor Sanniya.

POPULATION: 9,350,000 (estimate).

LANGUAGE: Arabic (official), Kurdish.

RELIGION: Muslim (official).

GOVERNMENT: Republic. **Head of state**—president. **International co-operation**—United Nations, Arab League.

CHIEF CITIES: Baghdad, Basra, Mosul, Kirkuk.

ECONOMY: Chief minerals—oil. **Chief agricultural products**—dates, wheat, barley, rice, livestock. **Industries and products**—oil refining, bricks, date packing, woolen and cotton textiles, cigarettes, cement. **Chief exports**—oil, dates, barley, cotton, wool, animal hides and skins. **Chief imports**—sugar, cotton and artificial textiles, motor vehicles, tea, timber.

MONETARY UNIT: Dinar.

NATIONAL HOLIDAY: July 14, Proclamation of the Republic.

NATIONAL ANTHEM: *Al-Salam al Jumhuriya* ("Salute of the Republic").

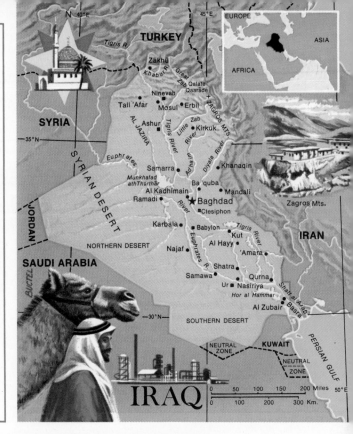

IRAQ

of increase is higher in cities and towns than in rural areas. Most of the people, who lived in tribal or semi-tribal groups before World War I, gradually abandoned a wandering life and began to settle in one place and raise crops.

One of the factors in the population increase and in the trend toward a settled life is the development of better health services. The number of hospitals in big towns and dispensaries in villages and the number of Iraqi doctors have been steadily increasing, especially since World War II. Social services have been extended to rural areas, living conditions have improved, and the high death rate of infants has fallen.

Before World War I education was confined to children of the well-to-do and of the ruling class. Less than 5 percent of the people could read and write. After 1921, when the modern state of Iraq was established, primary and secondary education were free wherever schools were available. Since 1959 primary education has been compulsory, and the number of schools has increased considerably, both at the primary and the secondary level. There are also many vocational and teacher training schools. The country's national universities emphasize graduate and professional training, but Iraq still depends to a large extent on schools abroad for graduate and technical education. Today between 20 and 25 percent of the people of Iraq are literate—that is, able to read and write.

ECONOMY

Oil is Iraq's greatest export and source of revenue, while farming and raising livestock are the two main occupations of the people. Dates are the country's major crop. Eighty percent of the world's date supply comes from Iraq. Barley, wheat, and rice are also important crops.

Before World War I a closed tribal economic system prevailed. Trade among different tribes and localities was limited and almost none existed with foreign countries. But trade with Western countries, which grew after the discovery of oil in northeastern Iraq in 1927, and technology have stimulated changes in the country's economic life. A system of dams and irrigation canals has made more land available for crops and has resulted in increased agricultural production. Machinery has been used recently in large-scale farming, and the government has introduced scientific methods supervised by experts.

Before 1958, when the government was overthrown in a revolution, the distribution of state lands among cultivators led to an increased number of wealthy landowners. After the revolution lands were taken from big landowners and distributed, together with state lands, among a greater number of tenant farmers and sharecroppers. The government has given them information about methods of cultivation and has encouraged the formation of farmers' co-operatives.

Before the revolution big industry was confined to oil, run by foreign companies. The principal Iraqi industries included flour milling, date packing, and the manufacture of wool and cotton textiles, bricks, and cement. After the revolution there was a demand for industrialization, and the government built a number of factories for the manufacture of clothing, chemical materials, and other products. But these domestic products cost more than foreign ones. In 1964 the government nationalized all banks and factories, although there has been a tendency recently to encourage a private sector of business and industry.

Oil revenues rose steadily from 1952 until 1965. Since that year Iraq has suffered some losses in oil production. One reason was Syria's closure of its transit pipeline in a dispute with the Iraq Petroleum Company in 1966. Another reason was Iraq's cutting off of oil supplies to Great Britain and the United States at the time of the Arab-Israeli War of 1967. However, the Iraqi National Oil Company now plans to increase oil production by developing new areas.

ANCIENT AND MEDIEVAL HISTORY

Excavations have shown that what may well have been the oldest of civilizations—the Sumerian—emerged in this country between 4000 and 3000 B.C. (Some archeologists maintain that an earlier civilization appeared in the Nile Valley.) The Sumerians, who were the first to put their language in writing, left clay tablets that indicate their achievements in literature, religion, and science. The Sumerians formed small independent states made up of a town and the villages surrounding it. Each state was ruled by a king who claimed divine authority, and rivalry among states was intense.

Sargon, King of Akkad, a region to the north, dominated the Sumerians during the 3rd millenium B.C. He unified the states into an empire, probably about 2300 B.C. (Historians are still uncertain about many ancient dates.) Sargon was a Semite, a member of one of the nomadic tribes whose original home was the Arabian Peninsula. After some 200 years Sargon's empire was dominated by another Semitic people who adopted the religion and literature of the Sumerians.

Two new empires, the Babylonian and Assyrian, were formed by other Semitic tribes migrating into the Tigris-Euphrates Valley. The

The ruins of the spiral minaret of the Great
Mosque at Samarra, built in the 9th century A.D.

Babylonians occupied the middle of the country, mainly around
Babylon, the capital. Their greatest period was under King Hammurabi,
who reigned during the 18th century B.C. (some scholars give a much
earlier date) and who is best-known for the code of laws that bears his
name. Babylon fell to invaders from the north and east, and for many
years no great empire emerged, until the Assyrians, who lived in the
north around the cities of Ashur and Nineveh, became dominant. The
Assyrian Empire fell in the 7th century to the Babylonians, who rose to
power for a second time under Nebuchadnezzar. During this second
period of Babylonian power, the Hanging Gardens, one of the wonders
of the ancient world, were built.

In the 6th century B.C. Cyrus, known as Cyrus the Great, who was
king of the Medes and Persians, conquered Babylon. The Persians were
succeeded in the 4th century by the Greeks under Alexander the Great.
After Alexander, the Parthians and then the Romans ruled Iraq until the
beginning of the 7th century A.D.

During this century the religion of Islam, founded in western Arabia
by the prophet Mohammed, rose as a world power. Muslims conquered
the area between the Mediterranean Sea and the Indian Ocean.
Mohammed was succeeded by caliphs who ruled the new Arab empire
first from Medina, then Damascus, and finally from Baghdad. The period
in which Baghdad was the capital came to be known as a golden period.

This was the age of the *Arabian Nights*. The empire reached a height of power, prestige, and learning, especially during the reign of Caliph Harun al-Rashid (ruled 786–809).

Weakness of the central authority became apparent in the 10th century, and Iraq entered one of its darkest periods. It was invaded first by the Mongols in the 13th century and then by the Turks. In the early 16th century Iraq became one of the provinces of the Ottoman Empire, remaining for the next 4 centuries an almost neglected part of that empire.

MODERN HISTORY

When World War I broke out, Iraq and other Arab countries joined the Allies to regain their independence. However, their hopes were not fulfilled after the war. The mandate system was devised by the Allies and applied to Iraq and some other former Ottoman possessions. Great Britain was given the mandate over Iraq at the San Remo Conference in 1920.

Iraq regarded the mandate as another form of colonial rule. Nationalist activities were fomented and led to a rebellion for independence. Great Britain suppressed the rebellion and decided to come to terms with nationalist leaders. In 1921 a provisional government was formed to prepare the country for a national regime. Faisal I, son of the Sharif of Mecca, an Arab leader, was chosen as ruler. Faisal was enthroned on August 23, 1921, and he ruled until his death in 1933. For the first time in 7 centuries, Iraq again had an Arab regime.

The country remained a mandate until 1932, when it was admitted into the League of Nations as an independent state. Lack of experience and the complexity of internal problems made Iraq's progress slow. The Army began to interfere in politics, and a series of military coups d'état took place between 1936 and 1941. During this time, Iraq became a center of Pan-Arab activities (those concerning all Arabs), and in 1945 it took an active part in the establishment of the Arab League.

After World War II Iraq began to turn its attention to reconstruction (including flood control, irrigation, and agricultural development) and to economic development. In 1950 the royalties from the oil industry, operated by the Iraq Petroleum Company (consisting of British, French, and American interests) were increased, and the bulk of the royalties went to reconstruction projects.

However, a nationalist upsurge, aroused by rival political groups and Pan-Arab sentiment, induced young men in both civilian and military ranks to seek the overthrow of the monarchy. A military uprising broke out on July 14, 1958, in which King Faisal II lost his life. The monarchy was replaced by a republican regime, with the government in the hands of the Army. The country is still governed by the military, and one set of army officers after another has come to power.

As a modern state Iraq is young. It has achieved modest progress in recent years, yet it still has much to do in the future. A stable regime, without political coups and counter-coups, could provide the country with the opportunity to realize its great potential in both human and natural resources.

M. KHADDURI, Director, Middle East Center, School of Advanced International Studies
Johns Hopkins University; author, *Republican Iraq*

TIGRIS AND EUPHRATES RIVERS

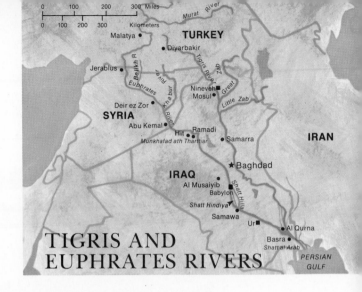

TIGRIS AND
EUPHRATES RIVERS

The story of civilized man probably began nearly 6,000 years ago in the region bounded by the Tigris and Euphrates rivers. According to tradition, the Garden of Eden was located in the Tigris-Euphrates Valley. It is also believed by some Biblical scholars that Noah's Ark came to rest on high ground just north of the valley. More definite is the role of the Tigris-Euphrates region—known to the ancient Greeks as Mesopotamia—in later history. For it was there that the Babylonian, the Assyrian, and the Islamic empires took root and expanded. Cities such as Nineveh, Samarra, Ur, and Baghdad were once flourishing cultural and commercial centers. Except for Baghdad, they have either lost their importance or are dusty ruins, and where there was once great wealth there is often poverty. But although empires have come and gone, the Tigris and Euphrates rivers still bring life to large areas of Turkey, Syria, and Iraq, which they serve as a means of transportation and irrigation.

The Tigris. From its source in the Kurdistan region in east central Turkey, the Tigris flows for nearly 1,150 miles (1,850 kilometers) down the length of Iraq into the Persian Gulf. The river first runs through rich farmland above Mosul, in northern Iraq, where apples and oranges are grown along with grain crops. Across the river from Mosul are the ruins of Nineveh, the proud capital of the Assyrian Empire until its destruction in the 7th century B.C. Between Mosul and the ruins of Samarra, an ancient Arab city, the Tigris is joined by its two main tributaries, the Great Zab and the Little Zab. The river then continues its winding course past Baghdad, the fabled city of the *Thousand and One Nights* and the capital of present-day Iraq. From there the river flows southeast to Al Qurna, where it combines with the Euphrates to form the Shatt al Arab. The Tigris is navigable by shallow draft steamers as far as Baghdad.

The Euphrates. The Euphrates is western Asia's longest river. It begins in the highlands of eastern Turkey, and its 1,700-mile course (2,700 km.) runs across the Syrian Desert and the plains of Iraq. Two principal tributaries, the Belikh and the Khabur, flow into the Euphrates in Syria. Between Hit and Ramadi in Iraq, the river enters its delta. At Al Musaiyib, the river divides into two branches, the Shatt Hindiya and Shatt Hilla. It merges again at Samawa and then proceeds southeast until it joins with the Tigris. The combined river, the Shatt al Arab, flows for about 120 miles (190 km.) into the Persian Gulf.

Reviewed by YASSIN EL-AYOUTY, St. John's University

The Elburz mountains loom over Teheran.

IRAN

Iran, better-known in the West until recently as Persia, is situated at one of the main crossroads linking Europe and the Middle East with Central Asia. The name Iran means "land of the Aryans"—a reference to the country's original settlers. More than 2,500 years ago, an Aryan people called the Persians united the country and founded a great empire. At its height, in the 6th and 5th centuries B.C., the Iranian (or Persian) Empire ruled nearly half of the ancient civilized world. Ancient Persia greatly influenced European and Asian political organization, science, art, and religion. But after many wars and invasions—by the Greeks, Arabs, Turks, and Mongols—the Iranian Empire collapsed, and the country entered a long period of decline. Since the early 1900's Iran has made a sustained drive to come abreast of Western material and technological standards. In this way, Iran hopes to be a strong nation again after centuries of backwardness.

THE LAND

Iran has an area of about 636,000 square miles (1,648,000 square kilometers). Dominating most of the country is the Iranian plateau, a somewhat triangular-shaped plateau fringed by mountains. The Elburz mountain range extends east and west—from Turkey to Afghanistan—along the northern frontier of Iran, which borders the Soviet Union and the Caspian Sea. Iran's highest mountain peak, Mount Demavend (18,600 feet; 5,670 meters), is located in this range. The Zagros range, Iran's other principal mountain chain, runs southeast along the western border with Turkey and Iraq. It ends at the Strait of Hormuz, which connects the Persian Gulf with the Gulf of Oman. These mountain walls separate the narrow coastal plains bordering the Caspian Sea and the Persian Gulf from the rocky highlands and great deserts of the interior.

Much of the interior plateau region—which varies in height from 1,000 to 6,000 feet (300–1,800 m.)—is dry and barren. Population density is therefore heaviest in the Caspian lowlands and in the northwestern mountain valleys where farming is possible. Roughly 25 percent of the country's land can be farmed, but only 10 percent is actually under cultivation. However, only about half of that 10 percent (about 20,000,000 acres; 8,000,000 hectares) can be farmed in any given year. The other half must be left unused each year to avoid exhausting the soil. Two fifths of the farmland is watered by irrigation. Since ancient times, there has been a system of underground channels, called *qanats,* for storing water. As part of modern development plans, many dams have been built—or are now being built—to provide irrigation for additional areas of land.

The principal river of Iran, the Karun, is located in the western part of the country. It is navigable for about 150 miles (240 kilometers) above its mouth at the port city of Khurramshahr. There are several smaller rivers in the north and southwest. In the interior, seasonal streams flow down from the mountains in the spring, when the snow melts. The water that they carry down generally evaporates quickly or is lost in the hot sands of the country's two great salt deserts—Dasht-i-Kavir and Dasht-i-Lut—which cover one third of Iran's land surface. A few small lakes dot the country, of which Lake Rizaiyeh (formerly Urmia) in the north is the largest.

Climate.   In the central plateau region, the climate is dry and the seasons are clearly defined. Throughout most of the year the days are clear and sunny. Winters can be quite cold and summers very warm. Along the Caspian coast, the climate is generally warm and humid. In the southern coastal plains, summers are extremely hot and the humidity is uncomfortably high. Except in the Caspian region, which is well-watered and fertile, the rainfall in Iran is scant. The annual average for the country as a whole is roughly 10 inches (25 centimeters). Heavy snows blanket the mountaintops in winter.

Natural Resources.   Petroleum is Iran's most important mineral resource, with production at over 140,000,000 tons a year. Ninety-five percent of the oil is extracted by an association of American, British, and other foreign oil companies (known as the Consortium), which operates under a royalty and profit-sharing agreement with the government of Iran. There is a large oil refinery at Abadan and smaller ones at Kermanshah and Teheran. Iran also has considerable quantities of iron ore and coal. Large copper deposits were discovered in 1967. Other minerals mined in Iran include lead, zinc, chromite, antimony, manganese, sulfur, barite, and gold.

Young Iranian boys are taught the art of decorating brass objects.

ECONOMY

By means of modern agricultural machinery, Iran is gradually changing from a subsistence-farming and handicraft economy to one of developing industries and cash-crop farming. Iran's petroleum industry provides employment for thousands of workers. Revenue from oil has increased steadily in the last 15 years and amounted to about $1,000,000,000 in 1970. This figure is expected to double within the next 10 years. Eighty percent of the money from oil sales is set aside each year to help finance the current reform and modernization program, which the present Shah (king) calls the White Revolution—as opposed to a Communist, or Red, revolution.

An outgrowth of Iran's oil industry has been the development of petrochemical factories, which produce fertilizers, plastics, artificial fibers, sulfur, and other chemicals. Other newly established factories include a steel mill, a machine tool plant, a tractor assembly factory, and automobile manufacture and assembly plants. In the agricultural sector, thousands of tractors are now in use on the farms. Crops have been increased, and efforts are underway to reduce food imports by building up local production of various staples, such as sugar.

Until 1967, the United States provided Iran with vast sums of money for development. American technical experts assisted Iranians in their efforts to learn modern skills. Although Iran must still borrow money to finance development, the country's economic relations with other countries are now on a more strictly commercial basis.

FACTS AND FIGURES

IRAN—Keshavré Shahanshahiyé Irân—is the official name of the country.

CAPITAL: Teheran.

LOCATION: Southwest Asia. **Latitude**—25° N to 40° N. **Longitude**—44° E to 63° E.

AREA: 636,294 sq. mi. (1,648,000 sq. km.).

PHYSICAL FEATURES: Highest point—Mount Demavend, (18,600 ft.; 5,669 m.). **Chief rivers**—Karun, Aras, Sefid, Atrek. **Major lake**—Rizaiyeh.

POPULATION: 28,000,000 (estimate).

LANGUAGES: Farsi, or Persian (official), Turki, Armenian, Arabic.

RELIGION: Muslim.

GOVERNMENT: Constitutional monarchy. **Head of state**—Shah. **Head of government**—prime minister. **Legislature**—parliament. **International co-operation**—United Nations, Central Treaty Organization (CENTO).

CHIEF CITIES: Teheran (Tehran), Isfahan, Tabriz, Meshed, Abadan.

ECONOMY: Chief minerals—oil, coal, lead, zinc, iron ore, barite. **Chief agricultural products**—wheat, rice, barley, sugar beets, cotton, tea, tobacco, vegetables, citrus fruits. **Industries and products**—food processing, textiles (cotton and silk), oil refining, carpets and rugs, glassware, cement, cigarettes, fishing (caviar), livestock products. **Chief exports**—oil, raw cotton, carpets, fruit, hides and leather, mineral ores, wool. **Chief imports**—machinery and electrical equipment, iron and steel (crude and manufactured), sugar, chemicals, pharmaceuticals, vehicles.

MONETARY UNIT: Rial.

NATIONAL HOLIDAY: October 26, Birthday of the Shah.

NATIONAL ANTHEM: Soroûde Shahanshahi ("Long Live our Shahanshah").

THE PEOPLE

There are more than 28,000,000 people now living in Iran. The majority of these people are descendants of the Medes, Persians, and Parthians, the most important of the Aryan peoples who settled in ancient Iran. There are also descendants of other ancient peoples, including the Scythians and Bactrians. A large number of Turkish people inhabit the northwestern section of the country. In addition, there are small groups of Jews, Arabs, and Armenians.

Language. Most Iranians speak Persian (Farsi), an Indo-European language that has changed very little over the centuries. Persian is the official language of the country and is taught in all the schools. It has been greatly influenced by Arabic, and it is written in Arabic script, from right to left. In northwestern Iran, a Turkish dialect is spoken. Certain segments of Persian society, including the Kurds, Lurs, Gilaks, and Mazandranis, speak different Persian dialects.

Religion. Nearly all Iranians are Muslims, mostly members of the Shi'a sect of the Islamic religion (as opposed to the Sunni branch in the Arab countries). There are some 20,000 Muslim mosques in the country, many of them ancient buildings of rare beauty. Small numbers of Christians, Jews, and Zoroastrians also practice their faiths. The Zoroastrians are followers of the religion that was founded in Iran and that was dominant there until the Arab conquest in the 7th century A.D. Zoroastrians believe that there are two competing spiritual forces in the universe; one representing truth and light, the other darkness and lies.

A view of the Madreseh Chahar Bagh Mosque in Isfahan, the former capital of Iran. Iran's mosques are noted for their beautiful design and tile work.

Way of Life. Although most Iranians live in rural areas, nearly 10,000,000 are now inhabitants of the cities. Because of new opportunities in industry, the movement from the countryside to the cities has grown rapidly in recent years. By 1976 it is expected that half of Iran's people will live in the cities. Before World War I, over 20 percent of the people were nomads who wandered about the desert in search of pastureland for their horses, sheep, and donkeys. But over the past 50 years most of the nomadic peoples have given up their traditional way of life to go to work in the cities.

The majority of Iranians earn their living in agriculture. Until very recently, most of the farmers dwelling in the 50,000 villages of Iran were sharecroppers. They rented their land from wealthy landlords, to whom they were obliged to give as much as 80 percent of their crops. But in 1951 the Shah began a major land reform program by distributing royal lands to the farmers. In the following years, laws were passed limiting the amount of land any one landlord could possess. The laws also required the big landholders to sell their surplus land to the government for resale to the tenant farmers. Eventually, about 2,500,000 families with 12,000,000 members received land. At the same time, farming co-operatives were formed, modern agricultural methods were introduced, and loans were made to support the farmers' efforts to increase crop production.

The Iranian family is a very close-knit unit. At its center is the father or another older male, who is known as the master (aga). He is absolute ruler in his home and commands total respect and obedience. Large families are the rule in Iran, and male children are particularly desired so that the family line can be continued. Where the old traditions are still followed, it is usual for sons to remain in the family home even after they are married.

Until recently, the status of women in Iran was an inferior one. In the last few decades changes have occurred, and Iranian women are gradually achieving a position of equality with men. Women now have the vote and have been elected to public office. They cannot be divorced

A group of Iranian women sit on scaffolding while they weave a rug. The manufacture of Persian rugs is Iran's most famous handicraft.

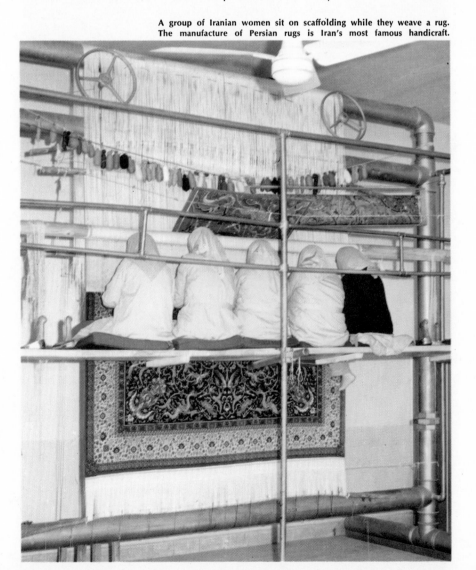

at will by their husbands, as was once the case. The Muslim practice of allowing men to have more than one wife (polygyny) is dying out. Before the 1966 census, women were not counted among the active members of the community, even though many were part of the labor force. Farm women, for example, not only are housewives, but also help their husbands plant crops and tend the flocks. More than two thirds of the carpet and textile weavers in Iran are women.

Iran has a serious housing shortage, and most existing housing is of poor quality. The vast majority of city families live in overcrowded one- or two-room houses. The situation in the farming villages is much worse. A typical farmhouse is made of sun-dried brick. It is usually small and sparsely furnished, with pillows often used in place of chairs. A high wall surrounds the house and encloses a courtyard that generally contains a pool of water for cleaning. Nomadic people live in round tents that are easily transportable. They eat and sleep on the ground, which is covered by rugs woven by the women.

Western-style clothing is rapidly replacing traditional Iranian dress both in the cities and the countryside. But many Iranian men continue to wear the old-fashioned cotton shirts and baggy trousers. It is still common to see women wearing the *chador* (*chawdar*), a long, scarflike cloth used as a head covering, wraparound, and occasionally as a veil. Mutton, chicken, rice, and cheese are the most common Iranian foods. The national drink is tea, which is grown on the hilly slopes around the Caspian Sea. A great delicacy is caviar, the eggs of the sturgeon, a fish that abounds in the Caspian Sea. But most caviar is sold abroad.

Leisure Time.   Iranians celebrate various Islamic religious holidays, including the birthday of the prophet Mohammed and Ramadan, the Muslim month of fasting. National holidays include the Shah's birthday and Constitution Day. The biggest festival time is the New Year, which in Iran falls on the first day of spring. The New Year is celebrated for 13 days and includes family gatherings, gift-giving, and country outings. Iranians are very fond of sports and games. Since ancient times, the people of Iran have been skilled hunters, polo players, and wrestlers. Iranians also enjoy playing chess, which originated in Iran and India.

Cities.   **Teheran** (also Tehran) is the country's largest city (population over 2,700,000) and its capital. It is situated at the foot of the Elburz mountains in the northern part of the country. A caravan town until the late 18th century, Teheran has since developed into a modern industrial and commercial city. Major railroads and highways converge here, and the city is also a center for artistic and educational activities.

Tabriz, near Lake Rizaiyeh in the northwest, is another important city (population over 400,000). It is noted for the production of Persian rugs of high quality and has a number of other industries. Other major urban centers include the former capital city of Isfahan (population about 425,000) and Meshed (population over 400,000).

Education.   The standard of living and the social condition of Iran's people are being improved through the efforts of the government to provide more and better education. As late as 1941, less than 5 percent of the people were literate. The well-to-do sent their children to private schools and then to foreign universities. But as a result of considerable government effort, the number of people who can read and write has greatly increased. To overcome the shortage of teachers, the Literacy Corps

A flock of sheep is led through Teheran's indoor gold bazaar.

Iran's capital city of Teheran is a modern, bustling metropolis.

was established in 1963. The Literacy Corps consists of educated army conscripts (draftees), who are given a brief teacher-training course and then sent into the villages to teach reading and writing to primary-school children and adults. Since the plan was put into effect, an average of 350,000 children and adults have been taught each year by the Literacy Corps.

The number of children in primary and high schools nearly doubled between 1962 and the end of the 1960's. Presently, some 60 percent of primary-school-age children and 20 percent of older children are attending schools. Iran has eight universities. The largest and most noted is the University of Teheran. There are a considerable number of teachers colleges and vocational schools.

Art and Literature.   Iranian literature, which has achieved worldwide recognition, reached its peak between the 10th and 16th centuries. Many significant works of history, philosophy, mathematics, medicine, astronomy, and poetry were written during this period—despite the fact that it was a time of wars, invasions, and internal conflict.

Among the most distinguished of the classical writers were Firdausi (A.D. 940?–?1020), the author of the *Book of Kings,* an epic poem of 60,000 couplets, and the poets Rumi (1207–73), Saadi (1184?–1291), and Hafiz (1324?–?88). The most famous Iranian poet was Omar Khayyám (?–1123), whose poem *The Rubáiyát* has been widely read in the Western world. Omar Khayyám was also a respected mathematician and astronomer. Most of the classical Iranian writers wrote in Persian, employing only a small number of Arabic words. However, since Arabic was the scientific language of the time, the most eminent Persian physicians and philosophers wrote in Arabic.

Ancient Persian architects pioneered in the development of the

The ruins of Persepolis, ceremonial capital of the ancient Persian Empire.

vault and dome and other architectural styles. The builders of that time solved the problem of building a round dome over a square base. They introduced the rectangular courtyard with four diagonal arches, later copied by the builders of mosques. Today only the ruins of the beautiful ancient Persian capitals of Persepolis and Susa remain, but the influence of Persian architecture can be seen in the mosques and other traditional buildings throughout the Middle East.

The art of Iran has been described as an art of decoration and detail. In earlier times, tiles of particular beauty were used to decorate the walls and domes of mosques. Painters excelled in creating miniature illustrations in manuscripts and on ivory and bone. Ceramics of a high quality were produced. These now are among the world's most treasured art works. Perhaps the best-known artistic creations of Iran—because they are eagerly sought after in the Western countries—are the famous hand-woven Persian rugs and carpets. Carpet weaving is a skill in which Iranians have excelled since ancient times, and it continues to be an important and profitable craft.

GOVERNMENT

Iran is a constitutional monarchy. Under the Constitution of 1906, the Shah is the head of state. There is an elected parliament composed of a lower house, called the Majlis, and an upper house, the Senate. Half of the senators are elected by popular vote, while the other half are appointed by the Shah. The head of government is the prime minister, who is appointed by the Shah with the approval of the Majlis. The prime minister heads the Cabinet, which includes the ministers in charge of the various government departments.

The position of the Shah is a strong and controlling one. He commands the armed forces and can dissolve the houses of parliament. The present Shah has used his power to promote his modernization program, which includes land reform, an improved educational system, a profit-sharing plan for industrial workers, and higher status for women.

HISTORY

Iran was settled some 3,500 to 4,000 years ago by Aryan people who migrated from the region around the Aral Sea, now a part of the Soviet Union. During the first 1,000 years of Iranian history, several kingdoms rose and fell. In the 6th century B.C., King Cyrus the Great, of the Persian Achaemenid dynasty, defeated the rival Medes in northern Iran and united the Iranian people. Under Cyrus' rule Persian armies conquered all of the present Middle East and Central Asia as far as what is now Pakistan. The great empire founded by Cyrus flourished for over 2 centuries until it was overthrown in the 4th century B.C. by Alexander the Great. Alexander and his successors established a new dynasty and extended Greek cultural influence throughout the area.

Greek rule lasted until the 2nd century B.C., when the Parthians (a people related to the Iranians) gained control. After 350 years of rule, the Parthians were defeated by the Sassanians, who established another Persian dynasty and turned back Roman invasions. The Sassanian period (A.D. 224–641) saw a revival of Persian culture and nationalism. In the 7th century Arab warriors swept across the Iranian plateau and toppled the Sassanid dynasty. The Arab conquerors introduced the Islamic religion—which replaced Zoroastrianism as the principal faith of the Iranians—and Arabic script.

Arab domination continued for several centuries, until the Islamic empire began to fall apart. The Arabs were followed by other foreign invaders. Over a period of 5 centuries (from about A.D. 1040 to 1500), Iran witnessed successive waves of intruders. Following a mainly peaceful movement into the area by Seljuk Turks, the Mongols under Genghis Khan and the Tatars under Tamerlane overran the country. Iran's cities were plundered and the countryside was left desolate. In the wake of these invasions, Iran suffered a period of political chaos and economic decline.

The Safawid dynasty (1499–1722) restored a measure of Persia's former greatness. Shah Abbas I (ruled 1587–1628), the most outstanding of the Safawid rulers, restored the cities, built new roads, and stabilized the government. Isfahan, the capital, was made into one of the world's most beautiful cities. Shah Abbas established the Shi'a sect of the Islamic faith as the state religion and used religious zeal as a force for unifying the nation.

The Safawid dynasty crumbled after an invasion by the Afghans in 1722. For a brief period under Nadir Shah (ruled 1736–47), Iran again became a military power. Persian armies conquered Afghanistan and reached as far as Delhi, in India. Among the treasures brought back after these conquests was the famed jewel-studded Peacock Throne. It is now preserved in a museum in Teheran, which became Iran's capital in 1788. Following Nadir Shah's death, Iran plunged into a civil war between two rival dynasties. In the 1790's, the Qajar dynasty emerged victorious and ruled Iran until after World War I.

During the centuries after Arab rule, Iran had little contact with the West and so fell behind in science and technology. By the 1800's the country had been greatly weakened by wars and invasions. Taking advantage of this weakness, Russia, Great Britain, and other Western powers became involved in Iranian economic and political affairs. Iran's geographical position made it important to Russia, which was seeking to expand southward to the Persian Gulf, while Britain wanted Iran as a buffer against Russian movement into India. After two successive wars with Iran (1801–13 and 1825–28), Russia gained control of all Iranian land north of the Aras River. In the early 1900's, Britain acquired commercial rights to the major oil fields of Iran. By this time the Iranian Government was virtually under the control of Britain and Russia, and the two powers divided the country into "zones of influence" in 1907.

To defend their threatened independence, the Iranian people began a movement to achieve internal reform. An important early step was the Constitution of 1906, which provided for representative government. Modernization efforts gained momentum after World War I, when Riza Shah Pahlavi—the founder of the present royal dynasty—started a major program of reform and development. During Riza Shah's reign the name of the country was changed from Persia to Iran. The present ruler, Mohammed Riza Pahlavi, who came to power in 1941, has continued his father's efforts to improve social and economic conditions within the country.

MEHDI VAKIL, Ambassador and Permanent Representative of Iran to the United Nations

AFGHANISTAN

Afghanistan is a landlocked country situated in the heart of Asia. At each of the four gateways to the country a historic city greets the traveler. The visitor from the north first arrives at Balkh, known to the ancient Greeks as Bactra. The modern city of Mazar-i-Sharif lies a few miles east of the site of the ancient city. Approaching Afghanistan from the west, the traveler soon reaches the city of Herat, capital of Herat province. The city was destroyed and rebuilt in the 4th century B.C. by Alexander the Great. From the south the road leads to Kandahar. This city was also built by Alexander on the ruins of a city he destroyed. From the east, through the winding Khyber Pass, one arrives at Jalalabad, the capital city of Ningrahar province. Beyond the next mountain range, and on the same route, lies the capital of Afghanistan, the historic city of Kabul, which has a population of about 456,000.

THE LAND

On the map, Afghanistan looks like a tortoise with an outstretched neck. The neck of the tortoise is the highest part of the country, the Wakhan panhandle. In this northeastern corner the ranges of the Hindu Kush climb to the Pamir, a great knot of mountains where Afghanistan, the Soviet Union, and China meet. The Hindu Kush slopes southwest, forming a broad network of highlands. North of the highlands flows the Amu Darya (formerly called the Oxus), one of the chief rivers of Asia. The eastern part of Afghanistan is a region of swift-flowing rivers, green and fertile valleys, and highlands. There are deserts in the south and southwest.

An arid country, Afghanistan's agricultural life depends to a large

Rugged mountains and barren deserts make up a large part of Afghanistan's landscape. But these Afghans are at home in their austere surroundings.

extent on the amount of snow stored on the mountain slopes from one year to the next. An intricate irrigation system covers almost the whole country. Afghanistan has a continental climate, marked by severe winters and hot summers and extreme changes in daily temperatures.

THE PEOPLE

All citizens of Afghanistan are called Afghans. (The name of the country means "land of the Afghans." The Persian suffix -*stan* means "land of.") Actually, the population—about 16,500,000—is made up of a variety of peoples, including Pushtuns, Tajiks, Hazaras, Uzbeks, and Turkomans.

The Pushtuns and Tajiks are the major ethnic groups. The Pushtuns speak Pushtu, a language related to Persian. The Tajiks and Hazaras speak Persian. The Hazaras are believed to be descendants of 13th-century Mongol invaders led by Genghis Khan. The Uzbeks and Turkomans are of Turkish origin. Pushtu is the tongue of the larger part of the population. Persian is commonly used in business.

Afghans are united, however, by a common religion, Islam. At various times in the past the country was a stronghold of the Zoroastrian and Buddhist religions. Zoroaster, the founder of Zoroastrianism, is said to have been born in Balkh, probably in the 7th century B.C. Islam reached the country in the 7th century A.D., but it was not until 1895 that the last group in Afghanistan accepted the religion.

The rugged nature of the land has produced a hardy, courageous people. Since its geographic location has made Afghanistan a crossroads for invading armies, the Afghans have also developed excellent fighting qualities. Although they may sometimes appear arrogant and fierce to outsiders, they are friendly and hospitable to strangers and have a keen sense of humor.

Its high ranges and isolated valleys have made Afghanistan a land of many villages and small towns. In recent years the cities have expanded greatly in population, but most of the people still follow age-old customs, and their living conditions have changed very little.

Their Way of Life

A typical house in an Afghan village is square or rectangular and made of mud or mud brick. It is divided into three or four rooms furnished with rugs, pillows, and mattresses. The flat roof is made of logs re-inforced with a thick layer of mud, often mixed with straw and chalk. Four walls 9 or 10 feet (about 3 meters) high are built above the roof, which then serves as a sitting room and sleeping porch in hot weather and as a place to dry fruits. More prosperous owners may erect towers on the four corners of their houses, or at least a single tower on one corner. Houses that have such towers are called *qalas*. A *qala* may house one or several family units.

Immediately inside the single large gate there is a room called the *hujra*, for housing guests and travelers. The *hujra* is also the center of social activities for the *qala*, especially on long winter nights, when storytelling and group singing are enjoyed. Here, too, *sohbat* is held—a potluck meal to which everyone contributes. Round loaves of unleavened bread are a staple of the diet, which also includes goat, mutton, beef, chicken, yogurt, rice, and fruits.

A tea shop along an old trade route offers travelers rest and refreshment.

The marketplace, where people meet and talk and buy everything from fruits to furs, is an important part of city life. Below, the grain market in Kabul.

Everyday life centers around the fields, gardens, and vineyards on which the people depend for a living. Men find evening fellowship in the *hujra,* especially in winter. When the weather turns warm, they turn to sports and dancing. Horse racing, *ghosai,* and *buz-kashi* are popular. *Ghosai* is a vigorous wrestling game in which a player tries to reach a goal while nine teammates run interference. Each player hops on one foot and with one hand holds the other foot behind his back. *Buz-kashi* is played by two teams on horseback.

The Afghan national dance is called the *attan.* The dancers, sometimes 100 or more, form a large circle with drummers in the center. At first the circle is small, but it spreads out as the pace quickens. When the dance is performed by men, guns and swords may come into play.

In the cities Afghans are gradually adopting Western dress, but in the villages people still wear traditional clothing. In general, an Afghan man wears a knee-length shirt, which may be embroidered, over full trousers. In the summer he wears a cotton shawl around his shoulders, and in the winter his coat is a long woolen or quilted *chapan.* The most popular headgear is a turban wound around an embroidered cap, or a karakul (lambskin) cap without the turban.

Afghan women generally wear long-sleeved, ankle-length dresses over full trousers with embroidered ankle bands. A large cloth, or *chawdar,* is draped over the hair. In the cities before 1959, women always appeared in public in a *chaderi,* a long, sacklike garment that covers the wearer from head to toe. A netlike piece over the eyes makes it possible to see. In 1959, however, the government authorized women to appear in public unveiled if they wished. Since that time a great change has taken place in women's clothing and in their way of life. Many women have given up the *chaderi* for Western dress. Increasing numbers of women are receiving a university education and are going to work. In 1966, for the first time, a woman was appointed to the Cabinet, and now there are women in Parliament.

A mosque in the mountain city of Kabul, capital of Muslim Afghanistan.

FACTS AND FIGURES

REPUBLIC OF AFGHANISTAN—Doulat Jamhory Afghanistan—is the official name of the country.

CAPITAL: Kabul.

LOCATION: Latitude—29° 22′ N to 38° 30′ N. **Longitude**—60° 28′ E to 74° 52′ E.

AREA: 250,000 sq. mi. (647,497 sq. km.).

PHYSICAL FEATURES: Highest point—peaks rising to more than 20,000 ft. (6,100 m.) in the Hindu Kush. **Lowest point**—Kham-i-ab (500 ft.; 152 m.) above sea level. **Chief rivers**—Helmand, Amu Darya, Kabul, Hari, Arghandab, Murgab, Farah.

POPULATION: 16,500,000 (estimate).

LANGUAGE: Pushtu, Persian (Dari).

RELIGION: Muslim.

GOVERNMENT: Republic. **Head of state**—president. **Head of government**—prime minister. **Legislature**—parliament (suspended as of 1973). **International co-operation**—United Nations, Colombo Plan.

CHIEF CITIES: Kabul, Kandahar, Herat, Jalalabad, Mazar-i-Sharif.

ECONOMY: Chief minerals—iron ore, natural gas, copper, coal, chrome, oil, lapis lazuli. **Chief agricultural products**—cotton, grains, sugar beets, fruits, nuts. **Industries and products**—cotton and wool textiles, carpets, cement, fruit processing, tanning, sugar. **Chief exports**—karakul skins, carpets, cotton, wool, fruits (fresh and dried), nuts. **Chief imports**—petroleum products, sugar, tea, textiles, motor vehicles.

MONETARY UNIT: Afghani.

NATIONAL HOLIDAY: May 27, Independence Day.

NATIONAL ANTHEM: *Loya Salami* ("National Anthem" or "Grand Salute").

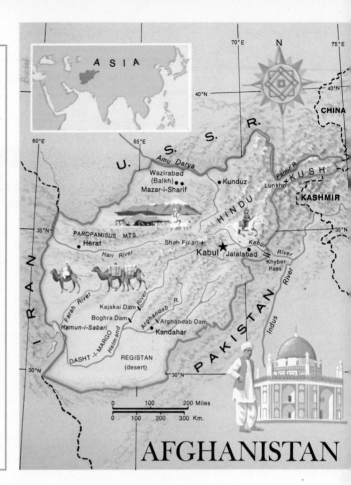

AFGHANISTAN

ECONOMY

In some respects Afghanistan is one of the richest nations of Asia. Its wealth is in the form of mineral deposits, forests, and agricultural products. The minerals include iron ore, natural gas, copper, coal, chrome, oil, lapis lazuli, and small amounts of gold, silver, and rubies. (Marble is also mined.) However, most of the mineral resources, except for coal, are still undeveloped.

Much of the country's forest wealth has been cut for timber, but there are many valuable nut trees, producing pistachios, pine nuts, walnuts, and almonds. Medicinal plants yield such products as asafetida and castor oil for local use and export.

Most important to the economy are the growing of crops and the raising of livestock. (More than 80 percent of the people are farmers and shepherds.) Practically all kinds of grains thrive in Afghanistan. Wheat is the major crop. Fruits, too, flourish. Melons, peaches, apricots, figs, apples, and pears are grown. Dried fruit is a major export. The vineyards of Herat, Kapisa, and Kandahar yield more than 30 varieties of grapes. Citrus fruits, rice, and cotton are also cultivated.

Most farmers today lead settled lives, but the herders of sheep, cattle, and goats still move from place to place in search of pasture. Karakul sheep are raised for the skins of the young lambs. These skins are prized as fur and are a leading export.

Modern industry is in its infancy in Afghanistan. Development has begun with the production of textiles, sugar, and construction materials.

On the other hand, cottage industries, or home crafts, have existed for centuries. The famous Afghan rugs and carpets are all made by people working at home.

HISTORY AND GOVERNMENT

In its earliest history Afghanistan was the home of nomadic tribes. Through the centuries they were overcome by a succession of conquerors —Alexander the Great in the 4th century B.C., Genghis Khan in the 13th century, Tamerlane (Timur) in the 14th, Baber in the 16th, and Nadir Shah of Persia in the 18th century.

In 1747 the Pushtuns unified the tribes into an independent nation. During the 19th century, however, the nation was caught between Great Britain and Russia as these two powers struggled for control of Central Asia. In 1879 Great Britain defeated Afghanistan and took control of the country's foreign policy. Its boundaries with Russia and India were agreed on later. During World War I Afghanistan was neutral. It regained control of its foreign affairs as a result of the third Anglo-Afghan War.

In World War II the country was again neutral. Trouble developed in 1947 between Afghanistan and Pakistan when an area inhabited by Pushtuns was claimed as part of Pakistan. The Afghans felt the area should be independent. The issue still remains a serious source of dis-agreement. However, in 1963 the two countries resumed diplomatic relations.

Since 1956 Afghanistan has been engaged in a program to improve social and economic conditions. Financial help has come from several nations. This aid has brought improvements in transportation, including a tunnel almost 2 miles (3 kilometers) long through the Hindu Kush. Both Kabul and Kandahar have international airports.

The government is making major efforts to develop agriculture by teaching farmers modern methods and by making more land available for farming. A vast system of dams, reservoirs, and canals is being built to provide irrigation and hydroelectricity. The government is also taking steps to improve health care and education. The high rate of illiteracy is a serious problem, but free education is available to everyone and many new schools are being built. After 6 years of elementary school, a student who wants to go to secondary school must go to Kabul or to the capital of a province. In Kabul there are vocational and technical schools and a university. There is another university in Jalalabad.

In 1963 a new constitution was proposed by King Mohammed Zahir Shah, who came to the throne in 1933. It was approved the following year, and provided for a prime minister appointed by the king and a legislature of two houses.

This constitution was set aside in 1973, however, when a military coup overthrew the monarchy. General Mohammad Daoud Khan, leader of the coup, proclaimed Afghanistan a republic. A new governing body, the Central Committee, was set up. General Daoud was elected president and prime minister. Daoud and thousands of his supporters were killed in 1978 when a new military regime seized power. The new government was headed by a revolutionary council.

ABDUR-RAHMAN PAZHWAK
Permanent Representative of Afghanistan to the United Nations

PAKISTAN

Pakistan is both a new country and a very old one. It did not exist as a nation until 1947. Yet within Pakistan are areas whose history dates back 500,000 years to the days when man first learned to make crude implements of stone. It is a land studded with the remains of ancient cultures and the monuments of past civilizations. These depict man's development from a primitive creature into a thoughtful being capable of creating philosophies and technologies. Perhaps the most famous of these ancient civilizations is that of the Indus Valley, which flourished between 2500 and 1500 B.C., and whose remains were found at Mohenjo-Daro and Harappa. A visitor interested in ancient cultures can see unfolding before his eyes a panorama of history that will take him from the Stone and Bronze ages through the Indus Valley, Hindu, Buddhist, and Muslim periods of Pakistan's past. He will also see the growth of modern Pakistan during the period of its recent emergence as an independent nation.

Pakistan's present culture bears the imprint of all its past civilizations. Among the many articles found at Mohenjo-Daro are toy carts that could be copies of the bullock carts still plying the dusty roads of the farmlands of Sind. Many utensils used by the village folk around Harappa in the Punjab are direct descendants, and sometimes exact reproductions, of the pots unearthed by archeologists. Excavations of these two sites have exposed orderly rows of well-built houses with straight, paved streets and

The Badshahi Masjid, or Imperial Mosque, in Lahore.

public sewers. The present-day dwellings in small towns still conform to this pattern, although the larger cities have adopted various forms of modern architecture. Buddhist temples, stupas (shrines), and monks in saffron (yellow-orange) robes remind the visitor of Pakistan's Buddhist past. The customs, places of worship, utensils, and even the furniture of the Hindu minority take one back to the time when Hinduism was dominant in the land. And since Pakistan is now a predominantly Muslim country, the traditional domes and minarets (towers) of the mosques are very much in evidence everywhere. Pakistani Muslims still respond to the daily calls to prayer just as their ancestors did many centuries ago.

With all this, the visitor will see streets crowded with automobiles; men and women rushing to their work in factories and offices; universities, colleges, schools full of students learning modern techniques of industry and scientific research; homes with modern conveniences in the cities and even in many smaller towns; and factories busy producing goods both for export and use at home. For Pakistan today is a society in rapid transition, maintaining many of its traditional features while adapting itself to the needs of modern industry and technology.

Pakistan has had an often stormy past. In the years since independence, its history has sometimes been no less turbulent. When Pakistan began life as a nation in 1947, it was composed of two distinct regions—West Pakistan and East Pakistan—situated almost 1,000 miles (1,600 kilometers) apart. Political antagonisms between the two regions eventually led to armed conflict and civil war, and to war between Pakistan and India. In 1971 the former East Pakistan broke away from Pakistan and declared itself the independent nation of Bangladesh. (A separate article on BANGLADESH appears in this volume.)

Shepherds tend their flocks in the northern mountains of Pakistan.

FACTS AND FIGURES

ISLAMIC REPUBLIC OF PAKISTAN is the official name of the country.
CAPITAL: Islamabad.
LOCATION: South Asia. **Latitude**—36° 55' N to 23° 41' N. **Longitude**—75° 23' E to 60° 52' E.
AREA: 310,402 sq. mi. (803,940 sq. km.).
PHYSICAL FEATURES: Highest point—Tirich Mir (25,263 ft.; 7,700 m.). **Lowest point**—sea level. **Chief river**—Indus.
POPULATION: 55,000,000 (estimate).
LANGUAGE: English (official), Urdu (national), plus various regional languages, including Punjabi, Pushtu, Sindhi.
RELIGION: Muslim (predominant religion), Hindu, Christian, Buddhist.
GOVERNMENT: Republic. **Head of state**—president. **Head of government**—prime minister. **Legislature**—parliament. **International co-operation**—United Nations, Central Treaty Organization (CENTO).
CHIEF CITIES: Karachi, Lahore, Lyallpur, Hyderabad, Rawalpindi, Multan, Peshawar, Islamabad.
ECONOMY: Chief minerals—chromite, gypsum, limestone, iron ore, coal, petroleum, natural gas. **Chief agricultural products**—wheat, cotton, millet, barley, maize (corn), rice, oilseeds, tobacco, sugarcane. **Industries and products**—cotton textiles, paper and board, cement, leather products, chemicals and pharmaceuticals, fertilizers, sugar, cigarettes, preserved food. **Chief exports**—cotton, cotton textiles, wool, hides and skins, fish. **Chief imports**—machinery, vehicles, electrical equipment, iron and steel, various consumer goods.
MONETARY UNIT: Rupee.
NATIONAL HOLIDAY: March 23, Pakistan Day.
NATIONAL ANTHEM: "The National Anthem."

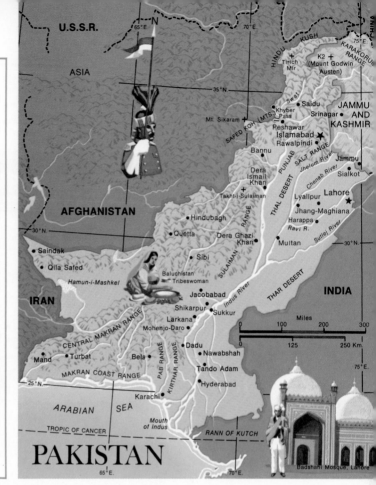

PAKISTAN

Badshahi Mosque, Lahore

THE LAND

Pakistan now comprises an area of 310,402 square miles (803,940 square kilometers). It is bounded by India in the east and southeast, Afghanistan in the northwest, Iran in the west, and the Arabian Sea in the south. Part of the northern boundary is covered by the disputed territory of Kashmir, which is claimed by both India and Pakistan. Pakistan consists of four provinces: the Punjab, the North-West Frontier Province, Sind, and Baluchistan.

The greater part of Pakistan is mountainous. In the north the mountains often reach great heights, and some are eternally covered with snow. Tirich Mir, the highest mountain peak in Pakistan, is 25,263 feet (7,700 meters).

The western areas of the region are part of the great Iranian plateau, which consists mainly of a series of plateaus of varying elevations surrounded by high mountain ridges that in Pakistan reach a maximum height of 11,290 feet (3,441 m.) at Takht-i-Sulaiman. The Iranian plateau is mostly arid and barren, although some valleys have water and are cultivated. The eastern part of the region is a plain, although in the north occasional hills jut into the otherwise level skyline.

The most important river in Pakistan is the Indus, which rises in Tibet and, after flowing through mountains and ravines, enters the plains of Pakistan. The Indus, together with its tributaries, is the mainstay of the agriculture of the country, which depends upon an extensive network of irrigation canals. There is limited rainfall, and mountain snow is an im-

A village in the Karakoram Mountains of northwest Pakistan.

A hillside in a mountainous region of Pakistan terraced for farming.

portant source of water for the rivers that feed the canals. There are two deserts, the Thal in the north and the Thar in the southeast. Adjoining the delta of the Indus River is a salt plain called the Rann of Kutch.

The northern mountains of Pakistan are well forested, but the western mountains are bare. In the plains the only natural forests are in the low-lying areas along the rivers. Temperatures vary widely. In the mountains the temperature depends a good deal on the elevation and ranges between Arctic and torrid. Temperatures on the plains also vary widely from season to season, ranging from 114 degrees Fahrenheit (46 degrees Celsius) in June to 40 degrees F. (4 degrees C.) in January.

The northern part of the region abounds in wildlife. It is the home of the majestic Marco Polo sheep, the blue sheep, the Siberian ibex and the markhor (both species of wild goat), the snow leopard, the black bear, and the brown bear. In the plains live deer, bears, jackals, wildcats, musk cats, hyenas, and various kinds of rodents and reptiles. There are over 100 varieties of birds.

THE PEOPLE

The population of Pakistan is estimated to be about 55,000,000. Muslims make up most of the population. Hindus form the largest minority religious group, and there are smaller numbers of Christians and Buddhists.

Since time immemorial, groups and tribes of peoples have migrated from Central Asia and settled in what is now Pakistan. These were mainly of Aryan or mixed Aryan and Mongol stock. The Aryan settlers spoke a language belonging to the Indo-European family, which includes most of the languages of modern Europe. There were similar migrations from western Asia, though on a much smaller scale. The people who built the Indus Valley civilization were Dravidians, who differed in physical appearance, customs, and language from the later-arriving Aryans. All these elements have mixed in varying degrees in the different parts of Pakistan, although the dominant racial strain is Indo-Aryan.

Way of Life

Pakistani Homes. Rich and upper-middle-class Pakistanis live in comfortable villas with all modern conveniences. In the larger cities the lower middle classes and sometimes the well-to-do live in apartments. A traditional Pakistani house in a small town is built around a courtyard. The homes of the well-to-do are often built of baked brick and may be divided into two parts: one is used by the men to receive visitors and conduct their affairs; the other is private and is reserved for women and children or for men when their presence is not needed outside. The number of such houses, however, is not large. In the villages the more typical houses are generally built of well-kneaded mud or unbaked bricks and usually have flat roofs. A peasant's house will have a fairly large enclosure where he keeps his cattle, his store of grain and fodder (feed for his cattle), and agricultural implements.

Food. Wheat bread is the staple food of the people of Pakistan. This is supplemented with meat and vegetables. The national drinks are tea and sour milk. *Dahi,* the Pakistani form of yogurt, is also eaten. Pakistani food is spicy, but it is not as hot as the food of, for example, South India or Mexico. On ceremonial occasions, pilaf (locally called *pulao*) is

These nomads live in a desolate part of Pakistan.

served. Pilaf is rice cooked with meat and broth. The more sophisticated dishes in the towns and the cities resemble Central Asian cooking. They consist of pilaf, various kinds of kebabs (broiled, baked, or fried meat, either cut into cubes and pieces or minced), and sautés or stews (called *qormas* and *salans*) made of meats, vegetables, and sometimes eggs and fish. Pakistani food achieves a wide range of flavors and subtlety because of the variety of ingredients and spices. Pakistani meat dishes mostly use lamb, chicken, or beef. Pork is not eaten, since its use is forbidden in Muslim religious law.

Dress.   A considerable number of Pakistani men living in the larger cities and towns wear European clothes, although on formal or festive occasions many put on national costume. This consists of a fur hat called a Jinnah cap, a long jacket called a *sherwani,* and either a *shalwar* (baggy trousers) or tight-fitting white breeches (*churidars*). The village folk wear the *shalwar* together with a rather long shirt, worn outside the trousers, and a turban.

Sophisticated women in Pakistan change the style of their clothes as rapidly as women in other countries. But generally speaking the most common dress for women is a *shalwar* and a shirtlike *qamis* or *kurta.* Women in the villages wear, in addition, a *dupatta,* a long scarf that covers the head and bosom. Fashionable women, however, wear the *dupatta* only to the chin and let it hang behind the back on two sides.

Festivals.   Pakistani festivals are mostly religious in nature. The greatest occasion is the Id al Fitr (the festival of the breaking of the fast), which follows Ramadan, the Muslim month of fasting. Early in the morning during this holiday the family will have a light breakfast of vermicelli (a thin spaghetti), sweet milk, pistachios, raisins, and dates. The mosques then fill with worshipers, all dressed in their best; children especially put on gay clothes. This is followed by visits to friends and relatives, and all visitors are offered sweets. The other Id, or festival, is the Id al Adha (the "feast of sacrifice"), which commemorates Abraham's willingness to sacrifice his son at God's command. It is celebrated similarly, except that on this occasion meat dishes predominate.

Language. English is the official language. However, it will eventually be replaced by Urdu, which now has the status of a national language. Urdu is the most commonly used language, but there are also several provincial or local languages, of which the most important are Punjabi, Pushtu, and Sindhi. These languages belong to the large Indo-European family.

Pakistan's Cities

Pakistan has a number of colorful cities. Of these the youngest is **Islamabad**, the new capital of the country. Many official buildings and houses for employees of the central government have been constructed,

The entrance to the old part of the city of Lahore.

A new housing project in the rapidly growing capital city of Islamabad.

Karachi, Pakistan's largest city.

and the population of over 50,000 is growing rapidly and is expected soon to triple. The city is surrounded by the green hills that are the foothills of the mighty Himalayas. The elevation of the city itself is close to 2,000 feet (610 m.) above sea level. All the buildings are modern, and expert landscaping has brought out the attractive natural features to their fullest advantage.

Islamabad lies just a few miles from the old city of **Rawalpindi**, which served as the interim capital of Pakistan while the new capital was being built.

Pakistan's largest city is **Karachi**. It is a comparatively young city, dating to the early 18th century when a fort was built there to protect the sea trade that was attracted by the excellent harbor. The harbor makes Karachi even today practically the only outlet to the sea for the whole of Pakistan and for neighboring Afghanistan. Since the creation of Pakistan the population of the city has increased many times and is now about 3,500,000. Its airport is of international importance, since it is located halfway between Europe and the Orient. Karachi has grown into a large industrial center, producing textiles and a great variety of consumer goods, some of which are exported. The city has a university, the University of Karachi, the Pakistan Institute of International Affairs, and a number of other educational, cultural, and research institutions. Karachi is spread out over a wide area, and most of it has a suburban look because of the many villas.

North of Karachi lies **Hyderabad**, a city of over 650,000, founded in 1782. The older part of the city is very picturesque. An interesting feature of the homes is a boxlike construction on the roofs designed to catch

Saidu, capital of the district of Swat in the North-West Frontier Province.

Pakistani children being taught in an outdoor school near Lahore.

the breezes coming from the sea about 100 miles (160 km.) away. The air is forced into the houses by its own pressure and cools the interiors during the hot summer months. The University of Hyderabad is located in a suburb, Jamshoro.

One of the oldest cities in Pakistan is **Multan**, which dates back to 320 B.C. and was in existence when Alexander the Great invaded the region. The city has an unbroken history and at one time was famous as a center of learning and culture. It is known for its tiles of medieval Persian design. It has about 560,000 people.

Lyallpur, with a population of almost 800,000, is a modern industrial city and a railroad junction. It is the home of the Pakistan University of Agriculture.

Lahore is the provincial capital of the Punjab and an industrial and trade center, with a population of about 1,750,000. It is the seat of the oldest university in Pakistan and is noted for its considerable educational and cultural activities. Lahore has many monuments from various historical periods because throughout its history the city has been the capital of either a province or a kingdom. It has several mosques of historical importance, the best-known of which is the Badshahi Masjid, or Imperial Mosque, built by the Mogul emperor Aurangzeb in the 17th century. The fort, built during this same period, contains several palaces and other interesting buildings.

Peshawar, which has about 285,000 people, is also a very old city, dating back many centuries to ancient Hindu and Buddhist times. It also has monuments from the period of the Moguls, a great Muslim dynasty that once ruled here. Peshawar has a university too. Both Lahore and Peshawar remind one strongly of the cities of Central Asia in their plans, general atmosphere, and feeling.

THE ECONOMY

The economy of Pakistan is predominantly agricultural. The main food crop is wheat, although some highly prized varieties of rice are grown that are in great demand for export. In recent years, through the

These boys are apprentice rug weavers.

Camels are a common sight in the arid regions of Pakistan.

A silk mill. The manufacture of textiles is one of Karachi's chief industries.

combined use of better seed and fertilizer, spectacular results have been achieved in raising the yield of wheat, so that Pakistan has become nearly self-sufficient in food. Other food grains include barley, maize (corn), and millet. Cotton is an important cash crop. The long-staple American varieties sustain the local textile industry and are also exported in large quantities. Tobacco, sugarcane, and oilseed also are grown on a large scale. Pakistan has many varieties of fruit and exports some of its citrus fruits. Poultry farming has made rapid strides, and fisheries are being developed scientifically. The coast of Pakistan is rich in seafood, and the freshwater lakes and rivers abound in fish.

In 1947 when Pakistan achieved independence there was hardly any industry, but during the last 2 decades the country has made rapid progress. The main industry is the production of cotton textiles, in which Pakistan has built up a sizable export trade. Other industries produce paper and board, cement, leather products, chemicals and pharmaceuticals, fertilizers, sugar, cigarettes, and preserved food. Plans are now being developed for establishing heavy industries. The discovery of large deposits of natural gas has helped industrialization considerably. Other mineral resources include deposits of coal, salt, gypsum, chromite, iron, limestone, brine, and clay. Some petroleum is also produced.

HISTORY

Ancient Times. The Indus Valley, along with some adjoining areas in what is now Pakistan, is one of the oldest cradles of civilization, com-

A spice seller.

parable to those of ancient Egypt, Mesopotamia, and China. In about 1500 B.C. there were massive migrations of Aryan peoples from the north. The Aryans brought with them a new language, Sanskrit, the ancestor of the Indo-European languages of modern Pakistan and India. They also brought with them their own customs and religious traditions, which, mixed with some of the earlier Dravidian culture, were to develop into Hinduism. Over the next several centuries numerous Indo-Aryan kingdoms spread across the northern part of the Pakistan-India subcontinent.

In the 6th century B.C. Pakistan came under the rule of the Persian Achaemenid dynasty. This was followed by Alexander the Great's invasion of the region in 327 B.C. The first empire to weld together almost all of the subcontinent was the Mauryan, which flourished from the 4th to the 2nd century B.C. Its greatest ruler, Asoka (who reigned during the 3rd century B.C.), became a convert to Buddhism

After the breakup of the Mauryan empire, Pakistan saw a succession of rulers establish themselves in different parts of the area. In the 1st century A.D., the Kushans, a people from Central Asia, established an empire that lasted until the 3rd century A.D. During the Kushan period Buddhism flourished in Pakistan and has left behind great architectural remains and sculpture. The Kushans were succeeded by the Sassanians (a dynasty that ruled the Persian empire) and by the Huns, a people who came from more easterly parts of Asia.

The Coming of Islam.   Hinduism had just begun a comeback in the area when, in the 8th century, the Arabs conquered Sind in southern Pakistan. The northern part of Pakistan came under Muslim sway in the beginning of the 10th century. Since then Islam has exercised great influence upon the life and culture of the people. Many of the long list of conquerors and rulers were from Central Asia, and the culture of modern Pakistan still retains strong traces of Central Asian influence.

The last Asian empire of the subcontinent was that of the Moguls, a Muslim dynasty related to earlier Turkish invaders. The Moguls flourished from the 16th to the 18th century, when control of the subcontinent passed into the hands of Great Britain.

In the beginning of the 20th century, nationalism began to sway the minds of the people, and anti-British feelings began to find means of expression. Following World War I, a mass movement was organized all over the subcontinent demanding self-government.

The Struggle for Independence.   After this there was continuous pressure upon the British to grant more self-government to the people. Two measures, the Government of India acts of 1919 and 1935, embodied British thinking on constitutional advance for their Indian empire. But both fell short of the expectations of the political leaders and the people. When World War II broke out, the demand for freedom grew more intense, and Great Britain also thought it advisable to gain the co-operation of the people in the war effort. At the end of the war the British decided to quit the subcontinent.

However, when the transfer of power seemed inevitable and near, the divisions among the peoples came to the surface. The deepest and the most important of these was between the Hindus and the Muslims. The nature of the difference between the two is still not properly understood outside the region. The Hindus and the Muslims do not form two religious denominations among the same people, as, for example, Christians and Jews do in many Western countries. Instead, they represent two distinct cultures and ways of life. The differences persist through all aspects of life—food, dress, utensils, literature, and thought patterns. Throughout history, since the coming of Islam to the subcontinent, the Muslims had made successful efforts to maintain their distinctive culture and to keep their communities intact. When the emergence of a numerically superior Hindu nation became imminent, the Muslims, through their national organization, the Muslim League, under the leadership of Mohammed Ali Jinnah, demanded a separate sovereign state in the areas where they were in an absolute majority. In 1947 an independent Pakistan came into existence.

Pakistan started as a parliamentary democracy in 1947. In 1958 power was taken over by Field Marshal Mohammed Ayub Khan, who established a presidential form of government. Under the Constitution proclaimed by him, the president and legislature were elected by an electoral college, which was elected on the basis of adult vote. Under that system Pakistan had a stable government for 10 years. In 1968, however, when the time for the second elections came near, the people demonstrated in favor of a return to parliamentary democracy and direct elections to the legislature. Widespread riots made it difficult to hold the elections.

In 1969 President Ayub Khan resigned and asked the Army to take over the country. The commander-in-chief of the Army, General Agha Mohammed Yahya Khan, then imposed martial law. Soon after, he assumed the office of president. In late 1970 elections were held to form a national assembly to write a new constitution.

The results of the election gave a majority of the seats to the East Pakistan Awami League, which favored local autonomy for the Eastern region. The Pakistan People's Party of West Pakistan, which supported a strong federal government, received the next largest number of votes. The conflict between political leaders of the East and West led President Yahya Khan, in 1971, to postpone the opening of the national assembly. The postponement, however, led to a general strike in East Pakistan and the threat of secession. The government responded by banning the Awami League and sending troops to the Eastern region.

War with India. The government, fearing a movement to break away from Pakistan, handled the disorders in the Eastern region with great severity. Many East Pakistanis were killed and millions fled across the border into India. In 1971, following incidents along the border, Indian troops invaded East Pakistan. In the brief war that followed, the West Pakistani forces were defeated, and East Pakistan declared its independence as Bangladesh. General Yahya Khan resigned and was succeeded as president by Zulfikar Ali Bhutto. In 1973 a new constitution came into effect under which Bhutto was elected prime minister.

The Present Government. Under the 1973 constitution Pakistan has a parliamentary form of government. The president serves as the ceremonial head of state. The head of government is the prime minister. The legislature consists of a parliament made up of two houses, the National Assembly and the Senate. The National Assembly, which is elected directly by the people, chooses the prime minister. The Senate is elected indirectly by provincial and tribal assemblies.

I. H. QURESHI, Vice-Chancellor, University of Karachi; author, *The Pakistani Way of Life*

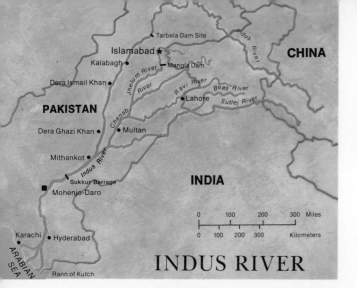

INDUS RIVER

INDUS RIVER

The Indus is one of the great rivers of Asia. From its starting point amid the jagged mountain peaks of Tibet, it flows across Kashmir and down the length of Pakistan into the Arabian Sea. To the people of ancient India, the Indus was known as "the king of rivers." Its name comes from the Sanskrit word *sindhu*—meaning "ocean" or "great water." It is from *sindhu* and Indus that India derives its name. Several thousand years ago, the Indus—like the Nile and the Tigris and Euphrates rivers—was the wellspring of one of the world's earliest civilizations. Between 2500 and 1500 B.C., a highly advanced culture thrived along the lower reaches of the Indus, centered around the city-states of Mohenjo-Daro and Harappa. It was also in the plains watered by the Indus and its tributaries that the Vedas, the most sacred texts of Hinduism, were composed. Sometimes called the gateway to India, the Indus has been important in history both for economic and military reasons. It served as a barrier against invaders and its waters nourished the soil of what was then northwest India and is now Pakistan.

The Indus begins its more than 1,900-mile (3,000 kilometers) course in the Trans-Himalayas of southwestern Tibet. The river first winds its way in a northwesterly direction through high-walled gorges and spectacular mountain valleys. Then it veers sharply to the southwest, finally breaking through the mountains to enter the Punjab plain near the city of Kalabagh. Below this point, the river flows parallel to a flat plain that is watered by the Indus and its tributaries. Near the town of Mithankot, the Indus is joined by its main tributaries (the "five waters"): the Jhelum, Chenab, Ravi, Beas, and Sutlej rivers. Along its route, the Indus passes several cities, including Sukkur and Hyderabad. South of Hyderabad, the river divides into several branches, which empty into the Arabian Sea.

During the rainy season, the Indus is dangerously unpredictable, and its swollen waters sometimes overflow its banks for many miles. Except for flood periods, however, the river is generally too shallow for large vessels. The river's low level is due to the extensive drawing off of its waters to irrigate the surrounding farming areas. The Indus Valley is a very fertile region, producing wheat, corn, rice, and a variety of fruits and vegetables. Because the waters of the Indus irrigate both Pakistani and Indian land, disputes have occasionally arisen between the two countries over the mutual use of the river water.

Reviewed by B. G. GOKHALE, Director, Asian Studies Program, Wake Forest University

INDIA: AN INTRODUCTION

by Chester BOWLES
Former United States Ambassador to India

Next to the control of nuclear weapons and the so-called population explosion, I believe that communication among people and nations is the most critically important question of our times.

Many of us find it difficult to understand the hopes and fears of our own fellow citizens. And when it comes to the people of other lands, with different languages, religions, living standards, and cultures, the barriers to genuine understanding are vastly more formidable.

For instance, when most Westerners think of India (which is not very often) the impression that comes to mind is of a faraway land with an overabundance of people, monkeys, cows, maharajas, cobras, polo players, dust, famine, dirt, and disease. Yet any thoughtful visitor who travels as long as 2 or 3 weeks in India can see with his own eyes that this stereotype no longer fits the facts. For this vast and complex land, with its ancient cultures and awakening masses of people, is in the process of profound change.

The nature of these changes, the kind of political and economic systems that finally emerge, and above all what the Indian people think and feel about their own future and their relations with other nations are critically important to the course of world peace.

There are three major reasons why this is so:

India, with its more than 500,000,000 people (over one seventh of mankind), is second only to China in population.

India is strategically located between East and West; a politically stable India with rising standards of living is essential to the peace of Asia.

India is one of the few developing nations genuinely committed to democratic methods in its efforts to provide a greater measure of security and economic and social justice for its people. The obstacles are formidable and it is often these with which foreign observers are most concerned. But side by side with the difficulties are some solid achievements. The following is an effort to strike a balance between the two.

In 1947 the British, after some 200 years of rule, left India in response to massive programs of nonviolent action, organized and led by Mohandas K. Gandhi and Jawaharlal Nehru. No doubt the British, with their superior military power, could have maintained their colonial rule a few years longer. But the cost in bloodshed and bitterness was a price that they wisely refused to pay.

The British left behind in India a competent administrative service, a significant amount of industry, one of the most extensive railroad systems in the world, and a genuine respect for democratic principles based on constitutional law.

On the negative side of the balance sheet, they also left some formidable problems. For instance, the Indian administrative services, although excellently run, were largely limited to communications, defense, tax collection, and law and order. Except in a very few places, the British made almost no effort to create an educational system, a public health system, or even a good road system in rural India. Agriculture under the British was old-fashioned, and the recurring famines were assumed to be unavoidable. (In the last major famine, which occurred in Bengal in 1943, it is estimated that some 3,000,000 people died.)

In addition, the conflicts between followers of the two major religions—Islam and Hinduism—often erupted in violence, and eventually led to the tragic partition of British India into India and Pakistan. The caste system was deeply rooted. Poverty and disease were widespread. The average life-span was short and the per capita income low. Railroads and industry were in a bad state of repair, largely as a result of the strain placed on them during World War II. Finally, and perhaps most important of all, India's long experience under foreign domination had created a deep sense of insecurity. Most Indians lacked confidence in themselves, their nation, and their future.

When we consider the magnitude of these difficulties, the economic, social, and political progress that India has made since independence is extraordinary.

The most publicized and perhaps the most important achievement is the recent progress in agriculture, which is referred to as the Green Revolution. This revolution was made possible by the massive introduction in the mid-1960's of new kinds of rice, wheat, and maize (corn) seeds; by a major expansion of irrigation; and by the more extensive use of fertilizers and pesticides.

There has also been a re-assuring expansion of industry. Modern Indian factories are now turning out, and even exporting, trucks, jeeps, electric generators, and railroad equipment such as locomotives and freight cars.

Educational facilities have been expanded greatly since independence. More than 2,000,000 young men and women are now attending Indian universities, several of which will stand comparison with the best in the United States and Western Europe. More than 85 new medical colleges have also been built since independence. Fifteen years ago 100,000,000 cases of malaria a year were accepted as inevitable; today the disease has been reduced to 50,000 cases a year. Agricultural schools are now graduating each year large numbers of able young Indian technicians who have the knowledge and the determination not only to keep the Green Revolution going but to expand it throughout India.

In respect to civil rights, Indian newspapers, with some occasional lapses, are free to publish the news as they see it, and freedom of speech is respected not only in Parliament but in the dialogue between the government and the people.

Although religious differences between Hindus and the 60,000,000

Muslims now living in India periodically erupt into violence, the leaders of India, supported by the vast majority of the people, deplore these conflicts and are earnestly striving to eliminate them; and there is some evidence of progress. For instance, in 1967 India elected a Muslim president; there are two Muslim members in the Cabinet, and many others hold high office in the government, the diplomatic service, and the Army.

But perhaps India's greatest achievement is the simple fact that it has survived as a unified, democratic, independent nation. Consider India's vastness and diversity. Its population is roughly equal to that of Europe without the Soviet Union. There are about the same number of Indian states and federal territories as there are European nations, and approximately the same number of languages. The physical and cultural differences between an Indian from the Punjab and one from Madras, or from Bengal or Gujarat, are as great as the differences between a Swede and a Greek, or between a Belgian and a Portuguese.

In the face of these difficulties, the new India has somehow succeeded in maintaining its political and economic unity, a unity that Europe has been seeking, with only slight success, for 2,000 years. After more than 20 years of independence, India still has the world's largest common market, a single prime minister, a freely elected national parliament, and an all-India constitution and court system.

So much for the favorable side of the ledger. Let us now consider the difficulties that still must be overcome.

India's political prospects are clouded by the fact that the Congress Party, one of the oldest democratic political organizations in the world, is now under heavy strain, and it is impossible to foresee what new political arrangements may develop. Among the most explosive political issues that Indian leaders will be called upon to face is the growing inequality of income in the rural areas. This, paradoxically, is a direct result of India's economic progress.

Three out of four Indians live in small villages, and most of them have for generations been poverty-stricken, insecure, apathetic, and committed to the old ways. When almost all families in a village are poor, there is little protest or bitterness—only apathy and a sense of resignation.

The Green Revolution has suddenly awakened tens of millions of these rural people to the possibilities of a better life for themselves and for their children, and they are eagerly pressing for their fair share. Although increased agricultural production and relatively high prices benefit almost everyone, the major share goes to those who own land. The landless laborers (who total 21 percent of the people of rural India) are, to be sure, receiving somewhat higher wages. But the gap between them and the landowners whose land they till has been steadily increasing.

There is also the question of population control, which is now recognized as India's number one problem. The objective of government planners is to lower the annual birthrate by one half. It is hoped that by 1985 this may stabilize India's population at about 750,000,000 people.

Once again, the outlook is mixed. On the plus side, birth control is accepted and supported by the major political parties. Cultural barriers are not a significant factor, and since 1965 the program has been given the very highest priority by the government. On the negative side is the hard fact that the birth control devices presently available are not well adapted to underdeveloped nations such as India. Only if a breakthrough in a more practical method of birth control occurs can India meet its 1985 target.

The efforts of the Indian Government are further complicated by the fact that rising living standards do not necessarily make for political stability in India—or anywhere else. On the contrary, expanding production may actually increase political unrest unless the masses are given a sense of personal involvement in the process of community and national development.

India's tense relations with Pakistan and China are another unpredictable factor. India has twice fought Pakistan over Kashmir, and any Indian government that agreed to withdraw from the Kashmir Valley at this stage would promptly be thrown out of office. The most likely solution may be an association between India and Kashmir comparable to the relationship between the United States and Puerto Rico. Such a development, if it comes, will take shape gradually.

India's relations with Communist China, once cordial, have been strained ever since the Chinese attacked India's northern frontier region, part of which China claims.

Finally there is the key question of how much economic assistance India can expect in the next few years from the United States and other Western nations. The help the United States has given India in the last 20 years has contributed significantly to India's economic progress. If the United States were to lend India each year for several years what it spent in Vietnam every 3 days during the height of the war, the prospects for a democratic, independent, stable India would be immeasurably improved, and economic assistance from foreign sources might no longer be required.

In closing let us consider India's position in an all-Asia framework. The costly war in Vietnam has demonstrated that an American-dominated alliance cannot maintain political stability in Asia, nor is an alliance dominated by the Soviet Union likely to be any more successful. The people of Asia are eager at long last to live their own lives in their own way.

Their future will be determined largely by the ability of the larger independent Asian nations, such as Japan, India, Indonesia, and Australia, to work together toward common objectives. Such an association could have a decisive political impact on the political balance not only in Asia but throughout the world. Eventually, it could be expected to bring about fundamental changes in the often difficult relations with China.

Whether or not this comes into being will depend in large measure on what happens in the next 10 years in India.

The Taj Mahal at Agra.

INDIA

The first Greek ambassador to India was a man named Megasthenes. Writing in the 4th century B.C. he said: "The Indians having abundant means of subsistence, exceed in consequence the ordinary stature, and are distinguished by their proud bearing. They are also found to be well-skilled in the arts. . . . And while the soil bears on its surface all kinds of fruits which are known to cultivation, it has also underground numerous veins of all sorts of metal . . . much gold and silver, and copper and iron in no small quantity, and even tin and other metals. . . ."

Some 16 centuries later, on September 28, 1919, Mohandas K. ("Mahatma") Gandhi, the great leader of the Indian independence movement, said: "India is a land of extreme poverty. Hundreds of thousands in India can get only one meal a day. . . . Indian peasants are destitute. . . ."

In 1952 Jawaharlal Nehru, the first prime minister of independent India, stated: "India is a mother country, which has influenced in the past vast sections of the human race in Asia. She still retains that storehouse of cultural vitality that has given her strength in the past, and at the same time she has natural resources, the scientific, technical, industrial and financial capacity, to make her a great nation in the modern sense of the word."

INDIA

U. S. S. R.

AFGHANISTAN

PAKISTAN

Disputed Boundary

K2 (Mt. Godwin Austen)

Cease-fire Line

Karakoram Pass

JAMMU

Islamabad ✪
Rawalpindi •

★ Srinagar

AND KASHMIR

CHINA

• Jammu

Amritsar •
Lahore •

PUNJAB

HIMACHAL
PRADESH

Indus River

Kamet •

T I B E T

• Lhasa

Mc Mahon Line

Mt. Everest •

Nanda Devi •

• Juliundur

Ludhiana •
Chandigarh-
★ Simla

Bhagirathi R.

Himalayas

NEPAL

ARUNACHAL PRADESH
(NORTH-EAST FRONTIER AGENCY)

★ CHANDIGARH

Katmandu ✪

SIKKIM
★ Thimbu

BHUTAN

Brahmaputra River

HARYANA

Gangtok ★
• Darjeeling

Naga Hills

Sutlej River

Delhi •
New Delhi ✪

• Meerut
• Moradabad

Jumna River

ASSAM

NAGALAND

Gauhati •

• Kohima

Bareilly •

MEGHALAYA

Shillong ★

MANIPUR

DELHI

Aligarh •

UTTAR

Cherrapunji •

Thar Desert

RAJASTHAN

Mathura •
★ Jaipur

• Agra

• Lucknow

BANGLADESH

• Imphal

Jodhpur •
• Ajmer

Chambal River

Kanpur •

PRADESH

Dacca ✪

Agartala ★

TRIPURA

Lushai
Hills

Indus River

Gwalior •

Allahabad •
Banaras •

Ganges River

• Patna

MIZORAM

BIHAR

WEST

• Hyderabad

BENGAL

Chin Hills

GUJARAT

Bhopal ★

MADHYA

Padma R.

BURMA

Ahmedabad •

Indore •

Vindhya Range

PRADESH

Chota Nagpur Plateau

Jamshedpur •

Howrah •
★ Calcutta

Rajkot •
Baroda •

Narbada River

Surat •

Tapti River

GOA, DAMAN, AND
DIU

Satpura Range

20°N

Nasik •

DADRA AND
NAGAR HAVELI

MAHARASHTRA

Aurangabad •

Nagpur •

ORISSA

Mahanadi River

Hooghly River

Bay of Bengal

Bombay ★

Poona •

Godavari River

Cuttack •

Bhubaneswar ★

Arabian
Sea

Sholapur •

Bhima River

ANDHRA PRADESH

• Vizagapatnam

Kolhapur •

Hyderabad •

Western Ghats

Panjim ★

KARNATAKA

Tungabhadra River

Kistna River

• Vijayavada

• Guntur

15°N

GOA, DAMAN, AND DIU

LACCADIVE, MINICOY AND
AMINDIVI ISLANDS

Deccan

Plateau

Eastern Ghats

Bangalore ★

Mysore •

Kozhikode •

KERALA

Madras ★

TAMIL NADU
(MADRAS)

• Salem

Nilgiri Hills

Tiruchirapalli •

PONDICHERRY

EUROPE

ASIA

15°N

Port
Blair ★

ANDAMAN AND NICOBAR ISLANDS

10°N

Cochin •

Cauvery (Kaveri) River

Palk Strait

INDIA

10°N

Tropic of Cancer

AFRICA

Coimbatore •

Madura •

10°N

Trivandrum ★

Gulf of
Mannar

Equator

Indian Ocean

AUSTRALIA

5°N

SRI LANKA ✪ Colombo

Indian Ocean

✪ *Country Capital*   ★ *State and Territorial Capital*

Map by J. Donovan

70°E   Maldives   80°E   90°E

A memorable past, a difficult present, and a promising future—all these are India. Since the dawn of history India has fascinated all who have come in contact with it. It is a land of gigantic contradictions and its contrasts are often overwhelming. Shaped like a giant pear, with its tapering end dipping into the Indian Ocean, India has been a bridge across which a long procession of invaders have marched and ideas and values have spread, both eastward and westward. India gave Buddhism to Asia, while to the West it gave its system of numerals, the so-called Arabic numerals. India's own traditions go back 5,000 years, and a list of its contributions to the civilizations of the world would be long and impressive.

India is a vast land. In area it ranks seventh among the nations of the world. With over 500,000,000 people, it is the second most populous country in the world (China is first). Perhaps nowhere else is the contrast between old and new so striking as in India. You may see a man reciting a 3,000-year-old prayer before beginning his work in an atomic energy plant. Equally sharp is the contrast between reality and promise. The per capita income of the average Indian is about $70 a year, one of the lowest in the world. India has some of the world's richest reserves of iron and coal, but its annual production of steel does not exceed 10,000,000 tons. It has great hydroelectric potential, but 75 percent of the people have to light their homes with oil lamps and use cow dung as domestic fuel. The Indian farmer has more than 2,000 years of experience behind him, but his yield per acre is one of the smallest in the world.

Such statistics, however, tell only a part of the story of India. Behind the figures of steel and agricultural production are the Indian people—industrious, long-suffering, patient, and hopeful. They are a product of their land and its history. Both exert a profound influence on their minds and lives. Both are a burden and a glory. The burden consists of the seemingly insurmountable problems of India; the glory lies in its past achievements and future possibilities.

FACTS AND FIGURES

INDIA—Bharat—is the official name of the country.

CAPITAL: New Delhi.

LOCATION: South Asia. **Latitude**—8° 04′ N to 37° 18′ N. **Longitude**—68° 07′ E to 97° 24′ E.

AREA: 1,261,813 sq. mi. (3,268,090 sq. km.).

PHYSICAL FEATURES: Highest point—Mt. Godwin Austen (K2) in Kashmir (28,250 ft.; 8,611 m.). **Lowest point**—sea level. **Chief rivers**—Indus, Ganges, Brahmaputra, Jumna (Jamuna), Godavari, Cauvery (Kaveri), Kistna (Krishna), Sutlej, Narbada, Tapti, Bhagirathi, Mahanadi.

POPULATION: 537,000,000 (estimate).

LANGUAGE: Hindi (official), English (second official language), plus various languages and dialects, including Gujarati, Bengali, Marathi, Oriya, Assamese, Sindhi, Punjabi, Urdu, Tamil, Telugu, Kannada (or Kanarese), Malayalam.

RELIGION: Hindu, Muslim, Christian, Sikh, Buddhist, Jain, Parsi, Jewish.

GOVERNMENT: Republic. **Head of state**—president. **Head of government**—prime minister. **Legislature**—parliament for the union. **International co-operation** —United Nations, Commonwealth of Nations, Colombo Plan.

CHIEF CITIES: Calcutta, Bombay, Delhi, Madras, Hyderabad, Ahmadabad, Kanpur, Bangalore.

ECONOMY: Chief minerals—iron ore, manganese, bauxite, gypsum, coal, copper, chromite, mica, salt, ilmenite, monazite, petroleum. **Chief agricultural products**—rice, millet, sorghum, wheat, peanuts, jute, tea, cotton, coffee, rapeseed, sugarcane, maize, rubber, pepper and other spices. **Industries and products**—textiles, cement, iron and steel, gasoline, food processing, chemicals, industrial machinery, handicrafts, copper and brass. **Chief exports**—tea, jute, iron ore, cotton goods and cotton yarn, leather, oilseed cake and meal, nuts, fish, tobacco, pepper and other spices, coffee. **Chief imports**—wheat, rice, raw cotton, machine tools and metalworking machinery, iron and steel, textile machinery.

MONETARY UNIT: Rupee.

NATIONAL HOLIDAY: January 26, Republic Day.

NATIONAL ANTHEM: *Jana-gana-mana* ("The Mind of the Multitude of the People").

THE LAND

The land of India is shaped by three major mountain systems, two great plains, six rivers, and the rain-bearing winds called the monsoons. In the north, standing like a giant wall, are the Himalayas, one of the world's great mountain ranges, stretching for some 1,500 miles (2,400 kilometers). Of the 146 Himalayan peaks, 40 rise above 25,000 feet (7,600 meters). The Himalayan landscape ranges all the way from the lush, tropical jungles of the mountain foothills to the snowy and windswept plateau bordering Tibet. In the folds of the Himalayas rise three of the great rivers of northern India, the Indus (now largely in Pakistan), the Ganges, and the Brahmaputra.

Almost in the center of the Indian subcontinent lie the Vindhya mountains, creating, in effect, two Indias. For centuries these mountains were a barrier to communications between north and south, and the two parts of the country were practically cut off from each other. To this day the people of northern and southern India are markedly different in many respects.

The south Indian peninsula is bounded by two mountain systems on its eastern and western flanks—the Eastern Ghats and the Western Ghats. The two systems meet in the Nilgiri Hills. In the extreme northeast are the Lushai, Naga, and Chin hills, which form the frontier between India, Burma, and China. These hills are the home of a number of tribal peoples. Because of the dense and often impenetrable jungle, they were isolated for a long period of time from the rest of India.

In spite of large-scale programs to modernize agricultural methods, many Indians still cultivate the land in centuries-old ways.

Rajasthan State in northwestern India is an arid region. These women are gathering drinking water.

The rivers of India are held in great reverence and affection by the Hindus, the major religious group in India. The holiest of Indian rivers is the Ganges. On its plains (the Gangetic plains) live over 160,000,000 people, who depend on the river for water for their crops and for transportation. Hinduism was born on the plains of the Ganges, and the place where the Ganges descends from the Himalayas onto the plains is especially holy to Hindus. About midway in the course of the river stands Banaras, the eternal city of Hinduism. Most of the famous temples of northern India are located along the Ganges and its important tributary, the Jumna (Jamuna), which flows by the great cities of Agra and Delhi. The Ganges is associated with the Hindu god Siva (Shiva). According to Hindu legend, it was Siva who first brought the river to earth from the heavens. The Jumna is associated with another great Hindu god, Krishna, for it is said that Krishna spent most of his childhood in the city of Mathura on the Jumna.

The three great rivers of the south are the Godavari, the Kistna (Krishna), and the Cauvery (Kaveri), which rise in the west and flow eastward to the Bay of Bengal. On their fertile plains and deltas live some 100,000,000 people. Unlike the northern rivers, which are fed by both the Himalayan snows and the monsoon rains, the rivers of the south are fed largely by the monsoon. Therefore, the volume of water in the southern rivers varies widely from season to season. Except during the summer monsoon, the south generally has little rainfall, and through the ages it has been necessary to irrigate the land to feed the people.

Upon these two great plains—the Gangetic plains of the north and the plains of the three southern rivers—rose the great kingdoms and empires of Indian history. Today 50 percent of the people of India live there. Year after year the rivers deposit rich mud on the plains, thus replenishing the fertility of the land. These lands have been farmed continuously for centuries, and only the rivers made it possible to continue to farm them without the use of fertilizers.

The Monsoon

The monsoon is as much a part of India as are the mountains, the plains, and the rivers. From March to June, the hot, dry season, the Indian sun beats down relentlessly on the land. Soil and vegetation become parched. The days are hot and the summer winds blow great quantities of dust over everything. The nights are often suffocatingly warm. Life becomes a great burden, and everybody, men and animals alike, is listless. Then toward the end of June the sky fills with black clouds, and the farmer looks up hopefully. Thunder and flashes of lightning light up the sky. Soon the skies open up and the rain pours steadily. Overnight, nature bounces back to life. Rivers that were once struggling ribbons of muddy water flow with great force through the hills and plains, spreading their waters across the land. The farmer is now busy with the work of planting and tending his crops. Through July, August, and September the rains continue and by October the crops dance in the wind. If the monsoon rains are good there is happiness everywhere; should they be late and wayward, it means disaster for a great many people. Some areas on the east coast may get some rainfall in the winter months, but for most regions the summer rains are practically the only source of water. The areas of heaviest precipitation are the Western Ghats and the northeastern part of the country. Cherrapunji in Assam State is one of the wettest spots on earth, with an annual rainfall of some 450 inches (1,100 centimeters).

Farmers transplanting rice during the monsoon season.

A work elephant hauling hardwood in a timber area of southern India.

Animal Life

India abounds in a variety of animal life. The foothills of the Himalayas teem with big game and the roar of the tiger is an almost familiar sound to many villagers. Assam has rhinoceroses and elephants, and the delta region of the Ganges is the home of the famous Bengal tiger. Leopards, wild boar, deer, and a great profusion of birds fill many of the forest areas. Peacocks and parrots are the most colorful, while the chief singing birds are the cuckoo and the mynah. Hawks are often trained to track game birds. The mountain eagle is regarded as sacred because it was the vehicle of the Hindu god Vishnu. India has hundreds of varieties of reptiles, of which the king cobra is the most majestic and the most dreadful. Along many jungle trails one can see the huge python, which can easily swallow a small goat or calf.

The elephant is both a royal animal and a beast of burden in the timber areas. In the past many maharajas, or native princes, kept elephants for ceremonial parades and processions. In the deserts of Rajasthan

Camels are a common sight in the city of Jaipur, capital of Rajasthan.

State the camel is the major form of transportation, and camel carts are sometimes seen trundling alongside modern automobiles in the streets of Delhi, Jaipur, and Ajmer.

The number of cattle in India is enormous. It is estimated that India may have nearly one fourth of the cattle population of the world. Hindus will not eat beef, for the cow is a sacred animal. It is associated with the god Krishna, who is called the Divine Cowherd. The cow is also the symbol of the sacredness of all life. In villages the work animal is the bullock, which pulls the plow and the carts. The bullock cart is the chief form of transportation in much of rural India, where roads become mud tracks during the rainy season. Each farm has at least a pair of bullocks and a cow. Each village has a number of dogs and cats, most of which are stray and make the rounds of homes for their food.

Monuments of the Past

The natural beauty of the Indian landscape is enhanced by countless monuments scattered across the length and breadth of the land. These include temples and mosques, palaces and tombs, cave temples and forts. Almost every dynasty in Indian history left great monuments behind, for their kings delighted in building great temples and palaces to reflect both their religious faith and their power. Each region developed its own particular style of art, and the treasures of the various regions adorn the country and museums the world over.

The great temples are in the south, especially in Tamil Nadu (formerly Madras State). In the Deccan region, near Aurangabad, are located two of the finest groups of cave temples. One, the Ajanta group, is Buddhist. An entire mountain has been scooped out to form a series of shrines and monasteries used by the Buddhist monks. The most impressive part of the Ajanta temples are the frescoes (wall paintings) in some of the caves. These were done between the 1st century B.C. and the 6th century A.D. and are world-renowned for their artistic beauty and richness of color. The themes of the paintings are taken from the life of Buddha and the world of man. The cave walls are crowded with scenes of men and women, gods and angels, birds and beasts. The second group, at Ellora, is chiefly Hindu. The most famous of the many temples at Ellora is the Kailasa temple. It was excavated from a hill and constructed from a single great rock. The walls are adorned with larger-than-life statues of the gods of the Hindu religion. The Ellora temples were built between the 7th and 9th centuries A.D. and are regarded as the greatest examples of the work of Indian craftsmen.

In the north the chief monuments are located in the cities of Agra and Delhi and in Rajasthan and Bihar states. The Taj Mahal at Agra is world-famous for its beauty. It is built of marble and was constructed as a tomb for Mumtaz Mahal, wife of the 17th-century Mogul emperor Shah Jahan. The building took some 20 years to complete and cost millions of rupees. The fort at Agra once housed the palaces of the Mogul emperors. Delhi has many renowned structures from different historical periods. Among these are the Qutb Minar, a minaret (tower used for calling Mus-

The Ajanta cave temples in the Deccan region of southern India.

The Kailasa temple at Ellora.

lims to prayer) some 238 feet (73 m.) high, and the tomb of Humayun, a 16th-century Mogul emperor. Rajasthan, the home of the warrior Rajputs, has many great palaces and forts that rank among the finest examples of Indian architecture.

INDIA—FROM ANCIENT TIMES TO INDEPENDENCE

The history of India is long and complex. The earliest known civilization is revealed in a few towns and cities that have been discovered by archeologists. The most important sites of this prehistoric period (about 2500–1500 B.C.) are Mohenjo-Daro in Sind and Harappa in the Punjab (now both in Pakistan). This earliest Indian civilization was contemporary with that of the Egypt of the pharoahs and was equally rich in its material life and sophistication. The cities were built according to well-laid plans and were busy commercial centers.

Around 1500 B.C. a group of people, the Indo-Aryans, migrated into India, probably from what is now southern Russia. The Indo-Aryans differed from the original inhabitants in their lighter skin color, language, social organization, and technology. Over the course of centuries they conquered various parts of northern India. The Indo-Aryans, who developed the religious beliefs that were to evolve into Hinduism, became

The Qutb Minar in Delhi is both a minaret and an emperor's tomb.

the dominant cultural element in India. Although they eventually became mixed with the original peoples of India, they have left their mark on many sections of the population.

In the 6th century B.C. two great religious movements—Buddhism and Jainism—arose in India. In the middle of the 6th century B.C. the Persians invaded the northwestern part of the country; they were followed by the Greeks under Alexander the Great in the 4th century B.C. The first great Indian empire—the Maurya Empire—appeared about 324 B.C. Its greatest ruler was King Asoka, who reigned from about 274 to 232 B.C. Asoka became a Buddhist and devoted his life to the spread of Buddhism in India and Ceylon. It was with Asoka that the monumental art of India really began. He had cave residences excavated for Buddhist monks and built stupas—semicircular mounds of brick that enshrine the relics of the Buddha. Asoka also had great pillars erected. Many of these are covered with inscriptions that tell us much about this period.

The next great empire of northern India (about A.D. 320–500) was that of the Guptas. This is considered the golden or classical age of ancient India. Under the Guptas, literature, the arts, science, and material prosperity reached peaks of greatness. During this period, too, Hinduism became firmly established as the religion of the masses of the people.

During the 5 centuries after the downfall of the Guptas, numerous small kingdoms fought against one another, and India was invaded by the Arabs and Turks. In 1206 the Turks established a kingdom in Delhi, and Muslim rulers controlled large parts of northern India. In 1526 the Moguls, who came from Central Asia, began to build a great empire, with capitals in Agra and Delhi. The Mogul Empire flourished until the beginning of the 18th century.

The European age of Asian history began when the Portuguese navigator Vasco da Gama reached India in 1498 by sailing around the Cape of Good Hope. In the 17th century the English East India Company set up trading stations in India. The English were followed by the French, and when the two nations were at war in Europe in the 18th century, their trading companies waged their own wars in India. By the second half of the 18th century the British emerged as the chief power in India. The last armed revolt against British rule by Indians, the Great Uprising (also called the Sepoy Mutiny) of 1857, was crushed. The following year responsibility for the administration of India was transferred from the East India Company to the British Parliament. From 1858 until 1947, when India achieved its independence, the country was ruled by a British governor-general, who also acted as viceroy in Great Britain's relations with the hundreds of separate Indian states.

One of the results of British rule was the rise of Indian nationalism. The Indian National Congress was organized in 1885, and after 1905 it became a mass-based militant organization. In 1920 Mahatma Gandhi assumed leadership of the congress, and in the 1920's and 1930's he led several passive resistance movements against the British. The British responded with a series of concessions that enlarged the membership of the Indian legislatures and increased their powers. In 1935 the provinces

The Golden Temple at Amritsar, center of the Sikh religion.

Bombay's Victoria Railway Station, a reminder of the British past in India.

were granted self-rule, and after World War II Britain began the withdrawal of its power from India.

Alongside the nationalism of the Indian National Congress there also arose a Muslim nationalistic movement led by Mohammed Ali Jinnah. Jinnah demanded the establishment of a separate sovereign state made up of areas with a majority of Muslims. This demand was granted in 1947. On August 14–15, 1947, the two independent nations of India and Pakistan were established and a historic epoch of Indian history came to an end.

The advent of independence roused great hopes among the people of India. Who are these people? What do they look like? How do they live?

THE PEOPLE

The people of India show great diversity in their ethnic makeup and cultural life. In skin color Indians range from very light to very dark, with an equally wide range in height, hair texture, and eye color. This is because India has been a melting pot of races since the dawn of the country's history. It is believed that the original inhabitants of India were black in color, short of stature, and broad-nosed. The Indo-Aryans, who are thought to have belonged to the Caucasian race, were light in complexion, long-nosed, and tall. Unlike the original inhabitants, who spoke a Dravidian language, the Indo-Aryans spoke a language belonging to the Indo-European family, which includes most of the European languages. These two groups mixed in the course of time. To this mixture were added other racial groups, such as the Mongoloid. It is therefore difficult, if not impossible, to speak of a single identifiable Indian ethnic type. If some generalizations are necessary, it may be said that the higher castes

A wedding feast. The people are Sikhs.

(the social and occupational divisions within Hindu society) tend to have lighter skin color. The people of the north also tend to be lighter-skinned than the people of the south. But there are many exceptions.

The people of northern and southern India differ markedly in a number of ways, including their choice of food and dress. In the south rice is the staple food, while in the north it is wheat. The southerner is a coffee-drinker; the northerner prefers tea. The southerner likes his food hot—liberally seasoned with peppers and red pepper powder; the northerner uses a lot of shortening in his cooking. Meat is eaten more frequently in the north than in the south, where the people tend to be orthodox Hindus and therefore vegetarians. The southerner usually dresses in a shirt and a long piece of cotton cloth draped around the lower part of his body; the northerner wears tight-fitting trousers, similar to riding breeches (jodhpurs), with a knee-length coat. Southerners are also more rigid in observing caste distinctions than northerners.

Indian Languages

There are 14 major languages and hundreds of dialects spoken in India. The two major language groups are the northern (which includes Hindi, Gujarati, Bengali, Marathi, Oriya, Assamese, Sindhi, Punjabi, and Urdu) and the southern (which includes Tamil, Telugu, Kannada or Kanarese, and Malayalam). English is an associate language of the Indian union. The northern Indian languages are offshoots of Sanskrit, the sacred language of Hinduism and one of the Indo-European family of languages. The south Indian languages belong to the Dravidian family and are pre-Sanskrit. Although the south Indian languages have been greatly influenced by Sanskrit, the two groups differ in vocabulary and grammar. Each language has its own script, which makes communication among the various peoples more difficult.

Village Life

India is primarily a land of villages. More than 70 percent of the people live in villages, which vary in size. Some are almost towns by Western standards, with as many as 5,000 people. Others may have no more than a few hundred people. The typical Indian village consists of a cluster of houses surrounded by farms and grazing areas, connected with the outside world by a mud path, which becomes practically impassable in the rainy season. Most of the houses are built of mud, plastered with cow dung, and thatched with the leaves of the palm and other trees. The living quarters are simple, usually one or two rooms that serve as living rooms during the day and bedrooms at night. The small kitchen is often dark and smoky, the common fuel being wood and cow dung. Some houses may have an additional room or two for the storing of food grains and tools. Close to the house is the cattle shed, sheltering a cow, a pair of bullocks, and a goat or two. Chickens run around the house and through the main street, and one may occasionally see a few pigs trotting through the narrow lanes.

The entrance to the house is through an open yard. In the middle of the yard is a sacred plant, the tulsi (also spelled "toolsey"), or holy basil, which is devoutly watered every day by the women of the house. The tulsi is held sacred as a symbol of the holiness of plant life and because of its medicinal qualities. It is used in the treatment of a variety of minor ailments. It also marks the presence of the divine near the household. The veranda at the end of the yard, on either side of the main door, is the part of the house most frequently used in dry weather. Here members of the family and their neighbors gather after work to exchange the latest news of the village and the surrounding areas. The village children

An Indian village.

A village on the Ganges plains, one of India's most densely populated areas.

play in the yard or in the streets and they are treated with a great deal of consideration.

Every village has a few wells that supply drinking water and water to wash clothes and pots and pans if there is no river nearby. These chores —the fetching of drinking water and the washing of clothes and kitchen utensils—are generally performed by the women. For them the village well is a convenient meeting place. As they do their work they gather all the news from each other and much of the news is later passed on to the men in the evening. Everyone thus knows what everyone else is doing or not doing. In the absence of newspapers (only 30 percent of the people of India can read or write) these channels of information keep people informed of the activities of their neighbors and the government.

If you step inside a village house you may see very little furniture in it. Most homes do not have chairs and tables, for people eat sitting on the floor. At night beds are spread on the floor, too. However, the head of the house may enjoy a charpoy, a four-poster bed with a wooden frame and hemp rope woven into a base. The pots and pans are generally made of clay, though the recent trend is to metalware. The hearth is made of three big rocks, which form an enclosure with one open side. In some homes there may be braziers, or grills, of brick and clay. The men are served their dinner first as they sit on the kitchen floor in a row or a semicircle. The food is eaten with the fingers. The meal generally consists of unleavened bread, baked fresh for every meal, rice and some curry, a vegetable, and red pepper powder or green pepper for seasoning. On special days there may be some meat (lamb or chicken) or fish. On holidays sweetened milk or rice and other sweets are served. The two big meals are the noon lunch (which the wife may carry to her husband at work in the fields) and the evening dinner. Many farmers also like a good breakfast of wheat cakes or unleavened bread before they start off for work. The menfolk drink tea or coffee.

The Caste System

The houses in the village are built close together and are grouped according to castes. Hindu society is divided into a number of castes in which membership is by birth. One belongs to a caste throughout one's life unless expelled for violations of its rules. Membership in a caste traditionally means attachment to a specific profession or occupation. Marriage is within the caste and is generally arranged by the parents. Caste membership also determines one's circle of social relationships, especially in the sharing of food in common.

The castes are arranged in a hierarchy, with the Brahman (priest) castes at the top and the Sudras (farmers, artisans, and laborers) at the bottom of the social ladder. There are hundreds of castes among Hindus all over India. They may generally be grouped under four broad categories: the Brahmans, Kshatriyas (warriors), Vaishyas (merchants and bankers), and Sudras. These are called the varnas (literally, "colors") and each varna is subdivided into a large number of castes and subcastes.

Essentially the caste system began with division of labor and occupational specialization, and many of the castes in traditional society still have this basis. A shoemaker, for instance, is obliged to render service to his village. In return he is given some land and payment in cash or its equivalent by each of his patron families. This ensures the continuation of essential services for the village as a whole and also a living for the craftsman. For centuries this kind of economic arrangement made for

A pottery merchant displays his wares in the town bazaar.

A wheelmaker in Agra. Cows, sacred to Hindus, are allowed to roam unmolested.

the smooth functioning of village life, and that is why the system worked so long. Each caste has its own occupational obligations and social and economic rights, which are carefully protected by two agencies of village government. These agencies are the caste council and the village council.

The Caste Council. Most castes have their own caste council (some have no formal council), with the right and duty of enforcing the caste rules. If a member violates the rules, the caste council tries him and orders various degrees of punishment. These range from a simple fine to expulsion from the caste. When a member is expelled from his caste he becomes an outcaste, and not only his own caste but also people of different castes will have nothing to do with him. An outcaste may redeem himself by suffering the prescribed punishment or penance. Otherwise, he must be prepared to lose caste or to migrate to an area where his past is not known. The caste council thus exercises strict control over the activities of its members. But it also protects the traditional rights of its members. In many cases a particular caste in an area may act as a clan or extended family. Its members often rely on the assistance of their fellow caste men in all social functions and when they are in economic difficulties.

The Untouchables. On the edge of the caste society live the untouchables. These are regarded as the lowest social groups, and their physical contact is considered to be ritually polluting. The origins of untouchability go back to the 2nd century B.C., when religious concepts developed among the Brahmans that required the avoidance of certain materials as unfit in Hindu ritual. People following occupations involving these materials (such as work in hides and skins and in disposal of the dead) were considered to be untouchable for ritual purposes. This ritual untouchability later developed (by the 7th century A.D.) into lifelong untouchability. Although untouchables are regarded as Hindus, they are not served by the Brahman priests, and their religious rites are markedly different from those of the caste Hindus. They are given the most menial of tasks, such as cleaning toilet facilities and sewers, disposing of dead animals, and other functions essential to the sanitation of the community. Their economic status is the lowest, and by tradition they were barred from any kind of formal education. Their life is the most degraded under the caste system. Every village has its own untouchable groups, and the total population of untouchables in India may number some 50,000,000. Both caste and untouchability have been declared abolished by the Indian Constitution of 1950. But although remarkable progress in improving the status of the untouchables has been made since independence, they are still the group most discriminated against in rural India.

The Village Council

The village council has functions somewhat different from those of the caste council. Representatives of all the households of the village are expected to take part in its activities. However, its executive functions are usually vested in a small group comprising the wealthy, powerful,

A sidewalk dentist in the city of Mathura treats his patient.

and leading men of the village. All matters concerning the life of the village are the concern of the village council. It settles disputes, punishes violators of rules of village behavior, and arranges for the worship of the village gods and the celebration of the village festivals. It may even ask for voluntary labor by the villagers on certain public work projects, such as keeping the village wells in good repair. The council is the representative body of the village, and its relationship with the Indian Government is channeled through the village headman and the accountant.

The office of the headman is generally hereditary and carries with it a great deal of prestige. He is expected to ensure law and order in the village and also to collect taxes for the government from the villagers. In this he is helped by the accountant, who maintains records of lands and taxes. Both officials have their own farmlands and are given additional payment for their work for the village. Their opinions are generally respected by both the people of the village and the government, and thus they function as a link between the two.

CITY LIFE

In spite of the fact that India is largely a country of villages, cities have always played a vital role in its history. As early as 2000 B.C. India had two large urban centers, Mohenjo-Daro and Harappa, each with an estimated population of over 40,000. Some of India's cities are as old as, if not older than, Rome. Banaras, for example, has had a continuous history since the 6th century B.C. India has over 100 cities with more than 10,000 people. Among these, Calcutta, Bombay, Delhi, Madras, Hyderabad, Ahmadabad, Kanpur, and Bangalore have 1,000,000 or more residents. Bombay, Madras, and Calcutta are among the largest cities in the world, and Delhi is one of the fastest growing cities in modern history.

While some cities, like Banaras, are very old, others, such as Bom-

These pavement-dwellers in Calcutta have no permanent homes.

Crawford Market in Bombay.

bay, Calcutta, and Madras, grew only during the British period of Indian history. Some of the Indian cities, such as Bombay, Calcutta, and Madras, are Westernized, with modern transportation and other civic services; others still retain their old character, typified by narrow winding lanes and crowded streets. But crowding is the general characteristic of all Indian cities, especially since 1947, because of migration from rural areas and the influx of refugees from Pakistan. The development of city services has seldom been able to keep pace with the increase in population; as a result, most cities are surrounded by large slum areas. In all the major cities a considerable number of people do not have homes, and most of the time they are forced to live on the sidewalks. Some families build makeshift houses of tin, bamboo, and gunny cloth (burlap). During the dry season one may see thousands of men, women, and children sleeping on the sidewalks. These unfortunate people do not have proper sanitary facilities and use the fire hydrants for their water supply. Calcutta is the worst in this respect, closely followed by Bombay and Madras. It is in these cities that the contrast between vast riches and grinding poverty is most glaring. Slums exist within the shadows of palatial homes, and the poor and the homeless wander through streets swarming with luxurious automobiles.

The Attraction of the Cities

These cities attract people because of the promise of jobs for the thousands of rural unemployed. Bombay, Madras, and Calcutta are important ports, have vast industrial complexes, and act as communications and transportation centers for their vast surrounding regions. They are also centers of India's large moviemaking industry. Trade and commerce are brisk, and since these cities are the headquarters of the state governments, this also adds to the influx of people. Bombay, Madras, and Calcutta have Western-style hotels and restaurants, nightclubs, movie and drama theaters, art shows and literary circles, and they function as pacesetters of India's cultural life. Their shopping areas are crowded with people buying a large variety of goods made in India as well as imported from the West. The movie theaters are always crowded, and long lines of people stand patiently at the ticket offices long before the beginning of the show. India is the second largest producer of movies in the world, and most of the Indian movies are made in the areas surrounding the great cities. The coming and going of movie stars also attracts large crowds. Most of these movies are based on religious or historical themes, and some are frank imitations of Hollywood.

Each city has its own public transport system, but they are sorely overburdened. The streets display all manner of vehicular movement, from electric suburban trains to carriages (rikishas) pulled by panting and scrawny men who earn barely a few rupees a day. Most of the cities are in urgent need of urban renewal, and in some this is currently taking place with the growth of satellite towns and suburbs. But the need always outpaces the availability of resources.

The social, political, and cultural importance of the cities cannot be overstated. They are symbols of prosperity and progress for millions of the rural people and act as magnets constantly attracting people from nearby areas. Here many of the Indian villagers are transformed into modern workingmen, who often live in state- or industry-subsidized housing. The most advanced people functioning in literature and the arts are to be found in the cities, and the cities have also played a leading part in the revival of the old Indian arts. The museums are packed with examples of ancient and modern Indian culture. The universities often become centers of intellectual movements, which gradually spread in ever-wider circles.

The cities also offer opportunities to make money. Since most of them are either state or district capitals, they function as centers of political life, too. From time to time the normal life of the cities is punctuated by intense political activity, with the numerous political parties leading demonstrations in support of a variety of causes. The trade unions are well organized and are often associated with political parties vying with each other for political power. The urban labor force, as a consequence, has become highly politicized, adding to the din and zest of city life.

Thus the Indian cities can be simultaneously charming and disturbing, throbbing with life or rebelling at stagnation, centers of dynamism and protest, of hope and aspirations.

The harbor of Calcutta, one of India's largest cities and major ports.

Calcutta

Calcutta, with a metropolitan population of over 5,000,000, is the capital of West Bengal State. Calcutta began as an English trading post in 1690. It grew in importance because of its location on the Hooghly River (a branch of the Ganges), which empties into the Bay of Bengal. In 1773 Calcutta became the capital of British India. Although the seat of government was moved to Delhi in 1912, Calcutta remains a major port and industrial city, producing paper, iron and steel, leather goods, and jute. It is also a center of Indian cultural and intellectual life. The University of Calcutta is India's largest university, and the National Library contains over 1,000,000 books.

Calcutta's problems—a severe housing shortage, slum conditions, and overcrowding—are the problems of many large cities, only magnified. The city itself provides a study in contrasts: wide thoroughfares lined with modern spacious buildings lie around the corner from narrow, cluttered side streets. Calcutta's main avenue, Chowringhee, is a hub of activity and entertainment, with many movie theaters and shops. The Maidan, a huge park in the center of Calcutta, offers areas for sports, picnicking, and just relaxing. In the Maidan is the Victoria Memorial, housing artifacts of Indian life and history. Calcutta's zoo, its botanical gardens (with the Great Banyan Tree that can provide shade for more than 200 people), and the Indian Museum are also popular attractions.

Bombay

Bombay, the capital of Maharashtra State, has a population of over 5,000,000. Like Calcutta, it is a modern, cosmopolitan city, with many of the same problems of that city. Like Calcutta too, Bombay grew from humble beginnings. Originally a fishing village, it was acquired in the 16th century by the Portuguese, who ceded it to the British in 1661. The development of railway lines and a textile industry in the 1860's began Bombay's major period of growth. Today it is India's major port on the west coast.

The port of Bombay, which opens into the Arabian Sea, lies on the eastern end of Bombay Island, on which the city itself is built. A huge stone arch, the Gateway to India, stands at the entrance to the harbor and is Bombay's best-known landmark.

One of Bombay's most striking sights is the Marine Drive, a four-lane highway; it is especially spectacular when its blazing lights are viewed at night from the crest of Malabar Hill, the city's most fashionable residential district. Other attractions include the various temples and shrines, the Prince of Wales and the Victoria and Albert museums, the zoo, and the city's famous Hanging Gardens.

Bombay's Marine Drive.

A less prosperous part of Bombay: squatters' shacks.

The section of the city known as the Fort is Bombay's cultural and commercial center. Its wide boulevards and large buildings give it the look of a European city. Here are located most of the city's business establishments, banks, theaters, the University of Bombay, state government buildings, and the Victoria Railway Station. To the north is the city's industrial area, with facilities for processing the great quantities of raw cotton brought in from the surrounding countryside.

Madras

Madras, the capital of Tamil Nadu (formerly Madras State) and the principal city of southern India, has a population of about 2,000,000. It is an important seaport as well as a major transportation center. The city has given its name to the colorful Madras cloth produced in local factories and exported to countries throughout the world. Besides textiles, the city's industries include iron foundries, bicycle factories, and engineering works. Madras came into being in 1639, when the East India Company acquired the area surrounding the present city from a local rajah. The British built a trading post, named Fort Saint George, around which a thriving commercial settlement gradually developed.

By the early 1800's Madras had become one of the administrative capitals of British India. Because of its location on the Bay of Bengal, the British developed Madras as a major port and began construction of an artificial harbor. The city presently extends for nearly 10 miles along the coast of the bay. Its magnificent beaches and the Marina, a splendid 4-mile-long (6 km.) shore drive, are major attractions. The Marina begins in the south at Saint Thomé Cathedral (where the remains of the Apostle

Thomas are said to be buried) and then continues past Madras University, state government buildings, and Fort Saint George until it reaches its northern end at the city's harbor. Madras is also noted for its beautiful Hindu temples, its Horticultural Gardens, and its Indian art collection in the National Art Gallery and Government Museum.

Delhi

A city with ancient roots, Delhi today is divided into two sharply contrasting sections. New Delhi, the capital of India since 1931, was built in the 20th century, while old Delhi was founded in the 12th century and was one of the principal cities of the Mogul Empire. The two parts of Delhi together form the largest city of northwestern India, with a population of about 2,500,000. Delhi is a bustling industrial and commercial center. The city stands on the west bank of the Jumna River, a tributary of the Ganges, and Hindus regard the place where these two rivers meet as a holy site.

Old Delhi is a mixture of beautiful landmarks and overcrowded slums. New factories have sprung up among the maze of narrow, winding streets, but the old city is still dominated by ancient buildings such as the Red Fort, a fortress erected in the mid-17th century by the Mogul emperor Shah Jahan. Not far from the Red Fort is the great mosque Jama Masjid, which rises 201 feet (61 m.). The main thoroughfare of old Delhi is Chandni Chauk (or Chowk)—the "street of silver." Today, as in ancient times, silversmiths still practice their craft along this street, but it is now filled with shops of every description. Between old Delhi and New Delhi lies the revered shrine Raj Ghat. Here Mahatma Gandhi, who led the fight for Indian independence, was cremated on January 31, 1948.

In contrast to the old section, New Delhi is a modern, planned city with wide boulevards, spacious public parks, and large government buildings. A long, tree-lined avenue called the Raj Path leads to the home of the president of India. The avenue is used for parades and official processions. The presidential residence, the Rashtrapati Bhavan, and the Parliament House are popular tourist attractions. New Delhi's main shopping center is Connaught Place, a circular street surrounding a park. The University of Delhi, the Central Agricultural Research Institute, museums, and theaters are among Delhi's many cultural institutions.

In addition to its commercial and governmental activities, Delhi is important also as the center of northwestern India's transportation and communications network. Industries include the manufacture of cotton and wool cloth, food processing, iron foundries, printing plants, and flour and sugar mills.

RELIGIONS OF INDIA

One of the most important elements in Indian life is religion. Hinduism is professed by about 85 percent of the people of India. Muslims number some 50,000,000—about 10 percent of the population. Christians account for about 2.5 percent of India's population and are evenly divided between Catholics and Protestants. There are small communities of Buddhists, Jains, Sikhs, and Jews. The Parsis, followers of the ancient Persian prophet Zoroaster, number 100,000, of whom half live in the city of Bombay. India also has a large tribal population (estimated at about 40,000,000), whose religion is primarily animism. They believe in the existence of powerful spirits or gods and that all natural objects have souls. Many of these tribal peoples have been converted to Christianity, especially in the northeastern areas. Others are gradually being absorbed into the structure of Hindu religious beliefs and practices.

Hinduism is like a vast sprawling mansion. Its ideas, beliefs, cults, and practices have ranged across the centuries and are drawn from many sources. It has no single founder, and within its generous folds may be found people close to spirit worship at one end and believers in a personal God at the other. The beginnings of Hinduism go back to at least 1500 B.C., when the composition of the Vedas, the sacred books of Hinduism, began. This early religion included belief in many gods who were personifications of natural phenomena—thunder, lightning, rain, the sun, and the moon—and who were worshiped through a cult of sacrifices. These beliefs were replaced, during the period 700–600 B.C., by the philosophical theories that laid down some of the fundamental principles of Hinduism. Essentially these are four in number.

The Hindus believe in God (or gods who are considered manifestations of a single god or universal spirit) as the creator and sustainer of this world. They believe in a soul that is eternal and indestructible and that merges into God at salvation. They believe in the moral responsibility of man for his actions (karma), since man is endowed with a will of his own by which he acts. These beliefs are similar, though not identical, to some of the major principles of the other great religions of the world. It

The Ganges at Banaras. The city and the river are holy to Hindus.

Hindu pilgrims en route to a holy site.

is in the fourth principle that the Hindus (and Buddhists and Jains) differ from others. This fourth principle is the belief in rebirth (reincarnation). Hindus hold that man must go through a series of births, deaths, and re-births to atone for his sins before he may work his way to salvation. The nature and form of the next birth are largely determined by one's action in an earlier life. Hence one has the capacity to shape one's future. Hinduism is a very tolerant religion, and Hindus generally do not try to convert people of other faiths. The caste system, though it arose independently as a social institution, soon became a part of Hinduism.

The Hindu mode of worship is largely familial. Each family has its own set of gods and every home has a small corner that is set aside as the family shrine. It is the duty of the head of the family to perform the daily ritual worship as well as the ceremonial worship on days especially sacred to the household gods. On special days the Hindus go to the temple to worship, again as a family group. The temples are dedicated to the great gods of classical Hinduism. These are Vishnu (who sustains the world), Siva (the destroyer and re-creator), Rama (the ideal man and king), and Krishna (the god of eternal youth and love). Together with Brahma (the creator), Vishnu and Siva comprise the Hindu trinity. There are also many other gods: Ganesha (the elephant-headed god of wealth and wisdom, and a son of Siva), Subramanya or Kartikeya or Kumara (another son of Siva and the general of the army of the gods), Kali, also called Durga (the goddess of destruction), Lakshmi or Laxmi (Vishnu's wife and goddess of wealth), and Hanuman (the monkey god, the devoted servant of Rama). In addition, each village has its own set of gods and goddesses who are regarded as protectors of the people, especially against dread epidemics of smallpox, plague, or cholera.

Hinduism operates on two distinct levels, which may be described as the classical and the folk levels. On the classical level the gods worshiped are the major gods—Rama, Krishna, Vishnu, and Siva—and the ceremonial service is performed by the Brahman priesthood. Its marriage rituals, fasts and holidays, pilgrimages, and sacraments relating to birth, marriage, and death come from the classical Sanskrit texts. On the folk level the dominant gods are the local gods and goddesses, who are regarded as local manifestations of the great gods. The folk ritual may not be performed by the Brahman priesthood; it is generally the responsibility of the non-Brahman priests and involves the use of animal sacrifices.

The classical ritual uses water, perfumes (usually sandalwood paste), flowers, leaves of trees considered sacred, oil lamps, coconut, and bells. Music is an inseparable part of Hindu worship. On religious holidays devotional songs are recited and events in the lives of the various gods are narrated in many of the great temples. The main part of the Hindu temple is the relatively small chamber containing the images of the gods. In the south the temples have magnificent gateways with richly sculptured scenes from the Hindu heavens.

The Hindu religion also lays down guides for individual and social behavior. Hinduism regulates family relationships, and the Hindu codes have governed the lives of millions of Hindus through the centuries. The discipline in the family is generally the concern of the eldest male mem-

A temple shop in Banaras with religious souvenirs for sale.

Muslims at prayer. They are the largest religious minority in India.

ber, whose orders are accepted by all members. He, in turn, is responsible for the happiness and welfare of all members of the family.

The Muslims believe in one God who is formless and must not be represented in any image. Their places of worship are the mosques. The mosque contains a place with a niche that points in the direction of Mecca in Saudi Arabia, the holy city of Islam. It also has an assembly hall where congregational prayers are offered every Friday at noon. The Muslims are also expected to offer prayers five times a day; to observe a strict fast during the holy month of Ramadan; to go on a pilgrimage to Mecca and Medina (also in Saudi Arabia); and to contribute a tithe (a portion of their income) to the community chest.

Both Hinduism and Islam have been great forces in the cultural history of India. They have been sources of spiritual strength and social cohesion for their followers. It is also true, at least in some aspects, that some of their practices have tended to obstruct social and economic modernization. The caste system and its institution of untouchability have divided Hindu society into hundreds of small, self-contained, and exclusive groups. This has made economic and social co-operation on a large scale difficult. The Hindu idea of the sacred cow has led to an excessive cattle population, which is a burden on the economy. Untouchability has been the most objectionable feature of Hindu society. It condemns millions of Indians to a permanent social inferiority and economic degradation from which there is little escape. Similarly, the generally inferior position of women in Muslim society has crippled their social and cultural development. This is especially true in the custom of veiling (purdah). Orthodox Hindus and Muslims have resisted attempts at rapid social change through legislation or social action, thus impeding economic progress. A part of the explanation of the poverty of the Hindu and Muslim masses, therefore, lies in their social structures, ritual concepts, and rigid traditions.

THE BEGINNINGS OF MODERNIZATION IN INDIA

The British period of Indian history saw the beginnings of modernization in many areas of Indian life. Large urban centers emerged during the 19th century, and these dominated vast regions surrounding them. Modern industries were concentrated in these regions and drew people from the rural areas. About 1856 modern communications (railways and the telegraph) began to unite the country in a manner never possible before. The British created a modern civil service and judiciary that were national in scope and spirit. The English language opened for Indians a window on the Western world, and Indian intellectuals began to absorb many modern concepts in philosophy, politics, and economics. One of the most dramatic results of this Western impact was the rise of Indian nationalism, which eventually led to the withdrawal of British power from India.

The process of modernization affected the ancient foundations of Indian life in many ways. The caste system began to lose much of its rigidity, especially in social relationships, rules of food and drink, residential patterns, and occupations. In the growing cities it became impossible for people to observe all these rules strictly. The occupational basis of the caste system became steadily more irrelevant with the great economic changes sweeping over the country. The old village agencies of control, such as the village and caste councils, broke down in the urban centers. The custom of untouchability also declined in the cities. Perhaps the most significant change took place as newer classes began to assume social and economic importance due to their leadership in trade and industry. The spread of Western education also sparked a number of social reform movements, especially in Hindu society.

But these changes were limited. Although the caste system became less rigid in some ways in the cities, it still continued to hold sway in the rural areas. The introduction of Western-style politics in some instances

A tractor factory in Delhi. Industry has expanded since independence.

Handicraft industries are important to the economy. These are workers in brass.

strengthened rather than weakened the caste system. The beginnings of industrialization were made, but the process was restricted. The modernization of India was thus only partial at best.

INDIA SINCE INDEPENDENCE

The major problem faced by the new nation in 1947 was the completion of the revolution in Indian life begun during the 19th century under British rule. The partition of the subcontinent into primarily Hindu India and Muslim Pakistan led to a mass migration of peoples unparalleled in history. Millions of Hindus, Sikhs, and Muslims were uprooted from their native soil during 1947 and 1948. These refugees placed impossible burdens on the Indian economy. The partition also created an economic imbalance, since the major food-growing areas remained in Pakistan while the main industrial areas remained in India. The influx of refugees, the growing population, and the economic dislocation depressed the already low standards of living. Economic development, therefore, became a major concern for the new government of India.

The young nation suffered a tragic blow with the assassination of Mahatma Gandhi in 1948. The leadership of the country passed to Jawaharlal Nehru, who resolutely began to seek solutions to the grave problems confronting India. The first need was to devise a constitution that would unify the country, create a workable political process, and give to the various regions a sense of national responsibility.

The Indian Constitution was adopted in 1950. It drew its inspiration from the United States Constitution and from British constitutional ideas and practices. It decreed India to be a union of states (to date, 22 in number) and some federally administered territories. Each state has a governor (appointed by the president of India), a cabinet of ministers, a legislature, and a judiciary of its own. The union, or federal, government is presided over by a president and a vice-president who are elected by an electoral college made up of members of the union and state legislatures.

The real executive power of the union government is vested in a cabinet of ministers led by the prime minister. The union legislature has two chambers—the Lok Sabha (House of the People), with some 500 members, and the Rajya Sabha (Council of States), with some 250 members. The members of the Lok Sabha are elected by the people every 5 years. The Rajya Sabha is elected by the members of the state legislatures. The union judiciary has a supreme court headed by a chief justice. Every Indian citizen over 21 years of age is eligible to vote.

Since independence and in spite of illiteracy voters have participated in their government in a responsible manner. There have been strains on the government, particularly in 1975, when Prime Minister Indira Gandhi declared a state of emergency and suspended some civil liberties, a move that aroused strong criticism. But in the 1977 elections she was defeated and democracy was restored.

TOWARD SOLVING INDIA'S PROBLEMS

The British impact on India was uneven in its intensity and resulted in the creation of regions of varying degrees of political and economic development. Compared with the provinces ruled directly by the British,

A brickyard in Agra. New building construction is urgently needed to keep pace with the population growth in the cities.

Women road sweepers. Many such jobs are still done by human labor.

the native states under the administration of the maharajas were generally backward. Upon independence these native states were merged into existing or new states. The political map of India is still in a state of change due to demands for the creation of separate states based on language or tribal populations.

Along with the people of the former native states, the untouchables and the tribal peoples suffered from isolation and backwardness. Special efforts, therefore, had to be made to bring them into the mainstream of Indian life. These efforts included educational concessions, reservation of seats in the legislatures, and government jobs. Equally pressing was the problem of social development. The Hindu social structure was out of date in many respects, and the new Hindu Codes of 1954–56 sought to change it. The new codes profoundly affected Hindu marriage customs and the organization of the Hindu family. Marriage was made a civil rather than a religious act through the registration of marriages; women were made eligible for inheritance rights; and divorce was permitted to Hindus. Other laws attempted to distribute power to the agencies of local and village government, and in many areas the new village councils have become politically vigorous. Considering the fact that many Indian institutions were centuries old, the changes brought about by the new legislation were truly revolutionary, though they were achieved by peaceful democratic means.

Economic Advances

At the time of independence the Indian economy was close to a complete breakdown. The steady rise in population due to better preventive medicine, public health and hygiene measures, and a drop in infant mortality added to the economic problem. The average annual rate of growth of the population is 2.5 percent. Thus every 2 years India adds some 25,000,000 people (or more than the total population of Canada) to its numbers. This population growth tends to outstrip economic advances, and consequently India has to run in order to stay in one place.

Three 5-year plans for economic development, begun under Prime Minister Nehru, have been undertaken in India so far, and a fourth was started in 1970. These have encouraged development in many areas. Giant hydroelectric stations have been constructed, increasing the amount of farmland under irrigation and supplying electric power to thousands of villages and towns that never had electricity before. Steel output has been raised since independence. Agricultural productivity has been increased, though it is still necessary for India to import additional grain to feed its people. Fertilizer production and the supply of improved varieties of seeds have also increased. Thousands of miles of new railway and bus transportation have been added, making the Indian population more mobile. Many new cities have sprung up, humming with industries and business. The average life expectancy has risen from 25 years in 1947 to 45 years today. Slowly but steadily, industrial and urban revolutions are sweeping across the land, creating new hope and new patterns of life for millions of people.

Overpopulation is a basic problem in India. This sign extols smaller families.

But these developments exacted a toll. Some of the efforts were too sudden and imposed unbearable burdens on the economy. There was a miscalculation in priorities. Steel production, for example, received precedence over the production of food. The emphasis on heavy industry led to a shortage of consumer goods and made the problem of inflation critical. For the first 10 years the government paid little or no attention to the problem of population control. The result was that the growing population wiped out most of the fruits of economic development.

These problems have been corrected recently, and India has now embarked on one of the most ambitious programs of population control in history. The production of chemical fertilizers has also been increased. Indian agronomists (agricultural scientists) have developed strains of rice and wheat seed that have doubled and even tripled the output per acre. If the monsoons continue to be good and the program is sustained, India may become self-sufficient in food supplies for its hungry millions.

Industry has also shown a steady growth. With its reserves of iron, coal, petroleum, and other minerals, India can become a great industrial nation in the not-too-distant future. Great strides have been taken, too, in the production of atomic energy. Indian scientists and technicians are capable of making atomic bombs if the country chooses to do so. But so far the Indian Government has insisted that India will not produce nuclear power except for peaceful purposes.

The old and the new in India: women pick aromatic plants, and in the distance is a factory producing cement, one of the country's important products.

A village elementary school. Education has been one of India's chief concerns.

Progress has been remarkable in other fields. Educational opportunities have been greatly widened. The number of children between the ages of 6 and 11 attending school has risen to 80 percent since 1947. There are some 70 universities in India today, compared with 17 in 1947. The isolation of the rural areas is being gradually eliminated through a network of roads and radios. Many villages that had neither a school nor medical facilities before now have both.

The Problems That Remain

But difficulties still lie ahead. It will take a long while before a balance between resources and population is established. The pace of economic growth is slow in comparison with the expectations of the people. The intake of calories per day is still too low and must be raised to at least 2,200 for the people to be well nourished. Schools will have to be provided for millions of children and jobs found for the growing numbers of college graduates and technicians. India is one of the most heavily taxed nations in the world, but it still has to depend on foreign aid (largely from the United States) to a great extent in carrying out its economic projects. The per capita income, although it has been raised from $40 a year to about $70, is still pitifully low.

INDIA'S INTERNATIONAL RELATIONS

In its international relations India has generally followed a policy of non-alignment. This has meant a reluctance to join any system of military pacts. India's main difficulties have been in relations with Pakistan

These two young women are learning to read and write.

and China. India has gone to war with Pakistan three times—in 1947, 1965, and 1971. The first two involved conflicting claims over Jammu and Kashmir. The 1971 war grew out of the civil war between East and West Pakistan. The flight of millions of East Pakistani refugees into India and border incidents led to the march of the Indian Army into East Pakistan. The result was a swift victory for India and the establishment of the nation of Bangladesh in the former East Pakistan. With the dismemberment of Pakistan, India emerged as the leading power in South Asia.

Relations with the People's Republic of China were originally friendly but turned hostile when the Chinese attacked Indian frontier areas in 1962. Since then relations between the two countries have been uneasy.

In 1971 India and the Soviet Union signed a 20-year treaty of peace, friendship, and co-operation. Indian leaders have insisted that this is not an alliance as is normally understood. But the treaty did have an effect during the 1971 war with Pakistan. India's relations with the United States have not been smooth recently. The massive economic aid from the United States was deeply appreciated in India, but friction developed over American military support for Pakistan.

THE FUTURE OF INDIA

Although much remains to be done, India has great hopes for the future. The Indian people have a remarkable capacity for patience and hard work, and they are quite willing to make sacrifices for their country. In the past they have shown qualities not only of survival but also of resurgence, and there is no reason why the same qualities should not sustain them through the coming crucial decades. After long subjection to foreign rule they are today a free nation, proud of their ancient heritage. They are creating a new way of life for a centuries-old civilization, but a way of life that will not destroy its ancient values. India today is an exciting land in an exciting age, one quarter of humanity engaged in carrying through a peaceful revolution.

B. G. GOKHALE, Director, Asian Studies Program, Wake Forest University
Author, *The Making of the Indian Nation*

GANGES RIVER

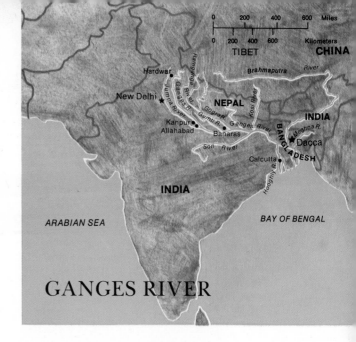

GANGES RIVER

The Ganges is India's principal river and one of the longest in the world. But to millions of Indian Hindus, the Ganges is more than just a great waterway. For them, Ganga Mai ("mother Ganges") is a sacred river, whose waters are believed capable of cleansing the soul of all sins and healing the body of all ills. Each year, tens of thousands of the Hindu faithful make pilgrimages in order to bathe in the Ganges.

Many Hindu temples line the banks of the Ganges, since the river is important in Hindu religious ritual. Flights of stairs (ghats) have been built along the river to provide pilgrims access to the holy waters and a place to sit and pray. Hindu funeral services are held on the ghats. The dead are cremated and the ashes scattered over the river. Besides Hinduism, two other major Indian religions—Buddhism and Jainism—were born and nurtured in the Gangetic lands. Buddhism began at Sarnath, near modern Banaras, where the Buddha preached his first sermon.

The Ganges begins in the Himalayas, some 10,000 feet (3,000 meters) above sea level. Fed by the melting snows and glaciers of the Himalayas, the Ganges winds its way down through deep mountain gorges until it enters the Ganges Plain at Hardwar. Then the river runs in a generally southeasterly arc across northern India and Bangladesh, finally emptying into the Bay of Bengal. During its nearly 1,600-mile (2,575 kilometers) course, the Ganges flows past Allahabad, Banaras, and Kanpur. In Bangladesh, the Ganges joins with the Brahmaputra to form a vast, swampy delta. The major tributaries of the Ganges are the Jumna, Son, Ramganga, Gogra, Gumti, and Kosi rivers.

Together with its branches, the Ganges waters an area that covers approximately one quarter of all India and a large part of Bangladesh. Water from the Ganges is used by millions of farmers to irrigate their fields of rice and wheat. For centuries the Ganges also has been an important artery of communication and transportation. The area watered by the Ganges and the Jumna was the economic heartland of two ancient Indian kingdoms—the Delhi Sultanate and the Mogul Empire. However, because the water level is often very low, most of the river can be navigated only by small boats.

Reviewed by B. G. GOKHALE, Director, Asian Studies Program, Wake Forest University

KASHMIR

KASHMIR

Srinagar

The outward calm and natural beauty of Kashmir are masks that conceal the true face of this troubled and battle-scarred land. For more than two decades, India and Pakistan have been involved in a constant and sometimes bloody struggle to control Kashmir. In 1948, 1965, and again in 1971, Indian and Pakistani troops clashed in Kashmir. Today, while tourists and vacationers enjoy Kashmir's magnificent valleys and lakes, there is tension along the cease-fire line that separates Pakistani-controlled Azad ("free") Kashmir from India's Jammu and Kashmir state.

The strategic importance of Kashmir becomes clear when one looks at a map. Kashmir's mountain passes form a natural gateway between China, India, and Pakistan. Moreover, water from several major rivers flowing through Kashmir is vital to the farmlands of Pakistan. However, the main cause of the conflict over Kashmir is religion. Most of Kashmir's people are Muslims, many of whom identify strongly with Pakistan. But there are also sizable Hindu and Buddhist minorities, and these people would prefer to have Kashmir become a part of India.

The Land

Kashmir nestles in the Himalaya mountains in the northeast corner of the Indian subcontinent. Its total area is approximately 86,000 square miles (228,000 square kilometers), which includes the Indian- and Pakistani-controlled sections and a portion ceded to China by Pakistan in 1963. Except for a narrow strip of the Punjab plains in the south, Kashmir is a mountainous country. North of the plains, elevations rise from 2,000 feet (600 meters) in the lower hills of the Outer Himalayas to over 25,000 feet (7,600 m.) in the Great Himalayas. These spectacular mountains are sacred to both Hindus and Buddhists. The tallest of Kashmir's peaks, second only to Mount Everest, is Mount Godwin Austen, or K2, which rises to an imposing 28,250 feet (8,611 m.). Several of Pakistan's major rivers, the Chenab, the Jhelum, and the Ravi, begin in the mountains of Kashmir. The Indus River begins in Tibet and flows through Kashmir.

The Vale of Kashmir, the region's most famous scenic wonder, lies in a broad, deep-set valley in the southwestern part of the territory. This beautiful valley has been an inspiration to poets and a resort for the kings and emperors of ancient times. It still draws many thousands of visitors, who marvel at its sparkling blue lakes, massive forests, and multicolored plants and flowers. Dal and Wular lakes are particularly popular with boating and fishing enthusiasts. Dal Lake, in the city of Srinagar, is noted also for its floating gardens (soil-covered rafts) of fruits and vegetables. Nearly half of Kashmir's people live in the Vale.

Climate. There is a great range in climate, which varies according to the region and the elevation of the land. In the southern plains the climate is subtropical, while in the higher mountains of the north it becomes subarctic and arctic. In the central valleys weather conditions are more moderate. Rainfall is heaviest during the summer monsoon period.

The People

A number of different ethnic and religious groups are represented among Kashmir's 4,500,000 people. In the Vale the people are mainly Muslims and are known as Kashmiris. South of the Vale, in the area around the city of Jammu, Hindu people called Dogras are dominant. The major language of the region is Kashmiri, an Indo-European language. A number of seminomadic tribal groups inhabit the northern and eastern mountain areas bordering China and Tibet. These people are

Shalimar Gardens, in Srinagar, a city in the beautiful Vale of Kashmir.

ethnically related to the Tibetans. They speak Tibetan dialects and follow the Buddhist faith.

Farming, sericulture (the raising of silkworms), and lumbering are the main occupations of the people. The richest farmland is in the Vale. Here are grown rice, wheat, and corn and a wide range of fruits and vegetables. In the mountain regions of Gilgit and Ladakh to the north and east, the people raise sheep, goats, ponies, and yaks (a shaggy-haired beast of burden). Farming in the highlands is accomplished by means of terraced fields dug into the mountain slopes. Many rural people are moving into the cities to work in factories and service industries.

Cities.   The leading city of Kashmir is **Srinagar**, with a population of over 300,000. Located in the Vale, Srinagar has long been noted for its handicrafts. The city's artisans produce jewelry, leather goods, silverware, wood carvings, woolen rugs, shawls, and the world-famous cashmere sweaters that bear the region's name. The Jhelum River flows through Srinagar, and wooden houseboats may be seen on the river and on the surrounding lakes. They are used as permanent or temporary homes by residents and tourists. A short distance from Srinagar are the ruins of ancient temples at Patan and Awanitpur. Kashmir's other city of importance is **Jammu**, near the border with Pakistan.

History and Government

For many centuries, parts of what is now Kashmir were ruled by Hindu princes. Then, in the 14th century A.D., Muslims conquered the southern parts of the region, and the Islamic faith was spread through the area. Kashmir was united politically in the 19th century by the Dogra Rajput family. Gulab Singh, a Dogra adventurer, built the kingdom through conquest. In 1846, the British confirmed him and his male heirs as the hereditary rulers of Kashmir. A century later, when India won its independence (1947), Kashmir became a state of the Indian republic.

The inclusion of Kashmir in the new Indian nation was the basis of the Kashmir dispute between India and Pakistan. When India was partitioned, in 1947, into separate Hindu and Muslim states—India and Pakistan—it was assumed that Kashmir's Muslim majority would choose to become part of Pakistan. But Kashmir's Hindu maharajah, faced with an invasion by Muslim tribesmen from Pakistan, sought union with India.

After the maharajah's decision, fighting broke out between Pakistani hill tribesmen and military forces, who invaded Kashmir, and Indian troops sent to defend the area. The fighting continued throughout 1948, until a truce was arranged by the United Nations on January 1, 1949. By the terms of the truce, India was permitted to control about two thirds of Kashmir, including the Vale. The northwestern section of the region is under Pakistani rule, and China occupies a section in the northeast. Following armed clashes in 1965, India and Pakistan agreed, in the Declaration of Tashkent (1966), to resolve the dispute peacefully. After the 1971 fighting the boundaries were altered slightly.

Both the Indian and Pakistani sections of Kashmir have local governments. Azad Kashmir, the Pakistani portion, is headed by a president and council of ministers based in the city of Muzaffarabad. The Indian state of Jammu and Kashmir is governed by a council of ministers and a two-house legislature. Srinagar and Jammu are both used as capital cities.

Reviewed by P. P. KARAN, University of Kentucky; author, *The Himalayan Kingdoms*

NEPAL

Nepal is a landlocked country located along the southern flank of the Himalayas. For centuries, this ancient and mysterious kingdom isolated itself behind mountains so high that they were believed to be "the home of the gods." Within the past few decades the kingdom has opened its doors to foreigners, but in life style it still remains apart from much of the modern world. Nepal's scenic wonders and centuries-old monuments have attracted tourists and other visitors from all parts of the globe. But while the ancient ways lure outsiders, the Nepalese Government continues its efforts to modernize the country.

THE LAND

Nepal is somewhat rectangular in shape, with a total area of over 54,000 square miles (140,000 square kilometers). The southern part of the country, known as the Terai, features broad plains and excellent farmland. There are also swamp and jungle regions, which are the home of elephants, tigers, rhinoceroses, and other wild beasts. The middle part of Nepal is mountainous, but has many fertile valleys. In the north are the Great Himalayas, a sparsely inhabited region of towering mountains. Mount Everest, the world's highest mountain at 29,028 feet (8,848 meters), is located there, along the border between Nepal and Chinese-occupied Tibet. Nepal's climate varies greatly according to altitude.

One of the many fertile valleys in the mountainous interior of Nepal.

A view of the main street of Katmandu, Nepal's capital and chief city.

Summers are hot in the Terai and the central hills, and cool in the higher mountain regions. Winters are mild in the south and severely cold in the higher mountains. There is considerable rainfall in some parts of the country during the summer months.

Economy. Farming and forestry are the chief occupations of the people. The crops and animals raised vary according to altitude. Rice, corn, and wheat are the major crops. Sheep and goats are grazed on the lower hillsides, while at the highest altitudes cattle, yak, and a cross-breed called the dzo, or *dzopkyo,* are raised. Nepalese factories produce glassware, pottery, paper, chemicals, and cement. The country's crafts-men are noted for their decorated woodwork and metalwork.

THE PEOPLE

Nepal's population of about 11,000,000 people is a varied mixture of related and unrelated ethnic, religious, and tribal groups. These fall into three main divisions. First are the Newars, a people who trace their ancestry back to Nepal's earliest known times. A second group consists of Hindus who came from India during and after the 13th century A. D. Third are the tribes that originated mainly in Tibet and Mongolia. They include the Tamang, Kiranti, Magar, Gurung, and Sherpa. Probably 40 percent of the people belong to various Hindu castes or social groups.

The Newars live throughout the extensive Katmandu Valley in cen-tral Nepal, and some 500,000 are farmers, tradesmen, and government employees. The tribal groups that came from the north long ago are primarily Buddhists and speak Tibeto-Burman languages. To the west of Katmandu are the Gurungs, who are known outside of Nepal as the Gurkhas—a term sometimes applied inaccurately to all Nepalese. The Gurkhas are famous as the tough mercenary soldiers who have served

FACTS AND FIGURES

KINGDOM OF NEPAL is the official name of the country.

CAPITAL: Katmandu (Kathmandu).

LOCATION: South central Asia. **Latitude**—26° 21' N to 30° 18' N. **Longitude**—80° 04' E to 88° 13' E.

AREA: 54,362 sq. mi. (140,797 sq. km.).

PHYSICAL FEATURES: Highest point—Mount Everest (29,028 ft.; 8,848 m.). **Lowest point**—150 ft. (46 m.). **Chief rivers**—Kosi, Gandak, Karnali.

POPULATION: 10,900,000 (estimate).

LANGUAGE: Nepali (official).

RELIGION: Hindu, Buddhist.

GOVERNMENT: Constitutional monarchy. **Head of government**—king. **Legislature**—National Panchayat. **International co-operation**—United Nations.

CHIEF CITIES: Katmandu, Patan, Biratnagar, Bhadgaon, Gurkha.

ECONOMY: Chief minerals—mica, coal, copper, graphite. **Chief agricultural products**—rice, maize, wheat, oilseeds, jute, sugarcane, livestock. **Industries and products**—textiles, glassware, ceramics, timber, chemical products, paper, cement, metalware, craftwork, **Chief exports**—food and live animals, jute, hides and skins, timber, oilseeds. **Chief imports**—cotton and silk, manufactured goods, minerals and fuels, iron and steel, chemicals, machinery.

MONETARY UNIT: Nepalese rupee.

NATIONAL HOLIDAY: February 18, National Day.

NATIONAL ANTHEM: "May Glory Crown Our Illustrious Sovereign."

Temple in Katmandu

in both the British and Indian armies for over a century. Equally well-known are the Sherpas, a sturdy mountaineering people whose villages are situated high in the Himalayas. Cattle and yak are raised by the Sherpas, who also serve as porters and guides for mountain climbers.

Religion. A distinctive part of Nepalese culture is the way in which Hinduism and Buddhism have mingled with each other. Over 95 percent of the people belong to these two religions. Many temples house both Hindu and Buddhist shrines, while religious ceremonies and festivals are shared by all.

Language and Education. Nepali, the official language, is an Indian tongue brought in by early Hindu immigrants. It is now spoken by well over half the population. However, many other languages and dialects of the Indo-Aryan and Tibeto-Burman families are also spoken. Most Nepalese are illiterate, and the percentage of children attending school is still relatively low. Because of the rugged terrain and difficult trails, villages are often isolated and this has hampered the spread of education.

Way of Life. The typical Nepalese village is made up of two-story houses of stone or mud brick, which are clustered together on hillsides above the flat valleys or irrigated, terraced fields. In most areas, rice is the staple food. Occasionally, mixed dishes of vegetables and fish or chicken are served. Tea is the most popular drink, and among the Sherpas it is taken with salt and butter. Clothing is varied and generally colorful. Blouses and long skirts are worn by the women, while the most common dress for men consists of a knee-length robe and tight pants.

For the Nepalese, who are primarily farmers, family life provides its own entertainment and excitement. Weddings feature elaborate celebrations, and each ethnic group has its special arrangements. There are also many other rituals. At harvesttime, men and women perform the traditional rice dance. In October, when the herds are brought down from the highest valleys for the duration of the winter, each male Gurung brings the head of a ram he has killed to a clan feast.

Many religious festivals are held during the year. One of the biggest is the week-long festival of Indra, the god of rain, which comes in autumn. A very young girl, a "living goddess" who is believed to be a reincarnation of the goddess Kumari, is carried through the streets of Katmandu. At festival times the temples are richly decorated, and there is feasting accompanied by music and dancing.

Cities. Within the Katmandu Valley are the country's three principal cities—Katmandu (Kathmandu), Patan, and Bhadgaon. They are close enough to each other to be considered one metropolitan area. **Katmandu**, Nepal's capital and economic and cultural heart, is a city of some 200,000 inhabitants. Its lively but uncrowded streets are flanked by numerous Hindu and Buddhist temples and statues, showing the elaborate woodwork and metalwork for which Nepalese craftsmen are famous. **Patan**, with a population of about 150,000, and **Bhadgaon**, with about 90,000 inhabitants, are also historic towns.

Government. Nepal is a constitutional monarchy in which the king is both head of state and head of government. The Constitution of 1962 assigns executive power to the king and the Council of Ministers, which he selects from among members of the National Panchayat (council), the legislative body. The National Panchayat consists of both elected and appointed members.

HISTORY

Authentic historical records of Nepal date from about the 4th century A.D. From that time until the mid-18th century there are accounts of a succession of Hindu dynasties and rulers controlling various parts of what is now Nepal. In 1769, Prithwi Narayan Shah, the ruler of the Gurkhas, established the modern nation of Nepal and became its first king. The Shah family still rules the country.

Nepalese expansion southward into India resulted in conflict with the British and the Anglo-Nepali War of 1814–15. Under the peace treaty of 1815, Nepal gave up some of its territory and agreed to allow a permanent British resident officer to be stationed in Katmandu. In 1846, a member of the Rana family forced the king to appoint him prime minister and to grant him absolute power. Thereafter, the premiership was held by the Ranas, who became the actual rulers of Nepal.

Eventually, there was dissatisfaction with the reactionary Rana family. An underground movement to restore the King to full powers began in the 1930's. The King escaped to India in 1950, and shortly afterward a revolt in Katmandu broke the power of the Ranas. King Tribhuwan returned to Nepal in 1951. He proclaimed a constitutional monarchy and opened up the country to the outside world.

King Tribhuwan died in 1955 and was succeeded by his son, King Mahendra. Until his death in 1972, Mahendra continued his father's modernization and reform program. The country's first elected parliament was installed in 1959, and a constitution was issued in 1962. With Chinese aid and supervision, a road linking Katmandu with Chinese-occupied Tibet was completed in 1967.

DONALD N. WILBER, Editor, *The Nations of Asia*

SIKKIM

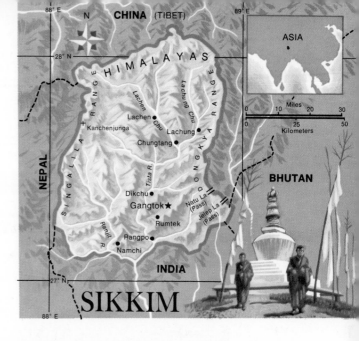

Sikkim lies high in the Himalaya mountains. It is a storybook land, a region of extremes. Skyscraping mountains alternate with low-lying green valleys, and the climate ranges from Arctic in the higher altitudes to subtropical in the lowlands. Its location along the shortest route linking Tibet with India has for centuries given Sikkim an importance out of all proportion to its size.

Sikkim is the smallest of the Himalayan lands. It has an area of 2,744 square miles (7,107 square kilometers) and a population estimated at 191,000. In this mountainous country land elevations start at about 700 feet (215 meters) above sea level and reach a high of 28,146 feet (8,579 m.) at the summit of Kanchenjunga, the world's third highest mountain. Several passes cut through the mountain barrier that surrounds nearly the entire region. The most important is Natu La, which channeled trade between India and Tibet for many generations.

Roughly 75 percent of Sikkim's population is Nepalese in origin, the descendants of people who moved into the area mainly in the early 1900's. The remaining Sikkimese are Lepchas, the original people of Sikkim, and Bhotias (Bhotes), who first migrated to the country in the 14th century. The three groups maintain their separate cultures, but intermarriage is now more common. Most Nepalese are Hindus, while the Lepchas and Bhotias are followers of Tibetan Buddhism (Lamaism), the official state religion. Several Himalayan dialects are spoken by the different peoples. Farming is the main occupation, and rice and corn are grown in terraced fields on the hillsides. Other crops include citrus fruits, wheat, and an herb called cardamom. Many Sikkimese also raise livestock. Sikkim's main town—**Gangtok**—has an urbanized area population of about 15,000 and is the capital and commercial center.

Sikkim became a separate nation in the mid-17th century, when the Tibetan Namgyal dynasty, or ruling family, was founded. Sikkim's Chogyals, or kings, were descended from this dynasty. In the 19th century Great Britain gained control of Sikkim and governed it as a protectorate. In 1950 it became a protectorate of India. In a referendum held in 1975 the people of Sikkim voted to become a state of India.

P. P. KARAN, University of Kentucky; author, *The Himalayan Kingdoms*

BHUTAN

Bhutan, a kingdom in the Himalayas, occupies a strategic position between the northern plains of India and the Chinese-occupied Tibetan plateau. A nation roughly the size of Switzerland, Bhutan has the potential to develop an economy that can adequately meet the needs of its people. But the resources of this mountain kingdom are underdeveloped, and so Bhutan remains the poorest of the Himalayan countries. It is nevertheless a richly scenic land. There are forested mountain ranges whose higher peaks are capped by snow; and there are broad, grassy valleys and thickly wooded jungle areas. Because of its remote location, Bhutan remains a strange and forbidding country. Its people call it Druk Yul, "the land of the dragon."

The Land.   Bhutan has an area of approximately 18,000 square miles (47,000 square kilometers). The kingdom is divided into three main geographic areas. Northern Bhutan lies within the Great Himalayas, where the mountains reach a height of more than 24,000 feet (7,300 meters). Spurs from the Great Himalayas radiate southward, forming watersheds for the country's principal rivers. The Middle Himalayan region of central Bhutan contains several fertile valleys, of which the Paro and Thimbu valleys are the best-known. Along the southern border of Bhutan lies the Duars plain, which extends into India. This is a hot, humid, and rainy area, consisting mainly of steaming jungle and malaria-infested swamps.

The People.   Only about 800,000 people live in this sparsely settled country. Because of the harsh physical conditions in the northern mountains and in the southern Duars, most of the people live in the valleys of central Bhutan. Most of the Bhutanese are Bhotias, a people of Tibetan

King's palace in Thimbu. The women in front are wearing traditional dress.

FACTS AND FIGURES

NAME: Kingdom of Bhutan.

CAPITAL: Thimbu.

LOCATION: South central Asia. **Latitude**—26° 45′ N to 28° N. **Longitude**—89° E to 92° E.

AREA: 18,147 sq. mi. (47,000 sq. km.).

POPULATION: 800,000 (estimate).

LANGUAGE: Bhutanese (Druk-Ke), Hindi.

RELIGION: Buddhist.

GOVERNMENT: Monarchy. **Head of government**—king. **Legislature**—national assembly. **International co-operation**—Colombo Plan, United Nations.

ECONOMY: Chief minerals—coal, gypsum, graphite. **Principal products**—rice, barley, fruit, textiles, timber, metalware, woodwork. **Chief exports**—timber, rice, coal, fruit. **Chief imports**—gasoline, fabrics, light equipment.

MONETARY UNIT: Indian rupee.

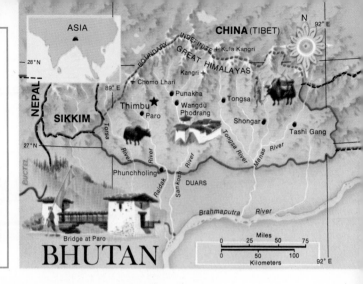

Bridge at Paro

BHUTAN

origin. About 20 percent of the population consists of Nepalese immigrants, who live mainly in the south. Several smaller tribal groups inhabit the eastern part of the country. The majority of the people are Tibetan (Lamaist) Buddhists.

Bhutan is almost entirely rural. Most of the people live in small, widely scattered villages. They are mainly farmers and herdsmen. A typical Bhutanese house is a two-story building constructed of stone or mud brick. The family occupies the upper floor, while the lower is used as a barn for the farm animals. Beds are made of straw and yak wool blankets. There are few towns. **Thimbu**, the capital, is merely a cluster of houses surrounding the Tashi Cho Dzong, an old fortress (recently rebuilt with Indian aid) that houses the government's offices. The king resides in the new capital, where a township is being developed.

Farming is the country's chief economic activity. A variety of crops are grown, depending on elevation and climate. Rice and buckwheat are grown up to an elevation of 5,000 feet (1,500 m.). At higher elevations, farmers alternate crops of barley and rice, while wheat is grown up to an altitude of 9,000 feet (2,700 m.). Many Bhutanese are expert handicraftsmen. They are particularly noted for embroidered wool and silk fabrics, bronze and silver ornaments, beautifully fashioned swords and daggers, and handsome woodwork. Examples of the fine wood carving may be seen in the ornate roofs and windows that adorn the old buildings in the country's many *dzongs* (castle monasteries). Bhutan's economic development has been hampered by the country's remoteness, as well as by a lack of transportation facilities, qualified technicians, and convenient markets. Coal is the only mineral mined.

History and Government.   Bhutan became a separate political state some 300 years ago, when a Tibetan lama (priest) named Sheptoon La-Pha proclaimed himself king. Subsequently, Bhutan was ruled by two leaders, a *dharma raja* for spiritual matters and a *deb raja* for political and administrative affairs. In 1907, aided by the British, the *penlop* (governor) of Tongsa in eastern Bhutan established a hereditary line of kings. An advisory council of civil servants and Buddhist leaders assists the king, and a national assembly, indirectly elected by the people, meets twice a year. Bhutan, whose foreign affairs have been handled by India, has been accepted into the United Nations and is establishing its own diplomatic relations.

P. P. KARAN, University of Kentucky; author, *The Himalayan Kingdoms*

Sherpas carry heavy loads high in the Himalayas.

HIMALAYAS

In May of 1953 two daring mountain climbers, Edmund Hillary of New Zealand and Tenzing Norkey, a Sherpa (one of the mountain peoples of Nepal), became the first men ever to climb to the top of the world's highest mountain—Mount Everest. When Hillary and Norkey reached the summit of this great mountain located on the Nepal-Tibet border, they were literally standing on top of the world. For Mount Everest soars nearly 5½ miles (8.8 kilometers) into the sky, the highest of the many towering peaks in the Himalayas, the world's tallest mountain chain.

Majestic and awe-inspiring, the Himalaya mountains form a natural wall, 100 to 150 miles (160–240 km.) in width, between Tibet in the north and India in the south. Located within the Himalayas are the Kingdom of Nepal and the smaller kingdoms of Bhutan and Sikkim. At their highest level the mountains rise like giant, snowcapped spires to well over 20,000 feet (6,000 meters). The name "Himalaya" means "the house of snow" and comes from the Sanskrit words *hima* ("snow") and *alaya* ("abode").

The People of the Himalayas. Except in the fertile valleys, few people live in the Himalayas. Most of the mountain people are ethnically related to the Chinese, Mongolian, and Indian peoples. They support themselves mainly as farmers and herders. The Sherpas in particular are known for their skills as mountaineers, and they often serve as guides for mountain-climbing expeditions. By means of terraced slopes, the Himalayan people are able to grow rice, corn, wheat, and barley at altitudes of up to 8,000 feet (2,400 m.), and sometimes higher. Herds of sheep, goats, yaks, and other livestock are grazed on the grassy slopes. Hunting, timber cutting, and tourism also provide income for the mountain people.

The Mountains. According to geologists, the Himalayas were formed during a violent upheaval of the earth in the Tertiary period, some 60,000,000 years ago. Parallel folds were created, which are now the three ranges of the Himalayas. The first of these is the Outer Himalayas. Heavily forested in some areas, this relatively low chain of mountains, with elevations up to 4,000 feet (1,200 m.), rims the northern Indian plains. Beyond the Outer Himalayas are the Middle Himalayas, which rise to heights of over 15,000 feet (4,600 m.). The last range is the Great Himalayas, whose peaks start at about 18,000 feet (5,500 m.) and reach a maximum elevation at Mount Everest, which looms above the clouds at 29,028 feet (8,848 m.). A fourth chain, the Karakoram range, is often associated with the Himalayas. It lies just north of the western fringe of the Great Himalayas. The world's second highest mountain, Mount Godwin Austen, or K2 (28,250 ft.; 8,610 m.), is in this range.

Rivers. Several major rivers run through the Himalayas. They include the Indus, which marks the western limit of the Himalayas, the Sutlej, and the Tsangpo, which runs eastward through Tibet until it becomes the Brahmaputra at the eastern edge of the mountains. The Ganges, India's most sacred river, also begins in the Himalayas. Glaciers high in the mountains supply water to these rivers. Over the centuries the glaciers and rivers have gouged out deep gorges and canyons.

Climate. On the lower slopes and in the valleys of the Outer Himalayas, extremely hot summers and mild winters—with temperatures above freezing—are the rule. In the upper regions of the mountains, summers are mild, and winter temperatures are much cooler. Snowfall usually begins at about 5,000 feet (1,500 m.), and above 16,000 feet (4,900 m.) there is a year-round blanket of snow. At the highest altitudes, temperatures plunge far below 0 degrees Fahrenheit, and blizzards are frequent. The air here has little oxygen, which makes exploration and mountain climbing very hazardous.

Animal and Plant Life. Dense jungles and swamps cover the valley floors of the Outer Himalayas, and various plants and trees are found up to about 12,000 feet (3,700 m.). Tigers and leopards prowl about the forests, and monkeys and elephants are also found there. The most useful animal in the Himalayas is the yak, a shaggy-haired, oxlike creature. Yaks are used as beasts of burden or are ridden like horses. Yak meat and milk are important foods for the Himalayan people, and the animal's hide and hair are used to make clothing and tentlike shelters.

The people of the Himalayas look upon the mountains with reverence. They believe that gods, spirits, and strange beings live among the tall peaks. A mysterious, apelike creature called the Abominable Snowman—or *yeti* in the Sherpa tongue—is said to inhabit Everest and other mountains. Many of the local people claim to have seen the *yeti*. But expeditions sent to find the creature have failed to discover anything except a few large footprints. No one can say for sure whether the footprints belong to the Abominable Snowman or to some large animal, and so the mystery remains.

Reviewed by P. P. KARAN, University of Kentucky; author, *The Himalayan Kingdoms*

BANGLADESH

In 1971, after 9 months of bloody civil war, the People's Republic of Bangladesh achieved its independence from Pakistan. Prior to that, Bangladesh was known as East Pakistan or East Bengal. Bangladesh is one of the world's youngest nations, although the region has a long history. With an estimated 75,000,000 people, it is also the world's eighth largest country in population.

THE LAND

Bangladesh—the name means "Bengal Nation"—has an area of 55,126 square miles (142,776 square kilometers). It lies astride the Bay of Bengal and is bounded on three sides by India and in the southeast by Burma. Bangladesh occupies the large delta formed by the mighty Ganges and Brahmaputra rivers. It is largely a country of lowlands crossed by innumerable waterways. In the southeast the Ganges and Brahmaputra join to form the Padma River. Further to the south the Padma is joined by the Meghna River. Here the rivers are actively building new deltas and islands by depositing huge quantities of sediment. It is a region of very fertile soil, most of which is devoted to the cultivation of rice, the country's staple food.

The hilly areas of Bangladesh are located in the Chittagong district to the east of the Bay of Bengal and in the Sylhet district. The hills of Chittagong are covered with tropical forests. The only river of importance here is the Karnaphuli, at the mouth of which lies the port of Chittagong, Bangladesh's outlet to the sea.

A family at work in a rice field. Rice is the main food crop.

Government and bank buildings line this street in Dacca.

CLIMATE

Bangladesh has a tropical monsoon climate with two principal seasons, a hot, wet summer and a cooler, drier winter. From April or May to October the onshore monsoon winds from the Bay of Bengal bring rain practically every day with high temperatures. The winter season begins in November. This is also the month of tropical storms, called cyclones, which often cause great destruction in the country. Some 500,000 rice farmers, fishermen, and boatmen and their families are reported to have been killed in the 1970 cyclone.

THE PEOPLE

Bangladesh is one of the world's most crowded nations. A density of over 1,000 persons per square mile is common over most of the country. The Chittagong district is one of the few sparsely settled areas. The great majority of the people live in small villages. There are only four cities with populations of over 100,000: Dacca, Chittagong, Narayanganj, and Khulna. The largest of these is Dacca, the capital, which has about 1,250,000 people, counting suburbs and surrounding towns.

The people of Bangladesh may be divided into two distinct ethnic groups—the majority Bengalis and the tribal peoples of the Chittagong Hills. Most of the population is Muslim. Hindus make up the largest minority group and there are relatively small numbers of Buddhists and Christians. Bengali is the language spoken by most of the people, although English is widely used in government, commerce, and higher education.

The typical Bengali is slim, brown-skinned, and rather short in stature, with glossy black hair and expressive features. The traditional dress for men is the lungi, which is worn wrapped around the waist and either stretching to the ankles or tucked up to the knees. Women usually wear the long, flowing sari.

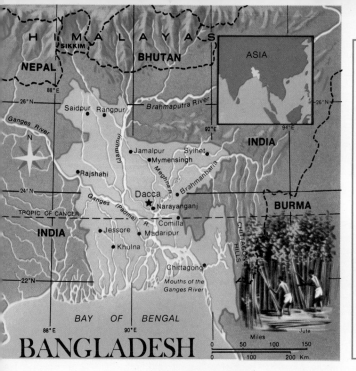

BANGLADESH

FACTS AND FIGURES

PEOPLE'S REPUBLIC OF BANGLADESH is the official name of the country.

CAPITAL: Dacca.

LOCATION: South Asia. **Latitude**—26° 38′ N to 20° 46′ N. **Longitude**—92° 41′ E to 88° 02′ E.

AREA: 55,126 sq. mi. (142,776 sq. km.).

PHYSICAL FEATURES: Highest point—Keokradong (4,034 ft.; 1,230 m.). **Lowest point**—sea level. **Chief rivers**—Ganges, Brahmaputra.

POPULATION: 75,000,000.

LANGUAGE: Bengali, English.

RELIGION: Islam (predominant).

GOVERNMENT: Republic. **Head of state**—president. **Head of government**—prime minister. **International co-operation**—Commonwealth, United Nations.

CHIEF CITIES: Dacca, Chittagong, Narayanganj, and Khulna.

ECONOMY: Agricultural products—rice, jute, tea, tobacco, oilseeds, mangoes, pineapples, coconuts, bananas. **Industries and products**—jute manufacturing, paper and newsprint, cement, fertilizer, handicrafts. **Chief exports**—raw jute, jute yarn and manufactures, fish, tea, tobacco, timber. **Chief imports**—metals, petroleum, coal, machinery, textiles.

MONETARY UNIT: Taka.

NATIONAL HOLIDAY: March 26, National Day.

NATIONAL ANTHEM: "My Golden Bengal."

THE ECONOMY

Bangladesh is one of the poorest nations in the world. It has a basically agricultural economy, with few mineral resources and little industry. Its income is mainly from exports of jute, tea, tobacco, fish, and timber. The chief agricultural products are jute and rice. Bangladesh is the world's leading producer of jute, a fiber used in making twine, sacks, and burlap. Rice is the basic food of the people.

The weather is of vital importance to the agricultural economy of Bangladesh. Floods, cyclones, and drought are a constant threat. Bangladesh is largely a nation of small farms, averaging only several acres each. The high rate of population growth impedes attempts to improve the standard of living of the people. And the recent devastation of the civil war has done much to disrupt the economic life of the country.

The importance of agriculture in the Bangladesh economy is shown in the number of industries based on agricultural raw materials: jute mills that process the raw jute fiber; paper and newsprint plants; and cement and fertilizer factories. A lack of adequate mineral and power resources, an underdeveloped transportation system, and a shortage of money are responsible for the low rate of industrialization. The extensive system of rivers has made the construction of railroads and highways very difficult. The rivers provide the largest single transportation network in the country and they are used chiefly by small boats.

HISTORY

The early history of Bengal is obscure. It is generally believed that about 1000 B.C. the Bang tribe, a Dravidian people, was pushed out of the upper Ganges valley by the advancing Indo-Aryans. The new territory occupied by the Bang later came to be known as Bengal. During the 3rd century B.C., the Maurya empire extended its domain over the area, and Buddhism spread under the rule of the emperor Asoka. Later Bengal came under the control of the Hindu Gupta empire.

During the 9th century A.D., the Pala dynasty came to power. The 3 centuries of rule by the Pala kings is regarded as the classical period of

Bengali history. There was a flowering of the arts and the shaping of a distinct Bengali culture. During the 6 centuries from the 13th through the 18th century Bengal was under Muslim rule. In this period Islam spread rapidly in Bengal, giving the region the Muslim preponderance it still has. But even today rural Bengali Muslims who are descended from Hindus continue the ancient Hindu religious practices. Most Muslims and Hindus even participate in each other's religious festivals.

In the 18th century Bengal came under the control of the British, who ruled it as part of their Indian empire until 1947. In that year Great Britain ended its rule in the Indian subcontinent. India received its independence, and at the insistence of the Muslim League, the separate nation of Pakistan was formed out of those parts of the subcontinent where the Muslims were a majority. East Bengal, which lay within the Bengal enclave of India, became the eastern wing—East Pakistan—of the new country. It was separated from the larger, western part of Pakistan by 1,000 miles of Indian territory.

CONTEMPORARY EVENTS

The only cement that Pakistan could rely on to hold the two distant parts of the country together was Islam. And in 25 years the Islamic faith proved unequal to the difficult task. In the 1950's Pakistan attempted to eliminate Hindu influence from the Bengali language and culture and mold it along Islamic lines by imposing Urdu, the main language of West Pakistan, as the official language of all of Pakistan. The attempt failed. A high percentage of the men who governed East Pakistan came from the West and did not speak the language of the Bengalis. West Pakistan tended to treat the Bengalis of East Pakistan with scorn and contempt.

The Bengali discontent was reflected by the growth of a powerful political movement called the Awami League, led by Sheikh Mujibur Rahman, one of whose goals was self-government for East Bengal. In the 1970 elections the Bengalis voted for Awami League candidates, who won a majority of the seats in the national assembly, the country's legislature. The West Pakistan government, however, prevented the successful candidates from taking their seats by postponing the opening of the national assembly. This led to riots and disorders in East Pakistan, which were put down with great harshness and bloodshed by West Pakistani troops. Mujib was imprisoned and an estimated 10,000,000 Bengalis fled across the border into India.

Burdened by the refugees, India began building up its own forces on the border between the two countries. The Indians also helped the Bengali guerrillas to resist the West Pakistan forces. Border incidents eventually led to a full-scale but short war between India and Pakistan. The West Pakistan forces surrendered and the new nation of Bangladesh was proclaimed. Sheikh Mujibur Rahman became its first prime minister and later its president.

Although independence found Bangladesh with a disrupted economy, the country has an excellent chance of developing into a stable and cohesive nation. It has a common language and a common culture and ethnic origins. But the problems are great ones. It must have effective leadership and efficient government. It must also solve the problems of overpopulation and food production if it is to maintain itself.

P. P. KARAN, Chairman, Department of Geography, University of Kentucky

A Hindu temple in Colombo (left) and a Buddhist dagoba, or shrine (right), at Anuradhapura. Buddhism and Hinduism are Sri Lanka's major religions.

SRI LANKA

Sri Lanka, formerly known as Ceylon, is a pear-shaped island separated from the southeast tip of India by just 22 miles (35 kilometers) of shallow water. But in spite of its closeness to India, Sri Lanka is a separate nation with its own identity. Few places are as rich and varied. Sri Lanka is rich in the beauty of its rivers and waterfalls, white beaches and thick forests, and in all that the land yields—tea, rubber, coconuts, spices, and gemstones. Sri Lanka is rich, too, in its history, which goes back 2,500 years, and in its large and varied population.

THE PEOPLE

The *Mahavansa,* the 6th-century epic of Sri Lanka, tells of a group of men from northern India who sailed to Sri Lanka in the 6th century B.C. They conquered the island's earliest inhabitants and set up a kingdom—the Sinhalese kingdom—that lasted more than 2,000 years. The *Mahavansa* also tells that in the 3rd century B.C. Buddhism was introduced into Sri Lanka by Mahinda, the son of the Indian ruler Asoka, and that the Sinhalese adopted the religion. In the centuries that followed, people from southern India came to Sri Lanka and eventually set up their own separate kingdom—the Tamil kingdom—in the north.

Today the Sinhalese make up about 70 percent of Sri Lanka's population. The majority live in the southern and western regions of the country and speak Sinhalese. Most are Buddhists. Tamils make up more than 20 percent of the population, and include Sri Lanka Tamils and Indian Tamils. (Sri Lanka Tamils are the descendants of Tamils who came to Sri Lanka in ancient times, and Indian Tamils are the descendants of people who were brought from India to work on plantations, starting in the latter part of the 19th century.) The Tamils live mainly in the north and east, speak Tamil, and are Hindus for the most part. The population also includes Muslims, some of them descendants of Arab traders who first came to Sri Lanka about the 8th century; people of mixed descent; and Europeans. A small group called Veddas, whose traditional home is the forests of Sri Lanka, may be the descendants of the island's original inhabitants.

The general mood of the people is cheerful and friendly. Even in the poorest villages, strangers are welcomed hospitably. The women are striking with their long black hair and erect, graceful walk. Women wear traditional saronglike ankle-length skirts with short jackets, or they wear saris. Many men wear Western clothes in the city, but traditional clothing is more common. It consists of a long piece of cloth, usually white, which is wrapped around the body and secured at the waist, and which is worn with a loose shirt or jacket.

Village houses are square or rectangular, with walls of dried mud or clay blocks, floors of beaten earth or cement, and roofs of coconut thatch or tile. Most houses in rural areas are fenced and have a small veranda. People find it cool to sleep on woven mats that are placed on the floor or on wooden frames.

People in Sri Lanka eat rice at every meal. Orthodox Hindus and

A street scene in Colombo, Sri Lanka's capital, largest city, and chief port.

most Buddhists are vegetarians and often serve rice with curries made of vegetables cooked in coconut milk and spices.

Boys and girls are required to go to school until the age of 14 and are taught in the language they speak at home: Sinhalese or Tamil. English is also taught. Schools are free from kindergarten through the university, and more than 70 percent of the people can read and write.

Artistic expression is important to the people of Sri Lanka and has been throughout their history. The sites of two ancient Sinhalese capitals, Anuradhapura and Polonnaruwa, contain a wealth of palaces, dagobas (domed Buddhist shrines), and statues of Buddha. The island's most famous paintings are a series of beautiful maidens painted in the 5th century on a wall of the granite fortress of Sigiriya.

The dance is an ancient art form that is still greatly respected. It is taught for 6 years in most government schools. Best-known of all the dancers are those of the lovely hill city of Kandy. Each August scores of whirling Kandyan dancers in lavish costumes take part in the Buddhist festival of Perahera, honoring a tooth of Buddha. The relic, which is housed in the Temple of the Tooth, is rarely seen. During the festival, a casket containing a replica of the tooth is carried on the back of a temple elephant in a procession.

Artisans' skills have been handed down for generations. Handicrafts include carved wooden masks used in ritual dances and folk plays, brasswork, handloomed cotton, tortoiseshell ware, and handmade lace.

THE LAND

Sri Lanka lies in the Indian Ocean, separated from India by Palk Strait. The island is 270 miles (430 km.) long, 140 miles (230 km.) across

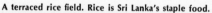
A terraced rice field. Rice is Sri Lanka's staple food.

FACTS AND FIGURES

REPUBLIC OF SRI LANKA is the official name of the country.

CAPITAL: Colombo.

LOCATION: Island in the Indian Ocean off the southeastern coast of India. **Latitude**—5° 55′ N to 9° 50′ N. **Longitude**—79° 42′ E to 81° 53′ E.

AREA: 25,332 sq. mi. (65,610 sq. km.).

PHYSICAL FEATURES: Highest point—Pidurutalagala (8,291 ft.; 2,527 m.). **Lowest point**—sea level. **Chief river**—Mahaweli Ganga.

POPULATION: 13,000,000 (estimate).

LANGUAGE: Sinhalese, Tamil, English.

RELIGION: Buddhist, Hindu, Christian, Muslim.

GOVERNMENT: Republic. **Head of state**—president. **Head of government**—premier. **Legislature**—parliament consisting of a House of Representatives. **International co-operation**—United Nations, Commonwealth of Nations, Colombo Plan.

CHIEF CITIES: Colombo, Jaffna, Kandy, Galle.

ECONOMY: Chief minerals—graphite, gemstones, ilmenite, monazite, quartz sand, limestone. **Chief agricultural products**—tea, rubber, coconuts, rice. **Industries and products**—cement, salt, gemstones, ilmenite. **Chief exports**—tea, rubber, coconut products, graphite. **Chief imports**—rice, sugar, wheat flour, fertilizers, milk products, fish, petroleum products, textiles.

MONETARY UNIT: Ceylon rupee.

NATIONAL HOLIDAY: February 4 (1948), Independence Day.

NATIONAL ANTHEM: *Namo, Namo, Matha* ("Hail, Hail, Motherland").

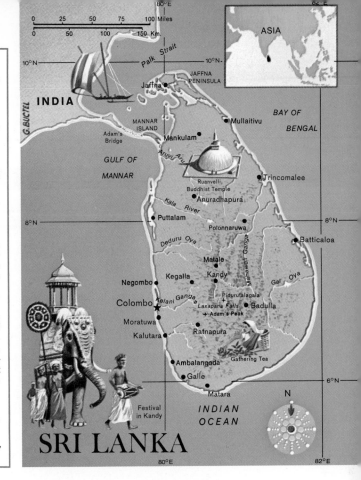

SRI LANKA

at its widest point, and has an area of 25,332 square miles (65,610 square kilometers). Much of the country is rolling lowland. The highest part is the mountainous south central region. The land rises dramatically from lowland rice fields plowed by water buffalo, through coconut and rubber plantations and highlands planted with tea, to grass-covered mountains 7,000 feet (2,100 meters) high. Many rivers flow down the mountains to the ocean. During the season of heavy rains, the water is stored in huge man-made lakes, or tanks. The average lowland temperature is 80 degrees Fahrenheit (25 degrees Celsius), but in the higher hills the climate is springlike. Rainfall varies from 25 inches (64 centimeters) a year in some places to 250 inches (640 cm.) a year in others. Rain falls during a summer and a winter monsoon, but each rainy season affects only one half of the island.

Sri Lanka is famous for the gemstones found in the southwest— sapphires, rubies, moonstones, topazes, and cat's-eyes. There are large deposits of graphite, the leading mineral export, in the southwest. The sands of the beaches yield ilmenite, a source of the metal titanium, and monazite, used in the production of nuclear energy. There are also deposits of quartz sand, used in making glass, and of limestone.

The flowering plants, trees, birds, and animals of Sri Lanka are a national treasure. Many kinds of plant and animal life are not found in any other part of the world. Among the loveliest flowers are orchids, jasmine, hibiscus, and jacaranda. The coast is fringed with coconut palms, while deep in the forests there are many hardwood trees, such as ebony, satinwood, ironwood, and halmilla, which has dark red wood. Nearly 400 different kinds of birds, including peacocks and flamingos, live in the forests and jungles, and there are more than 100 kinds of mammals, among them leopards, buffalo, deer, bears, and monkeys. Trained elephants help in construction work and in clearing forest land. Wild elephants are now strictly protected by law to save them from extinction. There are also crocodiles, lizards, and snakes, including the poisonous cobra and Russell's viper. To protect natural life the government has set aside land for national parks and bird sanctuaries.

ECONOMY

More than 50 percent of the people of Sri Lanka make their living by farming, working either on large tea or rubber plantations or on small farms where they raise a variety of crops, including rice, coconuts, fruits, vegetables, tobacco, and cinnamon. A major aim of the government is to increase rice production. Until recently rice was Sri Lanka's main import, but improved crops indicate that the country may soon grow most of the rice it needs. Industry is a growing part of the economy. There are cement factories, a steel rolling mill, and a tire factory. Textile mills are being built with foreign aid. The manufacture of salt by the evaporation of seawater is a government monopoly. Hydroelectric projects and the building of an oil refinery are under way. One of the colorful sights of the island is fishermen sailing in catamarans or perching on wooden stilts in the water, but efforts are being made to introduce modern methods and to develop fishing as an industry. Tourism is growing with the building of modern hotels.

The island's commercial center is the capital, **Colombo**. The docks of this major port are fragrant with tea and spices waiting to be shipped abroad. The cultural center of Sri Lanka is **Kandy**, once the seat of an ancient kingdom.

A tea plantation. Tea is the country's most valuable export crop.

HISTORY

Sri Lanka's location on the ocean route between East Africa and South Asia made it a natural stopping place for traders. Early Greeks and 8th-century Arabs knew the island, and Marco Polo visited it in 1293. But the first Europeans to occupy parts of Sri Lanka were the Portuguese, who landed in 1505, bringing Catholicism with them. In 1658 the Portuguese were driven out by the Dutch, who introduced Protestantism and Roman-Dutch law. At the end of the 18th century the Dutch were challenged by the British, who captured the ports of Trincomalee and Colombo. For a time Sri Lanka came under the rule of the British East India Company, and in 1802 it became a British crown colony. In 1815 the entire island came under British authority. Sri Lanka remained under the British, developing into a major grower and exporter of tea, until 1948, when it became independent.

Since independence Sri Lanka has gone through many economic and political crises. World prices for the country's basic crops—tea, rubber, and coconuts—have fallen, and Sri Lanka has had to turn to other countries for aid. In 1959 the Prime Minister, Solomon West Ridgeway Dias Bandaranaike, was assassinated. His widow, Mrs. Sirimavo Bandaranaike, succeeded him the following year, after a brief inter-regnum, becoming the world's first woman prime minister.

Sinhalese-Tamil relations have been a source of conflict, mainly because of language problems. Under the British, the country's official language was English. In 1956 a law was passed making Sinhalese the official language. At the same time another law was passed by Parliament providing for the use of Tamil for special purposes to be pre-scribed by regulation. Tamils resented the law, and periods of violence and national emergency followed. In 1966 the government, headed by Prime Minister Dudley Senanayake, passed regulations allowing Tamil to be used for certain official purposes.

Another problem was the nearly 1,000,000 descendants of Indians who came to Sri Lanka, starting in the late 19th century, to work on plantations. Most of these people were neither citizens of Sri Lanka nor citizens of India. Trying to decide how many of these persons of Indian origin would be accepted into India and granted Indian citizenship and how many would remain in Sri Lanka and be granted citizenship formed the subject of long and difficult negotiations until October, 1964, when Mrs. Bandaranaike and Lal Bahadur Shastri, the Prime Minister of India, signed an agreement. Sri Lanka granted citizenship to many of these people and India took back a large number. The status of the remainder was finally decided in 1974. Half went to India and half became citizens of Sri Lanka.

Government. The government of Sri Lanka is headed by a premier, or prime minister, who leads a council of ministers. The legislature consists of a single chamber, the House of Representatives. The president, who serves a term of 4 years, is head of state. All citizens of Sri Lanka over the age of 18 can vote. Following independence in 1948, Sri Lanka retained its allegiance to the British Crown, which was represented by a governor-general. Under a new constitution, adopted in 1972, Sri Lanka became a republic and ended its ties to the Crown, although it still remains a member of the Commonwealth of Nations.

Reviewed by H. S. AMERASINGHE, Permanent Mission of Sri Lanka to the United Nations

MALDIVES

If a Maldivian wants to rent his own island, he can choose from hundreds in the Republic of Maldives. For this archipelago, or chain of islands, in the Indian Ocean is made up of over 1,000 islands and atolls, and only about 200 are inhabited. The Maldivian atolls have long been praised for their beauty. In fact, the term "atoll," meaning a coral island (or group of islands) enclosing a lagoon, or pool of water, comes from a Maldivian word—*atolu*.

The Land.   The Maldive Islands lie several hundred miles southwest of Ceylon. None of the islands is larger than 5 square miles (13 square kilometers), and the entire archipelago is only 115 square miles (298 sq. km.) in area. The islands are low, rarely rising more than a few feet above sea level.

The climate is affected by the monsoons, strong winds that bring considerable rainfall. The weather is generally warm and humid throughout the year. Coconut palms and breadfruit trees in particular grow in abundance on the islands. The tropical waters abound with fish, and there are magnificent tortoises, whose distinctively marked shells are fashioned into jewelry and artistic objects by Maldivian craftsmen.

Transportation to and from the various islands is by boat. The bicycle is an important form of transportation on land. There are relatively few automobiles.

The People.   The origins of the Maldivian people are obscure. Their language, Divehi, is related to Sinhalese, a language spoken in Ceylon, and some scholars believe that the Maldivians are descendants of the Sinhalese of Ceylon, with a mixture of Arab peoples. Originally, the Maldivians were Buddhists, but since the 12th century they have practiced the religion of Islam.

About 108,000 people live in Maldives, some 12,000 of whom inhabit the capital, **Male**, located on Male Island. The Maldivians are

These Maldivian postage stamps commemorate the International Education Year.

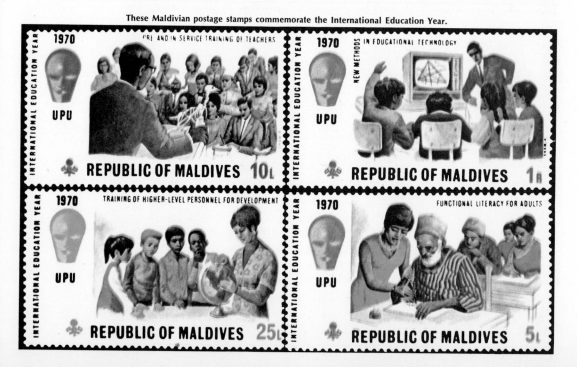

FACTS AND FIGURES

REPUBLIC OF MALDIVES is the official name of the country.

CAPITAL: Male.

LOCATION: Indian Ocean, southwest of Ceylon. **Latitude**—7° 6′ N to 0° 42′ S. **Longitude**—72° 31′ E to 73° 44′ E.

AREA: 115 sq. mi. (298 sq. km.).

POPULATION: 108,000 (estimate).

LANGUAGE: Divehi.

RELIGION: Muslim.

GOVERNMENT: Republic. **Head of state and government**—president. **Legislature**—Majlis. **International co-operation**—United Nations, Colombo Plan.

CHIEF CITY: Male.

ECONOMY: **Chief agricultural products**—coconuts, breadfruit, mangoes, papaw, plantains, pumpkins. **Industries and products**—fishing, coconut products (copra and coir), matmaking, lacemaking, cowrie shell and tortoiseshell objects. **Chief exports**—dried fish, cowrie shell, copra. **Chief imports**—rice, manufactured goods, fuel.

MONETARY UNIT: Maldivian rupee.

NATIONAL HOLIDAY: July 26.

NATIONAL ANTHEM: *Gowmee Salaam* ("Together in national unity, we salute you").

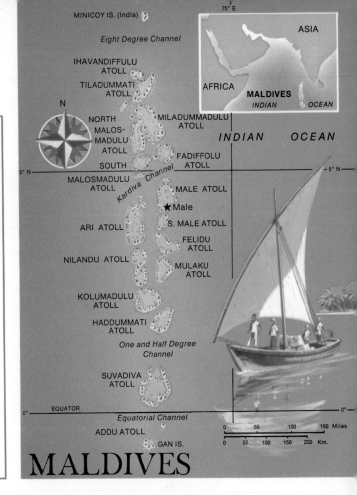

MALDIVES

skilled sailors, and fishing is the chief occupation as well as the country's most important industry. Dried fish (known as Maldive fish) is the major export, all of it exported to Ceylon. Rice, much of which must be imported, remains the staple food of the people.

The cultivation of coconuts and products derived from coconuts—copra (dried coconut meat) and coir (coconut fiber)—is the second most important industry. There are also some handicraft industries, such as lacquerwork and the making of mats and tortoiseshell objects. The sale of the country's interesting postage stamps is also a source of income for Maldives.

History and Government. According to legend an ancient prince of Ceylon, who with his bride was forced to anchor in a Maldivian lagoon because his ship was becalmed, became the country's first sultan. The Didi family, who governed Maldives as sultans for nearly 8 centuries, are said to be descended from this prince.

Contact with early Arab traders led to the acceptance of Islam. Later the islands came under the domination of the Portuguese and then the Dutch. In 1887 the islands became a British protectorate. Maldives attained complete independence in 1965.

In 1968, following a plebiscite, Maldives became a republic under a new constitution replacing the sultanate. The head of state and government is the president. The Maldivian legislature, the Majlis, is composed of 54 members, who are elected for 5-year terms.

Reviewed by EMBASSY OF THE REPUBLIC OF MALDIVES, Washington, D.C.

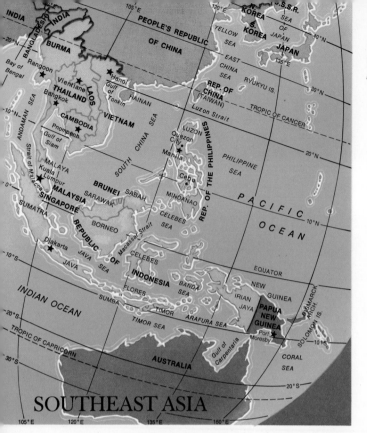

SOUTHEAST ASIA

SOUTHEAST ASIA

Southeast Asia is one of the world's great melting pots. Its diverse peoples moved into the region in search of a better life and greater security. This great population movement began about 4,000 years ago. Today Southeast Asia includes the independent nations of Indonesia, the Philippines, Malaysia, Singapore, Vietnam, Laos, Cambodia, Thailand, and Burma. Although Brunei is independent too, Britain is responsible for its foreign affairs. East Timor, which was formerly known as Portuguese Timor and was a Portuguese territory, is now part of Indonesia.

The original inhabitants of Southeast Asia were a very small, dark-skinned people, some of whose descendants still live in the highland regions of the Philippines, Indonesia, and Malaysia. Around 2500 B.C. the first major wave of migrating peoples entered the area. They were the Malays, or Indonesians, and it is their descendants who form the great majority of the populations of the Philippines and Indonesia today. The Malays formerly lived in what is now southern China, but pressure from the Chinese population in the north forced other peoples southward. These peoples in turn pressed upon the Malays, who moved through the mountain passes into mainland Southeast Asia, down the Malay Peninsula, and out into the Indonesian and Philippine islands. Skilled sailors, the Malays expanded eastward through these islands.

Other peoples followed, principally the Cambodians, the Vietnamese, the Burmese, and the Thai. They also moved south out of China, but settled in mainland Southeast Asia. The Thai were the last of the major peoples to settle in the area, establishing their first important kingdom in the 13th century.

These various peoples brought with them their own customs, cultures, and living patterns, but they were to be strongly influenced by still other peoples. Traders from India brought Indian ideas to Southeast Asia, especially the Hindu and Buddhist religions. Burma, Thailand, Laos, and Cambodia are today Buddhist countries as a result. Later, Muslim traders brought Islam to Malaysia and Indonesia, which are now predominantly Muslim. The culture and religion of Vietnam were influenced by China.

This process of infusion of both people and ideas has continued into modern times. The European powers began their colonization of the region (except for Thailand, which was never colonized) in the 16th century, bringing with them Western ideas, including Christianity. The Philippines, colonized by Spain, became largely Catholic in religion. In the late 18th and early 19th centuries large numbers of Chinese and Indians came to Southeast Asia to take advantage of the economic opportunities during the height of the European colonial period. During World War II the region was conquered by Japan. The years that followed World War II saw the appearance of the nation-states of modern Southeast Asia.

THE LAND

The nations and territories of Southeast Asia together cover an area of about 1,738,000 square miles (4,500,000 square kilometers) and are inhabited by approximately 280,000,000 people. Including the waters within it, Southeast Asia covers a portion of the globe as big as Europe from Ireland to Turkey. The largest country is Indonesia, whose approximately 120,000,000 people make it one of the six most populous nations in the world. Singapore is the smallest country of the region; its 2,000,000 people live on only 224 square miles (581 sq. km.).

China is Southeast Asia's neighbor to the north. To the east, west, and south the region is flanked by seas, principally the Pacific and Indian oceans. Burma, the westernmost of the Southeast Asian lands, borders India and Bangladesh. Southeast Asia thus is located very strategically— east of India and south of China. These are the two most heavily populated countries of the world. They are also lands with serious economic problems—problems that might be partly solved by access to the natural resources of the Southeast Asian countries.

Physically Southeast Asia is a much-divided land. Its mountain chains, which run in a north–south direction, historically have separated some of its peoples, such as the Burmese, Thai, and Vietnamese. Indonesia and the Philippines are also divided into literally thousands of islands, but contact was often easier among the islands than between coastal and interior regions. Southeast Asia's many divisions also made it very easy to conquer the region piece by piece. This is what the European powers did from the 16th through the 19th century and what Japan did in 1941–42.

The Climate. The climate in Southeast Asia is generally warm and very often wet. The seasons in most countries are alternately dry and rainy ones, although in Indonesia there is considerable rainfall throughout the year. Burma and the Philippines in particular have especially deadly monsoons—violent rainstorms—with hurricanelike winds and much flooding and loss of life. The temperature rarely falls below 68 degrees Fahrenheit (20 degrees Celsius) in most of Southeast Asia, except in the highlands, while the hot and dry season often brings temperatures of 90 degrees F. (32 degrees C.).

The Economy. The natural resources of the Southeast Asian coun-

These fertile rice fields are in Vietnam.

A causeway links Malaysia with the island nation of Singapore.

tries are considerable, but even today they are still largely underdeveloped. Compared with both China and India, Southeast Asia is well-endowed with resources and not densely populated (except for a few areas, such as the chief Indonesian island of Java), and it has a higher standard of living. This standard of living is low, on the other hand, when compared with that of the United States or Europe.

Southeast Asia's chief products are agricultural or mineral ones. Thailand is the world's leading exporter of rice, while Cambodia and southern Vietnam, in normal times, have also had substantial rice surpluses. The Philippines has recently become self-sufficient in its production of this vital grain, and Indonesia is expected to do so fairly soon. Malaysia is the world's leading exporter of rubber and one of its chief suppliers of tin. Indonesia and Burma both have substantial petroleum deposits, but coal is found only in northern Vietnam. Vietnam, Malaysia, and the Philippines also have some iron, although probably not enough to support significant iron-based industries without additional imports. The years since World War II have also seen the development of new products in many of the Southeast Asian countries—high-yielding corn in Thailand, for example. Southeast Asia's seas, lakes, and rivers, however, have never been adequately exploited commercially for their fish and other products.

THE PEOPLE

Almost all of Southeast Asia's peoples are Mongoloids, a majority of them being brown- rather than yellow-skinned. The area's major foreign minorities are the Chinese and Indians. Europeans have been relatively few in number, even during the colonial period.

Indonesia and the Philippines are probably the two least complex countries racially. More than 90 percent of their inhabitants are of the broad Malay or Indonesian ethnic grouping. Singapore—often referred to as a third China—also is a fairly homogeneous country: three out of four of its inhabitants are of Chinese descent. Cambodia's countryside is populated largely by Cambodians, but Chinese and Vietnamese minorities together outnumber native Cambodians in the capital city of Pnompenh.

The other countries have serious racial problems. Burma has four major minority groups—the Karens, Kachins, Chins, and Shans—some of whom have revolted against the Burman-dominated central government since independence came in 1948. Half of Malaysia's population are non-Malays; the number of Chinese, in fact, almost equals the number of Malays. Thailand has important, largely unassimilated Malay, Meo, Vietnamese, and Lao minorities. In Laos, the Lao, the dominant ethnic group, inhabit only the lowland area near the Mekong River. Vietnam was for many years torn by civil war and remained a divided country until 1976. Vietnam has four major religious groups, although their influence is expected to be severely limited under Communist rule. There are also differences between the majority Vietnamese and the minority Montagnards (mountain people).

Part of the problem is the artificiality of the borders drawn by the former colonial powers. There are Malays in Thailand and Singapore as well as in Malaysia; and Vietnamese live in Cambodia, Laos, and Thailand as well as in Vietnam. Burma's Shan minority is more closely related to

A woman of Laos.

Indonesian women from the island of Bali.

the majority Thai of Thailand than to the ruling Burmans of Burma. Thus there are sometimes stronger links between peoples divided by frontiers than among those living within the borders of a particular country.

Language. Southeast Asia's diverse peoples speak a variety of languages. The main languages of the insular, or Malay, countries—Indonesia, the Philippines, and Malaysia—are closely related, being of common Malay origin. But in the Philippines no one Malay dialect is predominant. English is more widely spoken than any single regional language, although the use of Pilipino, formed mostly from the Tagalog dialect, has grown considerably in recent years. A nationwide official Indonesian language has been largely the product of government stimulation since independence. Of the mainland Southeast Asian languages, only Thai and Lao are closely related. English is spoken by most educated persons in almost all of the countries. French is also spoken in the former French colonies of Cambodia, Laos, and Vietnam.

Customs and Beliefs. There are important differences in the ways the various Southeast Asian peoples look at life. Most people in Burma, Thailand, Laos, Cambodia, and Vietnam are Buddhists, although Vietnamese Buddhism is at least as different from the Buddhism of the other four countries as Roman Catholicism is from Protestantism. Nine out of 10 people in Indonesia are Muslims. The same percentage are Christians in the Philippines, but there are many Muslims in the southern Philippines. Thailand, too, has a Muslim minority of Malays in the south. Almost all the Malays of Malaysia are Muslims, but due to the large number of Chinese, less than half the total population is Muslim.

The Chinese follow a variety of religious beliefs—Confucianism, Buddhism, Taoism, Christianity, ancestor-worship, spirit worship, or often some combination of these beliefs. These Chinese do not believe that man must serve only one spiritual master. There are Christians in all of the Southeast Asian countries, though their numbers vary greatly. Only the Philippines has a Christian majority, the result of its conquest and conversion by Catholic Spain. About 10 percent of the people of Vietnam are Catholics; most live in the south. For many years they enjoyed a political and economic importance out of proportion to their numbers. Differences between Catholics and Buddhists probably weakened the government of South Vietnam in its struggle against the Communists.

Throughout the entire region, however, most people, especially the rural peasants (who account for 80 percent of the population), also believe in spirits, regardless of their formal faith. Thai and Laotian Buddhists, for example, honor the Buddha, but they also believe in *phi* (as the spirits are called). Such spirits inhabit all objects—rivers, trees, rocks, plants, and the like. They are believed, among other things, to be able to cause pain at night by pulling toes and to end the life of a loved one if not properly appeased with offerings of flowers and food.

Buddhists in the mainland Southeast Asian countries also believe in reincarnation—the process of being born over and over again. A man who is born again as a dog, for example, is thought to have lived a bad life in his previous existence; a dog reborn as a man, on the other hand, lived a good life. Such Buddhists do good deeds to acquire "merit" in order to be born as a higher being in their next incarnation.

There have always been at least two major ways of life in all the countries of Southeast Asia. Formerly a few people lived in and about the

A Buddhist monk strolls along temple grounds.

Interior of a Buddhist temple in Bangkok, Thailand.

Luneta Park in Manila in the Philippines.

Rush hour in Jogjakarta on the island of Java in Indonesia.

courts of the various kings, but probably over 95 percent lived in the countryside. Today, although most people still live in the rural areas, increasing numbers of them have moved to one or another of the several large cities that serve as the capitals and commercial centers.

The Major Cities.   The largest of Southeast Asia's cities is **Djakarta**, capital of Indonesia. Other important cities are **Bangkok**, the capital of Thailand; **Saigon**, Vietnam's largest city, and **Hanoi**, the Vietnamese capital; the city-state of **Singapore**; the Philippines' largest city and capital, **Manila**; Burma's capital, **Rangoon**; Cambodia's capital of **Pnompenh**; Malaysia's rapidly growing capital city of **Kuala Lumpur**; and the Laotian capital, **Vientiane**. In all of the big cities of Southeast Asia many young people arrived, attracted by the excitement of city life, before enough jobs were created for them. As a result, there is a relatively high level of unemployment.

A playground in a modern housing project in Singapore.

Meo children of northern Thailand.

Life in the Countryside.   The ordinary man in the countryside grows most of his own food. Increasingly, he has also been producing enough of a surplus to be able to supply city-dwellers with vital foodstuffs and to purchase such things as bicycles (an important means of transportation) and sewing machines. Many so-called small-holders also produce rubber and other export crops, while still other rural people work on big rubber and other plantations. Rice is the major export of mainland Southeast Asia, and most of the farmers here are engaged in the backbreaking labor of seeding, carefully transplanting, and harvesting the grain.

Most villages in Southeast Asia today have a school for children to attend, at least until they are old enough to go to high school. The high schools are usually located in a centrally situated town, to which the students walk or bicycle.

In many parts of Southeast Asia people still live in houses (usually thatched) built on stilts. This is especially true in those areas where heavy rains result in flooding. Building a house on stilts provides protection from wild animals as well as from burglars. It also provides a built-in place to keep the family work animal, usually a water buffalo.

Food and Dress.   Rice is the staple food in almost all parts of Southeast Asia, and it is frequently eaten with a very pungent sauce. Fish is the primary protein food. Meat is fairly scarce and comparatively costly. Buddhists are not supposed to eat the flesh of fish, fowl, or any other animal, but many do so. Muslims do not eat pork. Many of the favorite dishes of the Chinese, on the other hand, include pork as one of the prime ingredients. Vegetables and fruits are both plentiful, especially fruits, but vegetables form a surprisingly small part of the average Southeast Asian's regular diet.

Most people in the countryside throughout Southeast Asia—men as well as women—have traditionally worn skirtlike garments of one kind or another. In Burma they are called *longyis;* in Indonesia, sarongs (for men) and *kains* (for women). Traditional clothing is sometimes worn even in the major cities, particularly in Burma's capital of Rangoon, where almost everybody still wears the colorful, attractive *longyis*. Most city-dwellers, however, today wear Western-style clothing.

Life in even the conservative rural regions is changing today, although not as rapidly as in the cities. Many young men and women who go to the city and are unable to find jobs return to tell their relatives and friends how city people live. In almost all rural towns throughout Southeast Asia there are radios, which are located in some public place so that the people can listen to news and entertainment broadcasts. Some of the country people now own their own radios.

Traditional art forms are also changing, though slowly, under the impact of the outside world. The *wayang*, or shadow-puppet play, has historically been Indonesia's chief type of dramatic entertainment; the *pwe*, or folk opera, is the traditional Burmese popular art form. Both the *wayang* and the *pwe*, while still popular, are losing ground today to the motion picture, produced either locally or in the United States, India, or Japan.

HISTORY

European colonial rule began in Southeast Asia with the arrival of the Portuguese, who seized Malacca on the west coast of the Malay Peninsula in 1511. The Philippines was a Spanish colony for more than 3 centuries, from the second half of the 16th century until the Americans ousted Spain in 1898. The Dutch colonial conquest of Indonesia began in the 17th century but was not really completed until the early years of the 19th century. Burma, Malaya (as Malaysia used to be called), and Singapore became British colonies in the 19th century, while France took over Vietnam, Laos, and Cambodia (together known as French Indochina) during the same period.

The Europeans contributed much to the development of modern Southeast Asia, though not always intentionally. In 1830 there were probably only 10,000,000 people in all of Southeast Asia. The colonial powers ended local wars, which had been taking a high toll of the population, and they introduced improved health and sanitation measures, which also increased the population. The first universities in Southeast Asia were in the Philippines—Santo Tomás in Manila, and San Carlos in Cebu—and were established by the Spanish. Other universities and schools were built, and roads, railroads, and some industries were developed. New crops, such as rubber, were introduced; and old crops, such as rice, were made commercially profitable.

Southeast Asia, however, was still governed by Europeans in the interests of Europeans. Self-government was only grudgingly introduced, except in the Philippines after the Americans had taken over from the Spanish. The Western presence, however, foreign as it was, stimulated group consciousness, or nationalism, among the subject peoples. And in the 20th century increasing demands were heard for an end to colonial rule altogether. Following World War II there was an intensification of nationalist feeling and of the struggle for independence. During these years emerged the countries of Southeast Asia we know today.

RICHARD BUTWELL, The American University
Author, *Southeast Asia Today—And Tomorrow*

Sule Pagoda, Rangoon, one of Burma's many beautiful Buddhist temples.

BURMA

The Burmese people call their country the Golden Land. The name may come from the Burmese custom of decorating Buddhist pagodas with gold leaf, which makes them glitter in the sunlight. The term may also refer to the golden glow of Burma's bountiful rice crop just before harvesttime, for Burma grows so much rice—the basic food of much of Asia—that it exports large quantities of this grain. Or it may refer to the fact that the sun shines throughout many months of the year. Whatever the reason, it is an appropriate name. For Burma is a country that is well endowed with fertile land, great forests rich in valuable wood, and important mineral resources.

THE LAND

On a map Burma resembles a diamond-shaped kite complete with its tail. The country's long coastline fronts on the Bay of Bengal and the Andaman Sea, which form part of the Indian Ocean. Burma's neighbors include India and Bangladesh to the west and China, Thailand, and Laos to the north and east. Snow-covered Himalayan mountain peaks in the far north, rising to over 15,000 feet (4,600 meters), mark Burma's borders with India and China. Mountain ranges extend along the western and eastern sides of the country like the two arms of an inverted V. In the west, the Arakan mountain chain, extending in a series of ridges known as the Naga, Chin, and Lushai hills, forms the border with India. This area is thinly populated and has little land suitable for agriculture. To the west of the Arakan mountains, along the Bay of Bengal, is a narrow strip of land watered by rivers and streams flowing down from the mountains. This area contains rich farmland.

The Shan plateau (also called the Shan highlands), the eastern arm of the inverted V, extends southward into the Tenasserim mountain range, and the two together serve as a dividing line between Burma and Thailand. Between the Shan plateau and the Arakan mountains lies the great central basin of Burma. A lowland area, it is drained by the important north–south river systems of Burma—the Irrawaddy and its tributaries, including the Chindwin; the Sittang; and the lower reaches of the Salween. This vast delta is the center of Burma's major economic activity—rice production.

Burma's nearness to the equator and the influence of the southwest monsoon result in a tropical climate with three well-defined seasons. There is a wet, hot rainy season followed by a cool, or "winter," season (which is more like a warm spring) and then a dry, hot season. In the higher altitudes of Burma's northern mountains light frosts, freezing temperatures, and some snow occur. In the central part of the country temperatures range from about 65 to over 100 degrees Fahrenheit (16–

Burmese river life. People and boats of Twante, a small village of potters.

38 degrees Celsius). Along the coastal areas, temperatures are uniformly high. Rainfall varies from a high of about 180 to 216 inches (457–548 centimeters) annually along the coasts to a low of 21 to 40 inches (53–102 cm.) in the dry zone, which spreads across central Burma.

Economy. Agriculture is central to Burma's economy, employing seven out of 10 of the country's workers. In addition to rice, Burma's rich soil produces grains, cotton, sugarcane, tobacco, peanuts, sesame, and pulses (peas, beans, lentils). Forests blanket nearly three fifths of the land, and the production of wood, particularly teak, is a major industry. Burma is the world's leading exporter of teakwood. Mineral resources are plentiful, and the Burmese mine lead, zinc, tungsten, copper, gold and silver, tin, precious stones, and jade. A considerable amount of petroleum is also produced. Besides fishing, mining, and lumbering, industry is limited in Burma. Factories produce cement, textiles, fertilizers, tile, jute, pharmaceuticals, and steel, and there are a number of plants for processing foodstuffs, wood, and minerals. Burma's craftsmen are noted for their excellent wood and ivory carvings, silk weaving, and handmade jewelry.

THE PEOPLE

Burma's population of about 27,000,000 people—including some 350,000 Chinese residents—is primarily Mongoloid in origin. There are also over 500,000 Indian and Pakistani immigrants and a small number of Europeans. Nearly all of the people are Buddhists, although there are also Christians, Muslims, Hindus, and animists in Burma.

The great majority of the people speak Burmese. English is commonly spoken as a second language and is taught from the elementary

A bustling street in Rangoon. Men and women wear the traditional long skirt.

schools upward. About one fifth of the country's inhabitants speak other languages and dialects. Roughly 5 percent of Burma's people speak Chinese or Indian languages such as Hindi, Tamil, or Urdu. The Burmese alphabet is based on scripts taken from Indian languages, particularly Pali, which is used in writing Burmese Buddhist texts.

It is difficult to determine exactly who were the original inhabitants of Burma. Migrations into Burma from Central Asia began at least 2,000 years ago. First came the people of Mon-Khmer stock, who settled in the delta and Tenasserim areas and spread Buddhism and other elements of their culture throughout Burma. Early in the first centuries A.D. the Tibeto-Burman peoples—including the Pyu, Burman, Chin, and much later, the Kachin—began arriving in the area.

The third major group of immigrants was the Shan-Thai, who at one time dominated a South China kingdom called Nanchao. The main branch of Shan-Thai people—those who populated Thailand, Laos, and the Shan State of Burma—were driven from their kingdom in China during the Mongol conquests of the 12th and 13th centuries. Burma also has smaller groups of hill people, generally related to these three main Burmese ethnic families.

The various Burmese ethnic groups can be distinguished mainly by their dress and speech. The Burman men and women wear a *longyi,* or skirt, while a Shan male wears wide, cuffless trousers. There are also variations in shoes and head coverings, methods of tying knots, jewelry styles, and the colors and designs of textiles used by the different groups. A Burmese-speaking person from Tavoy or Mergui in the south, for example, has an accent quite different from someone living in Mandalay in the central part of the country.

Way of Life.   In their daily lives, most Burmese cling to traditional values and customs. In both the rural areas and the cities, the Burmese prefer to wear traditional garments rather than Western-style clothing. Modern household appliances, radios, and movie theaters are now found in the cities, but most Burmese live without these luxuries. Automobiles, trucks, bicycles, and motorbikes are found on all the roads, but the average rural Burmese either walks or travels by oxcart. Although some farmers work their fields with modern tractors, most rely on plows drawn by water buffalo and oxen. Elephants are often used to carry heavy loads, particularly in the lumber industry.

The great majority of Burma's people—about four out of five—live in small farming villages, mainly in the river valleys and delta floodplains of lower Burma. A typical Burmese family lives in a bamboo house elevated on stilts. Usually there is a long porch outside where the family eats and relaxes. Most men and women wear short jackets, skirts, and open sandals. Often the men wear colored headbands made of cloth. Rice is the basic food, and it is often taken with the fingers from a common bowl and dipped into dishes made with chicken, fish, or beef and spiced with curry powder and other seasoning.

Besides the cultivation of crops, life in the rural areas centers around the family, which includes uncles, aunts, and cousins. Burmese children are taught to show "respect," which in Burma is an important ritual involving correct manners and bearing and the use of proper forms of address. For example, one addresses an elderly person or one who has a high status as *U,* a title of respect meaning "mister" or

FACTS AND FIGURES

SOCIALIST REPUBLIC OF THE UNION OF BURMA—Pyee-Daung-Su Socialist Thamada Myanma-Naingan—is the official name of the country.

CAPITAL: Rangoon.

LOCATION: Southeast Asia. **Latitude**—9° 58′ N to 28° 29′ N. **Longitude**—92° 11′ E to 101° 10′ E.

AREA: 261,789 sq. mi. (678,033 sq. km.).

PHYSICAL FEATURES: Highest point—Hkakabo Razi (19,296 ft.; 5,881 m.). **Lowest point**—sea level. **Chief rivers**—Irrawaddy, Chindwin, Sittang, Salween. **Chief mountain peaks**—Saramati (12,553 ft.; 3,826 m.), Mount Victoria (10,018 ft.; 3,054 m.).

POPULATION: 27,000,000 (estimate).

LANGUAGE: Burmese (official), English, various local dialects.

RELIGION: Buddhist, Muslim, Hindu, Christian, animistic beliefs.

GOVERNMENT: Republic. **Head of state**—president. **Legislature**—People's Assembly. **International cooperation**—United Nations, Colombo Plan.

CHIEF CITIES: Rangoon, Mandalay, Moulmein, Bassein.

ECONOMY: Chief minerals—lead, zinc, silver, tin, petroleum, manganese, tungsten, gold, precious stones, jade. **Chief agricultural products**—rice, sugarcane, jute, pulses, sesame, cotton, tobacco, nuts, fruit. **Industries and products**—forestry, fishing, wood, rice, mineral processing, textiles, handicrafts. **Chief exports**—rice, wood (teak), petroleum products, minerals, cotton, rubber. **Chief imports**—machinery and transport equipment, textiles, iron and steel, pharmaceuticals, peanut oil, paper.

MONETARY UNIT: Kyat.

NATIONAL HOLIDAY: January 4, Independence Day.

NATIONAL ANTHEM: *Kaba Makye* ("Our Free Homeland").

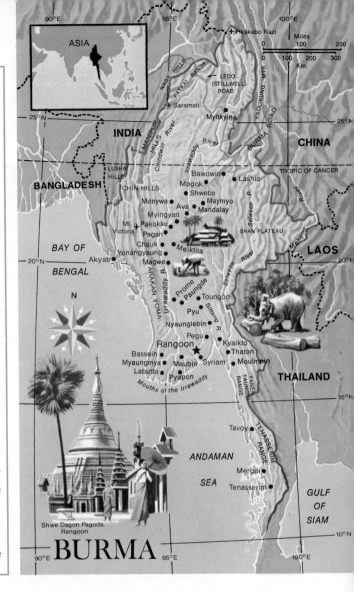

Shwe Dagon Pagoda, Rangoon

BURMA

"uncle" (as in the case of U Thant, former Secretary General of the United Nations). An important occasion in the life of every Buddhist male child is the Shinbyu ceremony, which signifies that the boy is ready to become a morally responsible Buddhist. After the ceremony, the boy enters the local monastery for a short stay.

Women have equal status with men in Burma. In the past women have ruled Burma as queens, and now they are active in politics and the professions. Women operate nearly all of the bazaar stalls and shops in local markets.

For the most part, Burmese Buddhists have no family names. Each person has a name based on the letters associated with the day of the week on which he was born. The *th* in U Thant's name indicates that he was born on a Friday. A Burmese may have only this one name or several. The others will generally represent hoped-for qualities of character and appearance. A boy may have in his name words like "honesty" (*yah*), "learned" (*thin*), or "big" (*gyi*). Girls are given names such as "lovable" (*khin*) or "pretty" (*hla*).

In their leisure time, the Burmese play at a variety of sports and games, including swimming, boating, and kite flying. A popular Burmese game called *chin-lon* is played by five people standing in a circle and tossing about a hollow ball—using only their feet and legs. The Burmese have many festivals, some of which are linked to religious holidays, or sabbath days (days of the month co-inciding with the four phases of the moon). In April the Burmese celebrate their New Year with a Water Festival (Thingyan). This is a 3-day period of merrymaking, in which the people give blessings to one another by sprinkling water on friends and passersby. Often the blessings develop into playful water fights. Another celebration, the Festival of Lights, is connected with the end of a 3-month period (beginning in July) of religious observance. Nearly every month of the year has a festival period. Country fairs are held during the festivals, and the Burmese attend open-air musical shows given by traveling theatrical groups known as *pwes*. Although Burma has artists who create modern paintings and sculpture, most art is inspired by traditional and religious themes.

Education.   Education is highly prized in Burma. One of the traditional duties of the monks is to study the Buddhist scriptures. Another duty is teaching children, and before Burmese independence in 1948 most elementary education was provided by Buddhist monastery schools. Since independence the Burmese Government has extended public education throughout the land. Approximately 60 percent of the people are literate. Low-cost college and professional education is available to all Burmese who can qualify on the basis of high school final examinations.

Cities.   Burma has about 50 cities, but only a few are major urban areas. **Rangoon**, the nation's capital, is the largest city, with a metropolitan population of about 1,600,000 (the city proper has about 900,000 people). The city sprawls along the left bank of the Rangoon River, 21 miles (34 kilometers) from where the river empties into the Andaman Sea. Rangoon is the country's main port, an important industrial and commercial center, and the hub of Burma's transportation and communications network. Besides having a number of imposing public buildings, Rangoon is noted for its historical monuments, including the great, gold-covered Shwe Dagon Pagoda, whose spires rise to 368 feet (112 m.). A number of educational institutions are located here, chief among them being the University of Rangoon.

Burma's second largest city is **Mandalay** (population about 317,000), which is situated on the Irrawaddy River. Mandalay is an active port and trading center, with several busy markets such as the Zegyo Bazaar. The old city contains many ancient buildings and Buddhist pagodas, though a large number of others were destroyed during World War II. Burma's other major cities include Moulmein (population about 157,000) and Bassein (population 125,000).

HISTORY AND GOVERNMENT

Burma's history as a unified nation began in the 11th century A.D., when King Anawratha (who ruled from 1044 to 1077) founded the first Burman dynasty at Pagan, a city on the Irrawaddy River. During the next 2 centuries, the Burmans conquered most of the other peoples in the area and absorbed their culture, which included Theravada Buddhism. Pagan became an architecturally magnificent city and a cultural center comparable to Athens during the Golden Age of Greece. Libraries and

Young Burmese pupils study Buddhist scriptures in monastery school.

The Engineering College on the modern campus of Rangoon University.

seminaries were established, beautiful Buddhist temples were built, and Burma developed into a firmly rooted Buddhist society, which it remains today.

At the end of the 13th century, the Mongol armies led by Kublai Khan invaded Burma and created political disorder. The Burman kingdom was split into warring factions and the Pagan dynasty collapsed. Burma's decline was only temporary, however. By the late 15th century, a new Burman dynasty known as Toungoo (the name of one of its capitals) had come to power. Under King Bayinnaung (who ruled from 1551 to 1581), the Burmans once again unified and even extended their kingdom. But King Bayinnaung's descendants exhausted themselves in costly wars, and the dynasty was finally brought down after an uprising of the Mons in 1740.

The Konbaung dynasty, the third and last, was founded by King Alaungpaya (ruled 1752–60), who decisively defeated the Mons and re-established royal authority over all of Burma. In the 1760's the new dynasty defeated the Siamese and repelled Chinese invasions. From 1782 to 1820 Burmese political and military power were at their height. The Burmese pushed westward to Assam and Manipur. But westward expansion brought Burma into conflict with the British in India, and three Anglo-Burmese wars were fought in the 1800's. After the last of these, in 1885, what remained of the kingdom of Burma was conquered by the British and annexed as a province of India. In 1937 Burma was made a separate colony within the British Empire.

During World War II, Japanese forces invaded and occupied Burma, and much of the country was devastated. After the Japanese were driven out in 1945, Burmese nationalists, led by Aung San, continued their efforts to gain independence for their country. The nationalists were successful, and on January 4, 1948, Great Britain granted Burma full independence. A new constitution, adopted 4 months earlier, provided for a parliamentary form of democracy, with a two-house legislature elected by the people, a prime minister as head of the government, and a president as titular head of state. The Constitution also contained provisions for national economic planning, state ownership of public utilities, land reform and distribution, and state or co-operative development of national resources. Under this Constitution, Burma developed into a democratic socialist welfare state.

Shortly after independence, the new Republic of the Union of Burma came under attack by organized Communist rebels. Although the Burmese Army eliminated many of the rebel bands, Communist guerrilla activity still continues. At times, the government has also had to face uprisings by Kachin, Karen, and Shan tribesmen, who have demanded independence from Burma. The continuing Communist rebellion resulted in political instability and was the major factor that led to a military take-over of the government in 1962, when General Ne Win seized power. General Ne Win set aside the Constitution of 1947 and established a moderate dictatorship, with himself as chief of state. In 1974 a new constitution was adopted. It called for a legislature, the People's Assembly, elected for a 4-year term; a council of ministers, headed by a prime minister; and a council of state, whose chairman is the president of Burma. Ne Win became president in the new government.

FRANK N. TRAGER, New York University; author, *Burma from Kingdom to Republic*

The splendor of Thailand—part of the grounds of the royal palace.

THAILAND

For many years, Thailand was known in the West as Siam, but, since 1939 (except for a brief period in the 1940's), the country has been called Thailand—"land of the free people." The title is justified, for Thailand is the only Southeast Asian country that has never been a colony of a European power. But although the Thai have been free of foreign domination, they have remained open to ideas from abroad. Visitors to Thailand who come with the mental image of a tradition-bound old Siam are often startled by their first contact with modern Thai society, which has borrowed freely from the West without losing its special Asian identity. Today, because of its strategic location in the midst of Southeast Asia, Thailand is fast becoming a regional hub of international activity.

THE LAND

Thailand on a map has been described as looking like a blooming flower on a stem—the stem being the Thai portion of the Malay Peninsula. This Southeast Asian country, which occupies a total area of nearly 200,000 square miles (about 520,000 square kilometers), shares borders with Burma in the north and west, Laos in the north and east, Cambodia in the southeast, and Malaysia in the extreme south. A short stretch of the Salween River separates Thailand from Burma, while the Mekong River serves as a dividing line between eastern Thailand and Laos.

There are four main geographic regions. The north, where several important rivers originate, is crisscrossed by teak-covered mountains and fertile valleys. The country's highest point, Inthanon Peak (8,452 feet; 2,576 meters), is located in this area. The most heavily populated region is the alluvial central plain, where the country's commerce, industry, and farming activities are mainly concentrated. A vast complex of canals and irrigation projects, fed by the Chao Phraya River system, waters the land. Northeastern Thailand, the third region, consists of a sandy, dry plateau that generally cannot retain sufficient water for irrigation. The heavily forested south, a 470-mile (750 kilometers) portion of the Malay Peninsula, contains much of Thailand's mineral wealth.

The country's major rivers, including the Chao Phraya and its tributaries, flow into the Gulf of Siam. The northeastern rivers, the Mun and the Chi, feed the Mekong.

Climate.   Thailand's climate, which is tropical and subtropical, is determined largely by the monsoons. The northeastern dry monsoon brings cooler weather from China during the cool season (November to February); while the southern wet monsoon carries rainfall up from the Indian Ocean during the rainy season (May to October). There is also a very hot season lasting from February to May. Regional variations in temperature occur, with the north generally being cooler than the rest of the country. But in the main, the weather is hot and humid. The country as a whole has about 60 inches (150 centimeters) of rain annually. The heaviest rainfall is in the south and southeast.

THE PEOPLE

Most of Thailand's population of about 35,000,000 belong to the Thai ethnic group. This includes the northern Thai (or Lao) who are related to the people of neighboring Laos. There are also over 3,000,000 people of Chinese background and more than 800,000 Malays in the country. Hill tribesmen, numbering some 300,000, are found mainly in the western and northern regions of the country.

Religion.   Approximately 90 percent of the people are Buddhists. Roughly 4 percent are Muslims, while Confucianists, Christians, and other religious groups constitute the remainder. Every morning Buddhist monks carrying small bowls and wearing saffron robes go among the people to receive food. On holy days and during festivals food is also brought to them in their temple-monasteries, or wats. Hundreds of these wats dot the countryside. There are nearly 400 in Bangkok and its neighbor Thonburi. The most elaborate among them is Wat Phra Keo (Temple of the Emerald Buddha), situated within the confines of the Grand Palace in Bangkok.

Language and Education.   Thai is the national language, and the official dialect is Krung Thep, or Bangkok. However, a number of regional dialects are spoken in the northern and southern provinces. English is generally taught as a second language, and many Thai speak it fluently. Primary school is compulsory, and children attend for 7 years. Smaller numbers go on to a 5-year secondary school program and then to one of the nine state-run universities.

Way of Life.   Thailand is still largely a nation of small villages, most of which are located along the coast or near the rivers. Rivers and canals play an important role in Thai daily life, although many canals in Bangkok have been filled up in order to widen traffic-congested streets. A number

FACTS AND FIGURES

KINGDOM OF THAILAND—Raj-Ana-Jak-Thai—is the official name of the country.
CAPITAL: Bangkok.
LOCATION: Southeast Asia. **Latitude**—5° 32′ N to 20° 28′ N. **Longitude**—97° 21′ E to 105° 38′ E.
AREA: 198,456 sq. mi. (514,000 sq. km.).
PHYSICAL FEATURES: Highest point—Inthanon Peak (8,452 ft.; 2,576 m.). **Lowest point**—sea level. **Chief rivers**—Chao Phraya (Menam), Mekong, Mun.
POPULATION: 34,700,000.
LANGUAGE: Thai (official), English.
RELIGION: Buddhist (official state), Muslim, Confucian, Christian.
GOVERNMENT: Constitutional monarchy. **Head of state**—king. **Head of government**—prime minister. **Legislature**—national assembly (suspended, 1976). **International co-operation**—United Nations, Colombo Plan.
CHIEF CITIES: Bangkok, Thonburi, Nakhon Ratchasima, Ubonratchthani, Chiangmai.
ECONOMY: Chief minerals—tin, iron ore, lignite, tungsten, manganese. **Chief agricultural products**—rice, rubber, sugarcane, maize, cotton, cassava, peanuts, tobacco. **Industries and products**—forestry and wood products, food processing, fishing, cement, paper, handicrafts. **Chief exports**—rice, rubber, tin, kenaf and jute, timber, maize, tapioca flour. **Chief imports**—machinery, motor vehicles and parts, fuels and lubricants, iron and steel, clothing and household goods, chemicals, dairy products, medicines.
MONETARY UNIT: Baht.
NATIONAL HOLIDAY: December 5, The King's Birthday.
NATIONAL ANTHEM: *Pleng Chard Thai* ("Thai National Anthem").

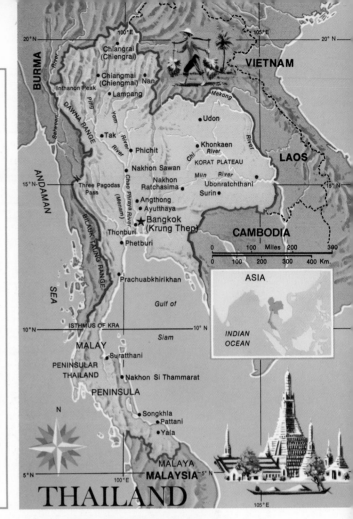

THAILAND

of people still live in floating houses moored to the riverbanks. They earn their living by selling various products from their floating stores.

Traditional-style houses in the villages are made of wood or bamboo. The roofs are generally made of thatch or sometimes corrugated iron, and, when located near the water, the houses are usually built on stilts as a protection against floods. In the homes of wealthier people, floors are paved with tiles, and there are more modern conveniences and furnishings. New public buildings, as well as private homes, successfully preserve the grace of traditional architectural design while using modern construction materials and techniques.

In spite of the increasing popularity of Western clothes, traditional costume is still worn by many women on formal occasions and by older people, especially those living in the countryside. An informal and less elaborate version of this costume is worn at home. It consists of a loose-fitting collarless shirt and a sarong or a loincloth for men and a blouse and sarong for women.

Boiled or steamed rice is the principal food. It is eaten together with fish, pork, chicken, meat, and vegetables, which are often fried and heavily spiced. At home, as well as in restaurants, the Thai delight in eating a wide variety of traditional Thai and Chinese dishes. Fish sauce and hot peppers are indispensable ingredients in Thai cooking. Fruits such as mango, mangosteen, rambutan (a Malayan fruit related to the litchi nut), and pineapple are found in abundance.

Boats, buses, cars, taxis, and other vehicles are used as means of transportation by the Thai. Three-wheeled motorbikes and cars are a common sight. Motorboats have been in use for some time on the rivers. The different regions of Thailand are served by a network of road, rail, water, and air transport. The construction of modern highways has also linked the country with other parts of Asia. Thailand is co-operating in the construction of the Asian Highway, an international roadway that will eventually permit car travel between Europe and Asia. Bangkok, in particular, has become a center for international air travel. Its airport, currently used by some 30 airlines, can handle the largest passenger jets.

A number of sports are popular in Thailand. They include traditional Thai boxing (in which the hands, feet, elbows, and knees may be used), wicker ball, kite flying, and cockfighting. Western sports such as rugby, soccer, and basketball have recently gained favor with the Thai. Among Thai festivals, perhaps the most colorful is the Surin Round Up, which takes place every October at Surin, in the eastern part of the country. Thousands of people come to watch the elephant races and the parade of the "war elephants," a re-creation of the pageantry of ancient times when elephants were used in battle.

ECONOMY

Historically, Thailand has been a farming country, although currently other sectors of the economy are being expanded. Agricultural output accounts for nearly 30 percent of the gross domestic product, with rice production making up about half of this. About 80 percent of the working population is involved in some form of agricultural activity. Most of

Kenaf is soaked, beaten, and washed before it is used to make rope and sacks.

these people are engaged in the cultivation of rice. Wet rice methods are used, and there is a heavy concentration of paddy fields along the alluvial coastal areas of the Gulf of Siam. Thailand produces enough rice so that a substantial amount can be exported. However, prices have fluctuated unfavorably in recent years, and as a result, rice farmers are among the poorest people in the country.

Other important agricultural products are rubber, maize, and cassava, which is sold abroad in the form of tapioca flour. Kenaf (a plant whose fiber is used to make rope) and sugarcane are also grown and exported, but both prices and production levels vary from year to year. Thailand's vast forests provide teak and other wood products. Although Thailand is one of the world's leading producers of tin, it is generally lacking in mineral wealth. Tungsten, iron ore, manganese, and lignite are mined in small amounts.

Industrialization is underway, and gradual progress is being made. Manufacturing is still on a small scale, but there are factories producing cement, sugar, gunnysacks, paper, and petroleum products. In addition, there are rice and cotton mills and sawmills and a number of rubber factories. Industrial development generally has been left in the hands of private enterprise. The state's role has been limited to providing incentives and assistance, partly through its Board of Investment.

Thailand's exports consist mainly of rice, tin, rubber, maize, jute and kenaf, tapioca, teak, and kapok (a silky fiber). Imports, on the other hand, are composed largely of heavy machinery and other goods for development, raw materials, and manufactured consumer products. Since merchandise imports have been rising faster than merchandise exports, a

Busy canals, such as this one in Bangkok, serve as floating marketplaces.

A busy street in the business section of Bangkok.

trade deficit has resulted. However, due to foreign aid and other capital income, including the tourist trade, a crisis in Thailand's balance of payments has so far been averted.

Cities. Thailand is divided into 71 administrative areas, within which are more than 42,000 towns and villages. The most important of the urban areas are the two cities of **Bangkok** (Krung Thep) and **Thonburi**, which are separated only by the Chao Phraya River. They have a combined population of over 2,500,000. While the two cities take up less than 1 percent of Thailand's area, they dominate the entire country economically and politically. Bangkok (meaning village of olives) was only a fishing village and fort centuries ago, when Ayutthaya was the capital city. Thonburi was founded as Thailand's capital after the fall of Ayutthaya in the 18th century. Both cities are crisscrossed by numerous canals. But while Bangkok, which is filling in many of its canals, is fast losing its claim to the title "Venice of the East," Thonburi is keeping its more extensive canal network as an essential means of transport and trade.

Other major cities are **Nakhon Ratchasima** and **Ubonratchthani** in the east. **Chiangmai** (Chiengmai), with its natural beauty and cool winter season typical of the north, ranks next in importance to Bangkok and Thonburi. Located in a major agricultural district, Chiangmai consists of an old walled city on the right bank of the Ping River and a modern city on the left bank.

HISTORY

Historians believe that the Thai originated in northwestern Szechwan in China about 4,500 years ago. Because of conflict with the Chinese, the Thai were forced to migrate south. In the 7th century A.D. they established a kingdom called Nanchao in south China. But trouble with the Chinese and Tibetans continued, and in 1253 the Thai kingdom crumbled under the attack of Kublai Khan and his powerful army.

Continuing their southern migration, the Thai moved onto the Indochina peninsula and drove the Lao and Khmer people from the region around the Chao Phraya River. A series of kingdoms were founded, beginning with the Sukhothai Kingdom in the 13th century. During the period of the Ayutthaya Kingdom (1350–1767), Thailand expanded its frontiers and became the dominant nation of mainland Southeast Asia. It also established contact with European trading powers such as Holland, Portugal, and Great Britain.

Following the short-lived Thonburi Kingdom (1767–1782), Thailand began modernization efforts. The two kings most responsible for introducing extensive reforms were King Mongkut, or Rama IV (ruled 1851–1868), and Chulalongkorn, or Rama V (ruled 1868–1910). Slavery was abolished, outmoded royal customs were done away with, and the power of the aristocracy was limited. For the most part, however, only the top level of Thai society was changed. Life for the average Thai remained much the way it had been. It was not until the mid-20th century, particularly after World War II, that political, economic, and social reforms affecting all levels of Thai society were begun.

GOVERNMENT

Thailand continued into the 20th century as an absolute monarchy, and, as a result, there was pressure for political change. In 1932, a bloodless coup transformed Thailand from an absolute to a limited monarchy. Today the king has considerable authority but, under law, very little political power. The Constitution of 1968 placed executive power in the hands of a prime minister, who was assisted by a cabinet, the Council of Ministers. The parliament, called the National Assembly, included an upper house, the Senate, whose members were appointed by the king, and a popularly elected lower house, the House of Representatives.

In 1971 the prime minister, Thanom Kittikachorn, declared martial law. Parliament was dissolved, the constitution set aside, and the Council of Ministers ousted. The prime minister governed by decree. In 1973, however, following student riots, Thanom Kittikachorn was forced to resign, and a new civilian government was formed. In 1974 a new constitution was proclaimed that restored a parliamentary form of government.

Political stability was short-lived, however. Economic and political unrest led to disturbances during the elections of 1976. The situation grew worse when former prime minister Thanom returned to Thailand from exile. Students demonstrated against allowing the former Thai leader to remain in the country. The demonstrations were put down by the military, who insisted that they were Communist-inspired. A military junta, called the Administrative Reform Committee, was established. The constitution was set aside and the National Assembly suspended. A civilian prime minister was appointed by the King.

PRACHOOM CHOMCHAI, Chulalongkorn University, Bangkok

LAOS: AN INTRODUCTION
by H.R.H. Prince SOUVANNA PHOUMA
Former Prime Minister of Laos

Like many ancient Asian countries, Laos has a legendary beginning. The first historical evidence for the existence of Laos, however, dates from the late 13th century, and the first kingdom—known as the kingdom of One Million Elephants—from the 14th century. In its quiet way, Laos, assimilating various influences, assumed its own national character, culture, and civilization. It began to acquire a harmonious personality that continues to this day. Laos's unity as a nation is real, despite its diversity of peoples and dialects. Its personality has been profoundly marked by Buddhism, which permeates all aspects of our lives.

Like many of the Southeast Asian countries, Laos has had a colonial period. It regained its independence in 1953 after having been involved against its will in the Indochina War. Unfortunately, the history of Laos after its independence was marked principally by the continuation of this war and by the aggressiveness of its neighbor North Vietnam, which used the territory of Laos not only for conveying men, armaments, and supplies to South Vietnam, but also for political objectives. The first attempt to end the conflict in Laos was made in 1962 when 13 countries, including North Vietnam, signed agreements in Geneva recognizing and guaranteeing neutrality and independence for Laos. A tripartite government was created, of which the presidency was given to me. The government included three political factions: neutralist, conservative, and Communist. These accords were welcomed by the people of Laos. Unfortunately, the problems were never completely settled, and fighting resumed. The three-member International Control Commission was unable to fulfill its mission of upholding the terms of the Geneva accords, largely because of the intervention of the international policies of some of the great powers. The result was the violation of our territory and occupation of part of our country by North Vietnamese troops. One can imagine how disastrous were the consequences of such a situation: 20 years of guerrilla warfare, 700,000 refugees, and delays in the sorely needed economic development of our country.

If the problems in Vietnam had remained purely Vietnamese, the fate of Laos would have been different. We would have had, as many countries do, a Communist or Marxist-Leninist party in the legal opposition; this is what we agreed to in Geneva in 1962. Our Constitution allowed for complete freedom of political thought; unfortunately, the Neo Lao Haksat, the Lao Marxist party, preferred a revolutionary fight on the side of the North Vietnamese. We wish to have peace in Laos. Our country's economic needs are great. Surrounded by countries more powerful and populous than we are, our only reasonable status is neutrality. This would enable us to develop Laos's potential natural wealth.

Prince Souvanna Phouma resigned as prime minister in 1975 following the fall of the coalition government and the establishment of a Communist regime in Laos—The Editors.

The former Royal Palace in Luang Prabang overlooking the Mekong River.

LAOS

A small nation, still largely undeveloped, Laos is a land of towering mountains and dense tropical forests. In the 1300's Laos was a powerful kingdom ruling over parts of the present countries of China, Cambodia, Burma, Thailand, and Vietnam. After several centuries the empire collapsed. In the late 19th century Laos became a colony of France, and did not win its independence until 1953. During the war in Indochina, Laos found itself reluctantly drawn into the conflict, largely because of its geographical position. It was also torn by civil war. Laos was a kingdom until 1975, when a Communist government was established and the monarchy was abolished.

THE LAND

Laos is situated in the rugged, mountainous interior of the Indochinese Peninsula, an extension of the vast Asian landmass to the north. The eastern part of Laos, bordering on Vietnam, lies along the high ridges of the Annamese Cordillera. Historically, these mountains have been a natural barrier to travel and a major reason for the separation of the ethnically related Lao and Thai to the west from the Vietnamese to the east.

The country's climate is generally warm, though subject to seasonal changes. Temperatures range from about 82 degrees Fahrenheit (28 degrees Celsius) in summer to the 60's and 70's F. (15–26 C.) from November through February. March and April are usually dry, hot months. From May to October, southwest monsoon winds deposit about 10 inches (25 centimeters) of rain per month. During the dry season, from November to April, the average is less than 1 inch (2.5 cm.).

Laos's natural resources, largely untapped, include teak, tin, lead, silver, and gold. There is considerable hydroelectric potential along the numerous rivers, especially the Mekong, one of the largest river systems in eastern Asia. Over the centuries, water transport has been the chief means of moving people and goods within Laos. The Mekong, which marks the western boundary with Thailand, provides a main artery of communication between northern and southern Laos. Tributaries of the Mekong afford natural avenues leading into the mountainous interior.

Villagers from surrounding areas crowd city markets during the dry season.

FACTS AND FIGURES

PEOPLE'S DEMOCRATIC REPUBLIC OF LAOS is the official name of the country.

CAPITAL: Vientiane.

LOCATION: Southeast Asia. **Latitude**—13° 55′ N to 22° 30′ N. **Longitude**—100° 05′ E to 107° 44′ E.

AREA: 91,429 sq. mi. (236,800 sq. km.).

PHYSICAL FEATURES: Highest point—Phu Bia (9,250 ft.; 2,820 m.). **Lowest point**—594 ft. (181 m.). **Chief river**—Mekong.

POPULATION: 3,000,000 (estimate).

LANGUAGES: Lao (official), French.

RELIGION: Buddhist.

GOVERNMENT: Communist republic. **Head of state**—president. **Head of government**—prime minister. **Legislature**—national assembly. **International cooperation**—United Nations, Colombo Plan.

CHIEF CITIES: Vientiane, Luang Prabang, Savannakhet, Pakse.

ECONOMY: Chief minerals—Tin and rock salt are mined. Unexploited minerals include iron, copper, gold, gypsum, lead, sulfur, coal. **Chief agricultural products**—rice, corn, tobacco, cotton, citrus fruits. **Industries and products**—teak, bamboo, silk weaving, pottery, leathercraft, silverwork. **Chief exports**—cardamom, tin, wood, leather, hides. **Chief imports**—motor vehicles, tractors, bicycles, machinery, electrical equipment, cotton, steel, other metals.

MONETARY UNIT: Kip.

NATIONAL HOLIDAY: December 2, establishment of the republic.

NATIONAL ANTHEM: *Pheng Xat Lao* ("Lao National Anthem").

THE PEOPLE

The broad floodplains of the Mekong Valley are suitable for growing wet rice in irrigated paddy fields. Most of the 1,000,000 Lao—the dominant ethnic group in this nation of over 3,000,000 people—live within these plains and valleys. The Lao are related by language to the Thai people, and many Lao are to be found across the border in northern and northeastern Thailand. Generally, the Lao prefer to live near water, surrounded by irrigated paddy fields and the lush vegetation of a tropical river valley. Lao villages are small, normally containing 300 to 500 inhabitants. Houses are usually constructed of bamboo and are raised above the ground on wooden piles. The space beneath the house is used to store tools and to secure livestock at night. An elevated granary stands a short distance from the family's living quarters.

Settlements are often quite isolated during the height of the monsoon rains in July and August. Oxcarts are then the only form of vehicular transport possible in rural areas. This is a period of rest, since it follows the arduous plowing and transplanting of the rice seedlings in May and June. Buddhist monks and older villagers retire to the village monasteries for meditation. Men, women, and children enjoy fishing in the canals and flooded paddy fields.

The dry season, which comes after the rice harvest in November, is the time for traveling and for visiting friends and relatives. Village monasteries hold colorful celebrations and sponsor fairs at which touring theater groups perform. Traders go up into the hills and many of the hill tribesmen come down to the village markets, as well as to the shops and bazaars of larger towns. By March the countryside is dry, hot, and dusty again, and the people eagerly await the return of the rains and the beginning of another rice-growing season. Such is the unchanging annual cycle that regulates the lives of the rice-farming people of Laos and other countries of Southeast Asia.

The Lao are Buddhists of the Theravada school (one of Buddhism's two branches), like the Thai, Burmese, and Cambodians across their borders. Each Lao village has its Buddhist temple, which serves also as a cultural and social center. The monastery compound, with its preaching hall and living quarters for monks, is the scene of frequent temple ceremonies and fairs. A young man's initiation into the monkhood is a great occasion in the lives of the novice and his family. Until recently, the training period for the monkhood was the only education available to most Lao boys. Even now, with improved public education, most Lao villagers receive their limited schooling from Buddhist monks. But despite the importance of Buddhism, many of the people still maintain a strong belief in local spirits (or *phi*), which may be good or evil.

The mountains of Laos are inhabited by ethnically diverse peoples, many of whom are related to upland tribes in the surrounding areas of Thailand, Burma, China, and Vietnam. These hill tribesmen live in small, scattered hamlets, practicing slash-and-burn agriculture on the steep slopes near their homes. In this primitive form of agriculture, the farmers hack away wood and brush to make a clearing for their crop. The wood is then burned and the ash used as a fertilizer.

Among the mountain peoples are the Meo and Yao tribes, who emigrated from South China during the past century. There are also numerous Kha tribes who are descendants of the original inhabitants of the Indochinese Peninsula. The Kha people are related to the Montagnard (hill tribesmen) of Laos and Vietnam.

Life in the hills is rigorous, and the tribesmen have few possessions. Families work long hours to clear and burn the brush and then to plant and harvest dry (or hill) rice, the staple crop. The tribesmen take time off from farming to make ceremonial animal sacrifices to ancestral spirits. Occasionally they journey to lowland towns to trade. Although the upland peoples make up almost one half of the population of Laos, they remain isolated from the lowlanders. The hill tribesmen are suspicious and fearful of lowlanders, who exploited and enslaved the upland tribes in past centuries. Differences between these two groups add to the difficulties of forging a modern nation-state in Laos.

Cities.   Laos has only a few cities of importance, chief among them being **Vientiane** and **Luang Prabang**, both on the Mekong River. Vientiane (population 163,000) is the largest city and the nation's capital and leading commercial center. The country's main airport is located here. One hundred and thirty miles (210 kilometers) northwest of Vientiane is Luang Prabang, the former royal capital (population 22,000). It is primarily a market town, where farmers, fishermen, and lumbermen come to sell their goods.

HISTORY

Originally, the Lao, the Siamese, and other speakers of the Thai languages were inhabitants of southern China. There they developed a characteristic way of life centered around the growing of wet rice in valley bottoms. Sometime during the first centuries A.D. these peoples began moving south and west. Eventually they reached as far west as Burma and as far south as the great Menam floodplain of Thailand.

Organized along semi-feudal lines, with a military and governing elite, the newcomers conquered, absorbed, or enslaved the local peoples.

In the area of what is now Laos, these ancestors of the Lao established small independent princely states. In A.D. 1353 Fa Ngoun, the first historic Lao ruler, united these principalities into the kingdom of Lan Chang (Lan Xang)—the kingdom of "one million elephants." By the early 1700's feuding among rival states led to the kingdom's being split into three sections. In the 19th century most of Laos was taken over by Siam (now Thailand). Siamese control continued until 1893, when the French, seeking to expand their colonial power in Indochina, ousted the Siamese from the Lao states.

Laos became a protectorate within what was to become French Indochina—including also the present-day nations of Cambodia and Vietnam. The ruling family of Luang Prabang was declared the royal family, and the French governed the Lao states indirectly through the king of Luang Prabang.

Laos was granted virtual independence as a self-governing state within the French Union in 1949 and complete independence in 1953. Under the terms of the 1954 Geneva agreements, which ended the French Indochina War, Laos was recognized as a unified, independent, and neutral country. But as a result of fighting between Communist Pathet Lao forces, supported by North Vietnam, and royal government troops, Laos remained divided. In 1960 neutralist army officers led an uprising against the government, and their forces occupied a strategic area of the country. In 1961–62 a multi-nation conference was held in Geneva to try to resolve Laos's problems. The resulting 1962 Geneva agreements offered international guarantees of Laos's neutrality and independence. A coalition government was formed, made up of Laos's three political factions (neutralists, conservatives, and Communists), with Prince Souvanna Phouma as prime minister.

During the Vietnam War, Laos tried to maintain a position of strict neutrality. But this proved difficult. The so-called Ho Chi Minh Trail, which runs through the mountains of eastern Laos, was used by the North Vietnamese to move troops and supplies to Communist strongholds in South Vietnam. At the same time, the struggle between Communist and non-Communist forces in Laos continued. In 1973 a peace treaty was signed between the Laotian Government and the Pathet Lao. At the time of the treaty, Pathet Lao and North Vietnamese forces controlled most of the country's three northern provinces. A new coalition government was formed in 1974, with Souvanna Phouma again serving as prime minister. The new government did not last long, however. In 1975, following Communist victories in Vietnam and Cambodia, the Pathet Lao took complete control of Laos.

GOVERNMENT

Laos was a constitutional monarchy until 1975, when King Savang Vatthana gave up the throne and a Communist people's republic was established. Prince Souvanna Phouma resigned as prime minister. A people's congress elected Prince Souphanouvong, a half brother of Souvanna Phouma and a longtime leader of the Pathet Lao, as president. A new prime minister and cabinet were also chosen.

FRANK M. LeBAR, Co-editor, *Laos, Its People, Its Society, Its Culture*

Monks walk in courtyard of 12th-century Buddhist temple.

CAMBODIA (KAMPUCHEA)

For more than a decade, Cambodia was a kingdom without a king. When the last King died in 1960, his son, Prince Norodom Sihanouk, refused to take over the 2,000-year-old throne. Instead, he chose to become chief of state—a title he had created for himself so that he could become more active in Cambodian politics. Under Prince Sihanouk's leadership, Cambodia appeared to be the most stable and democratic of the former Indochina states. But in 1970, while the Prince was in Europe, a group of high-ranking army officers and government officials seized power. The new government abolished the monarchy and proclaimed Cambodia a republic.

Until he was overthrown, Prince Sihanouk was the centerpiece of Cambodian life. A colorful figure, he seemed to enjoy the respect and affection of most Cambodians. Outside Cambodia, he was watched with great interest and fascination—and sometimes bewilderment—as he maneuvered to protect his small country from rebellion within and the spreading Indochina War without. After the successful coup against him, Prince Sihanouk established a government-in-exile in China. In the meantime, Cambodia was increasingly torn by war. When the government that had ousted him itself fell to Communist rebel forces in 1975, Prince Sihanouk was again named chief of state and returned to Cambodia. However, he remained in the new government only briefly, resigning from his post in 1976.

THE LAND

Cambodia today is what remains of the ancient Khmer Empire, which dominated most of Southeast Asia from the 10th to the 14th century. Cambodia occupies an area of nearly 70,000 square miles (180,000 square kilometers) and is more than half the size of its eastern neighbor, Vietnam. About half of the land is heavily forested, and another 10 percent is water, including the great Mekong River, which flows through Laos, Cambodia, and Vietnam into the South China Sea. The remaining 40 percent is excellent farmland, but only a fourth of it is cultivated. The central portion of the country is a level, fertile plain bounded by several mountain chains. Here the floodwaters of the Mekong River back up into the gigantic lake Tonle Sap (Great Lake) during the monsoon season, when the river channel clogs with silt. Then the lake expands to about eight times its dry-season size and irrigates the surrounding rice fields. Since Cambodia is mainly a nation of rice farmers, most Cambodians live in this area.

Cambodia has a mild, tropical climate, with temperatures ranging from about 70 to 95 degrees Fahrenheit (21–35 degrees Celsius). As in other countries affected by the monsoon, there is a dry season, which lasts from November to April, followed by 6 months of very heavy rainfall.

THE PEOPLE

The Khmer (or Cambodian) people, whose ancestors built the Khmer Empire more than 1,000 years ago, are the major ethnic group in Cambodia. More than 85 percent of the country's people are Khmer, and their chief occupation is farming. The better-educated Khmer hold most of the country's government and religious posts.

Chinese and Vietnamese minority groups each represent about 5 percent of the total population. The rest of the population consists mostly of primitive hill tribesmen and a sprinkling of people from other Asian countries. The Chinese live mainly in the cities, where they own and operate most of the country's small businesses and shops. Most moneylenders are also Chinese. The Vietnamese work as laborers on rubber plantations or in fisheries and also as artisans and merchants. In recent years, the government has tried to attract more Khmer into commercial fields. Because most Khmer lack funds or skills for such activity, the government took over the banks, the insurance companies, the export-import businesses, and other key economic functions in order to "Cambodianize" them.

A common language, religion, and historical tradition, shared by the Khmer, bind the nation together and make Cambodia a basically stable country. Khmer (also known as Cambodian) is the language most commonly spoken, although Chinese and Vietnamese are also heard in city business districts, where most commercial enterprises are owned by Chinese and Vietnamese. Since Cambodia was formerly ruled by France, French is widely used by government officials, professional people, and intellectuals. English is becoming increasingly important among government officials and the Buddhist clergy.

Education.   In the recent past, the Cambodian Government made great efforts to improve the country's educational system and to eliminate illiteracy. Although the number of public schools was increased, Buddhist monks (bonzes) traditionally conducted classes in the small villages. Since 1975, however, when the Communist (Khmer Rouge) government came to power, both education and religion have been discouraged, and the monks have been sent to work in the rice fields.

Religion.   Buddhism of the Theravada school, which came originally from India, is the chief religion, and nearly 90 percent of the people are Buddhists. Most Cambodians, particularly those in rural areas, are very religious. A majority of Cambodian men traditionally received religious training as bonzes for periods ranging from a few months to several years. Buddhism builds loyalty to the country itself, for it is in their own country that devout Buddhists hope to be reborn (reincarnated) into a better life because of the good things they did in their previous lives. Their ultimate goal is Nirvana, the Buddhist version of heaven. Religious observance also includes spirit worship. Many Cambodians believe that there are good and bad spirits that influence events. A decision to do something on a certain day is often determined by how the people feel the spirits will react. A sick person's family will often make ceremonial offerings to the good spirits to drive away the evil ones.

Way of Life.   Eight out of 10 Cambodians are farmers, most of whom live a simple existence in villages of only a few hundred people. Before the Communist takeover in 1975, everyday life revolved around the home, the plot of land the family cultivated, and the local Buddhist pagoda (temple).

Rice farming is the main source of livelihood for the people. There is an old Cambodian proverb that says, "To destroy growing rice is as serious as to insult one's mother and father." The pace and pattern of the farmer's life are influenced by the seasonal schedule of rice planting and harvesting. Planting is done mainly in May and June, and the rice crop is harvested from November to January.

The Royal Palace in Pnompenh was formerly the residence of Cambodia's kings.

FACTS AND FIGURES

DEMOCRATIC KAMPUCHEA is the official name of the country.

CAPITAL: Pnompenh.

LOCATION: Southeast Asia. **Latitude**—10° 24′ N to 14° 26′ N. **Longitude**—102° 21′ E to 107° 38′ E.

AREA: 69,898 sq. mi. (181,035 sq. km.).

PHYSICAL FEATURES: Highest point—Mt. Aural (5,948 ft.; 1,813 m.). **Lowest point**—sea level.

POPULATION: 6,700,000 (estimate).

LANGUAGE: Khmer (Cambodian), French.

RELIGION: Buddhist.

GOVERNMENT: Communist republic. **Head of state**—chairman of the presidium. **Head of government**—prime minister. **Legislature**—People's Representative Assembly. **International co-operation**—United Nations.

CHIEF CITIES: Pnompenh, Battambang.

ECONOMY: Chief minerals—iron ore, coal, copper, gold, salt. **Chief agricultural products**—rice, rubber, corn, sugar, beans, fish, livestock. **Industries and products**—rice milling, fish and wood processing, sugar refining, handicrafts, cement. **Chief exports**—rice, rubber, corn, wood products, livestock.

MONETARY UNIT: Riel.

NATIONAL HOLIDAY: April 17 (Victory of the Khmer Rouge).

NATIONAL ANTHEM: "April 17, the Great Victory."

CAMBODIA (KAMPUCHEA)

Ballet Dancer

After the farming season is over, time is spent repairing equipment and the thatch and bamboo houses, which are built on stilts as protection against the monsoon floodwaters. Traditionally, when the season was over, some farmers would go to the cities to look for work. Festive celebrations are held during this period, including the 3-day Water Festival celebrating the return flow of the floodwaters from the Tonle Sap to the Mekong. Cambodians look forward eagerly to this event, which features a dramatic race of long boats made from hollowed-out tree trunks. Cambodians will travel long distances to attend such events as well as family gatherings celebrating births or marriages or honoring the dead.

Mealtime is also an occasion for ceremony, though the food served is very simple. The basic diet is rice and fish. Fish sauce and a paste made of dried fish are very popular. Soft drinks, ice cream, and other foods have been introduced from foreign countries.

Music and poetry are an important part of Cambodian culture. Often Cambodians use traditional melodies as musical background for lyrics describing current events. Popular poetry, on the other hand, retells ancient stories and legends. Cambodians enjoy theatrical performances, particularly the folk dramas featuring pantomime, singing, and dancing. The Cambodian ballet, in which the Cambodians have traditionally taken great pride, is widely acclaimed.

Economy. Although Cambodia is mainly an agricultural country—with rice, rubber, corn, sugar, and pepper its chief products—industries are increasing in number. These include rice mills, fish-processing plants, a sugar refinery, and factories producing lumber, textiles, paper, cement, jute, and cotton.

Cities. Cambodia's largest and most important city is **Pnompenh**, the capital. Situated on the Mekong and Tonle Sap rivers, Pnompenh is the country's government, communications, and trade center. All of the country's major roads and waterways converge on Pnompenh, which has a population of about 450,000. Pnompenh used to be Cambodia's

Relics of ancient glory. Statues along causeway in the ruins of Angkor.

only port for oceangoing ships, and the Mekong River was once its only outlet to the sea. The Cambodians did not like being so dependent on Vietnam, and in the early 1960's the port city of Sihanoukville (Kompong Som) was established on the Gulf of Siam. A road and railroad connect the port with the capital. Except for Pnompenh, no Cambodian city had a population of more than 45,000. Battambang was the second largest city and Kompong Cham the third largest. Since the takeover of Cambodia by the Khmer Rouge, there have been reports of the widespread dispersion of people from the cities, especially the capital, into the countryside.

HISTORY

Cambodia's history began some 4,000 years ago, when the Khmer and other tribes migrated to this area from the northwest. Along this route also came Hindu doctrines and practices from India. After centuries of tribal warfare, the Khmer won out and established their own kingdom. Like other Asian peoples, today's Khmer take great pride in their ancient origins. The name Cambodia derives from the legend of Kambu Svayamb-

Produce grown in the countryside is on display at a central market.

huva, who is said to have founded the Khmer dynasty sometime in the 6th or 7th century A.D.

During the next 6 centuries, the Khmer extended their rule over an area that now includes most of the countries of Vietnam, Laos, and Thailand. This was a time of great social improvement and cultural achievement in Cambodia. Art and architecture flourished. Angkor, the capital of the Khmer Empire, was a city of magnificent palaces and temples. Today the ruins of this once great city—including the famous temple Angkor Wat (or Vat)—still claim a special place in the hearts and minds of the Cambodian people.

The Khmer Empire began to decline after the 14th century. By the 19th century, as a result of wars with the Thai and Vietnamese, Cambodia had been reduced to approximately its present size. In 1846 Vietnam and Thailand arranged to place a king they felt they could control on the Cambodian throne. King Ang Duong was chosen, beginning the royal line from which Prince Sihanouk is descended. However, King Ang Duong sought independence from both the Thai and the Vietnamese and called upon France for protection. A French protectorate over Cambodia was established in 1863. In 1941 the French placed Norodom Sihanouk, then only 18 years old, on the throne, hoping to control him as they had previous kings. Instead, the new King became the leader of a Cambodian movement for independence, which was finally granted by France on November 9, 1953, and formally recognized by the Geneva agreements of 1954.

King Sihanouk turned over the throne to his father in 1955 to begin an active political career. He had the position of chief of state created in 1960, upon the death of his father, so that he might assume the duties of king without giving up his political power. Prince Sihanouk founded his own political movement, the People's Socialist Community (known as Sangkum), and until 1970 he dominated both the party and Cambodian politics. Following the 1970 coup, the new government, headed by Marshal Lon Nol, ended royal rule and set up The Khmer Republic.

In foreign affairs, Cambodia had declared itself a neutral, interested in friendship with all nations. However, the Vietnam War placed this policy in jeopardy. During the course of the Vietnam War, American and South Vietnamese forces frequently chased Vietcong and North Vietnamese troops across the Cambodian border. The Vietcong and their North Vietnamese allies obtained food and medical supplies from Cambodians. South Vietnamese and American officials charged that military supplies also reached the enemy via Cambodia and that enemy sanctuaries existed on the Cambodian side of the border. Cambodia at first denied the charges and claimed in turn that United States air and land forces were violating its border with South Vietnam. Later the presence of Vietcong and North Vietnamese troops in Cambodia was acknowledged, and the Cambodian Government urged their withdrawal. In the spring of 1970, following the ouster of Prince Sihanouk, United States and South Vietnamese troops entered Cambodia for the stated purpose of destroying these Communist sanctuaries. Cambodia's new government supported this action, and government forces battled against those of the Cambodian Communists (the Khmer Rouge). Between 1973 and 1975 there was almost continuous fighting. War devastated this once-peaceful land as the Communist forces cut deeper and deeper into government-held territory. In the spring of 1975 the last government forces surrendered and the Khmer Rouge took control of Cambodia.

GOVERNMENT

The Communist leadership of Cambodia continues to hold firm military control of the country. Late in 1975 the Khmer Rouge announced a new constitution for Cambodia, which went into effect in early 1976. Under the constitution the country was renamed Democratic Kampuchea, and a legislative assembly called the People's Representative Assembly was set up. In the spring of 1976 elections for this 250-member body were held throughout Cambodia. Of the 250 seats, 150 were allocated to farmers, 50 to soldiers, and 50 to workers. The assembly meets once a year, day-to-day business being conducted by a standing committee. Members of the assembly serve 5 years. A prime minister heads the government. The head of state of Democratic Kampuchea is the chairman of the presidium.

DAVID J. STEINBERG, Senior author, *Cambodia: Its People, Its Society, Its Culture*

Fieldworkers near Dalat in Vietnam.

VIETNAM

In Vietnam, four symbolic animals are engraved, painted, or otherwise represented in homes and on public buildings: the unicorn, the tortoise, the phoenix, and the dragon. Of the four, the most important is the dragon. In spite of their frightening appearance, dragons do not represent evil to the Vietnamese. Rather, they are symbols of nobility and power. As such, dragons became the chief attributes of the Vietnamese emperors. Because of a similar role played by the dragon in the mythology, history, and art of Vietnam's huge neighbor, China, Vietnam is frequently referred to as "the country of the smaller dragon."

The relationship between the smaller and larger dragons is not accidental. In its long history, Vietnam has been strongly influenced by Chinese culture and civilization and for many centuries was ruled by its northern neighbor. Later, Vietnam was governed by France as part of that country's Indochinese empire. When the French departed, Vietnam gained its independence. But there were years of war and a division between North and South until the two parts of the country were united in 1976.

THE LAND

Vietnam occupies the eastern part of the Indochinese peninsula. In the north Vietnam borders on China and in the west on Cambodia and

Laos. Vietnam's long coastline faces the South China Sea in the east, and in the south touches the Gulf of Siam. The total area of Vietnam is over 128,000 square miles (332,000 square kilometers), or slightly larger than the area of Italy.

The two most important regions of the country are the Red River delta in the north and the Mekong River delta in the south. A narrow strip of flatland and smaller deltas connects the two great deltas, which are Vietnam's fertile rice-growing regions. The country's length and its narrow width between the two deltas give Vietnam its peculiar shape, which has been described as a bamboo pole with a basket attached to each end.

The two great deltas of Vietnam were formed by the country's two great rivers—the Red and the Mekong. The Red River in the north flows from China in a southeasterly direction and empties into the Gulf of Tonkin. The volume of water in the Red River varies widely. High dams to contain the waters have been built, but these have not always been able to prevent catastrophic floods. Less erratic than the Red River is the Mekong. One of the world's great rivers, the Mekong originates in China. It flows through southwestern China, then forms the border between Burma and Thailand and Thailand and Laos, and after passing through Cambodia, enters southern Vietnam. There the river splits into several branches before reaching its mouth in the South China Sea. The Mekong River irrigates the large southern delta, whose fertile lands are increased each year by sediment carried by the river. Both delta regions are crossed by many smaller rivers and canals, Vietnam's most important routes of transportation. (A separate article on the MEKONG RIVER appears in this volume.)

All the rest of Vietnam—more than two thirds of its area—is made up of mountains and high plateaus. These comprise the Annamite chain, or Annamese Cordillera. Some of the branches of the Annamite chain extend directly into the sea. The highest mountain, Fansipan peak in northern Vietnam, reaches 10,308 feet (3,142 meters).

Climate. Vietnam has a tropical climate both in the north and the south, except at high altitudes. The climate is governed by the seasonal winds known as the monsoons, which produce a dry and a wet season. Summer temperatures differ very little between north and south, and temperatures well over 100 degrees Fahrenheit (37 degrees Celsius) are quite frequent in both parts of the country. The average winter temperatures, however, are lower in the north. In the mountains of the north and the high plateaus of the south temperatures are considerably lower all year round. Rainfall is heavy but irregular. During the few rainy months, an average of 72 inches (183 centimeters) falls in the cities of Hanoi and Saigon.

Animal Life. Domesticated animals found in Vietnam include such work animals as elephants, water buffalo, and (very rarely) horses. The water buffalo is the most common beast of burden. Animals used for food include goats, pigs, and various kinds of fowl. Silkworms are also raised. Among the many species of wild animals are tigers, panthers, wild oxen, boar, bears, various small game birds, monkeys, a great variety of snakes, turtles, rodents (including rats, some of which are as large as house cats), and among the many kinds of insects, the ever-present mosquitoes.

These parading Vietnamese women wear the traditional dress, the "ao dai."

THE PEOPLE

Over 85 percent of the people of Vietnam are ethnic Vietnamese. Earlier, they were usually called Annamese (or Annamites). This name came from Annam, the name of the country when it was ruled by China. The chief minority groups are aborigines, who inhabited the country before the arrival of the Vietnamese, and immigrants from China. There are also large numbers of Khmers (Cambodians), who are gradually being assimilated by the Vietnamese, and a smaller number of Chams. The Chams are the descendants of the people who made up the ancient kingdom of Champa in central Vietnam.

Approximately two thirds of the Chinese immigrants live in Cholon, the twin city of Saigon. The Vietnamese inhabit the two large delta regions and the long narrow chain of land between them. The aboriginal peoples are found in the mountainous regions of the country, both in the north and the south. They form the largest ethnic minority. They are a mixture of racial groups, divided into numerous tribes and speaking different languages and dialects. In the south they are generally referred to by the French term "montagnards" (mountain people). Most southern tribes live a seminomadic existence. When the land, which they clear by burning down jungle woods, becomes exhausted, the whole village moves on and settles on new land. The religion of these people consists largely of magic, superstition, and mythology.

In contrast to the great variety of the minority peoples, the Vietnamese have a remarkable racial and cultural unity. There is little difference in language and customs, even between the people of north and south, except for differences in local dialects. The Vietnamese also vary little in physical appearance. The dominant features are straight, black hair, skin that ranges in color from a very light to a medium brown, and dark eyes with the distinctive Mongolian single fold of the eyelids (often incorrectly called "slant" eyes). The average male is short and slight. Vietnamese women are even smaller and as a rule quite slender. They usually wear their jet black hair long, and they move very gracefully in their traditional flowing garment, the *ao dai,* which most women, even in the cities, still wear. The *ao dai* consists of black or white satin trousers with a long-sleeved dress, fitted at the waist, from which two long panels of material, one in back and one in front, reach to the heels. Vietnamese women can wear the *ao dai* gracefully even while riding a bicycle, a common form of transportation.

Their Way of Life.   The Vietnamese staple foods are rice and fish. The fish is often in the form of a pungent sauce called *nuoc mam.* For the great mass of the people, meat, such as pork, chicken, and buffalo, is only an occasional addition to their rice and fish diet. Some variety is provided by soups, usually with noodles, vegetables, fruits, dishes prepared from animal organs (intestines, heart, stomach), the use of spices, and by sweets made of rice and coconut. Meals are served on low tables and eaten with chopsticks.

During their history most of the people have lived in villages, subdivided into smaller hamlets. They grow rice in irrigated paddies, raise some chickens and pigs, and plant vegetables. If the village is near one of the country's many rivers or near the sea, they catch fish to supplement their diet.

Farms always have been small. But those in the Red River delta in northern Vietnam were smaller and more intensively cultivated than farms

Once the war ended, new buildings began to spring up in Hanoi.

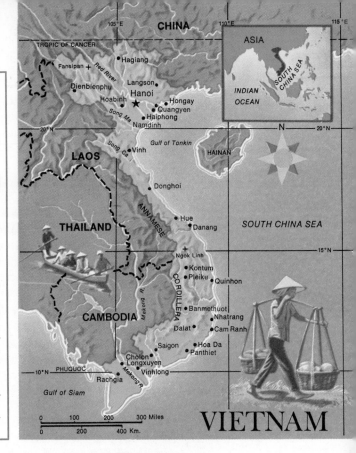

FACTS AND FIGURES

SOCIALIST REPUBLIC OF VIETNAM is the official name of the country.

CAPITAL: Hanoi.

LOCATION: Southeast Asia. **Latitude**—8° 34' N to 23° 23' N. **Longitude**—102° 09' E to 109° 28' E.

AREA: 128,402 sq. mi. (332,559 sq. km.).

PHYSICAL FEATURES: Highest point—Fansipan peak (10,308 ft.; 3,142 m.). **Lowest point**—sea level. **Chief rivers**—Mekong, Red.

POPULATION: 43,000,000 (estimate).

LANGUAGE: Vietnamese, French, Chinese, English.

RELIGION: Buddhist, Christian, Taoist, Confucian.

GOVERNMENT: Communist republic. **Head of state**—president. **Head of government**—prime minister. **Legislature**—national assembly.

CHIEF CITIES: Saigon, Hanoi, Haiphong, Danang, Hue.

ECONOMY: Chief minerals—coal, tin, gold, iron ore, antimony, chrome, phosphates. **Chief agricultural products**—rice, rubber, sugarcane, maize (corn), cotton, tea, tobacco, coffee. **Industries and products**—food processing, cement, mining, textiles, paper, consumer goods. **Chief exports**—rubber, coal, rice. **Chief imports**—fuel, machinery, food, metals.

MONETARY UNIT: Dong.

NATIONAL HOLIDAY: Tet (New Year; January/February).

NATIONAL ANTHEM: *Tien Quan Ca* ("Forward, Soldiers").

in other parts of the country. When the Communists came to power in northern Vietnam, the government began vast land reform programs there. Most of the farms were grouped in co-operatives, where the land and equipment are owned by the community. Land reforms and long-range agricultural programs were planned for southern Vietnam after the war ended and the two parts of the country were united under a Communist government.

Houses in the villages are mostly built of mud and bamboo with roofs of palm leaves or grass. Houses along the rivers are built on stilts. In the cities most of the dwellings are one-family houses of wood or concrete. Many of them have electricity and modern plumbing.

In the agricultural areas the older children usually work with their parents in the rice paddies. The youngest children remain in the village, where they are cared for by a sister or brother who is not quite old enough to work or by other villagers.

Men and women in the country wear the same kind of everyday clothes—a shirtlike garment over cotton trousers and a wide conical hat or a cloth formed into a turban for protection against sun and rain. Western-style clothes are common only in the cities.

The Vietnamese people enjoy many holidays and festivals. Among the dozen nationally celebrated holidays, Tet is the most important. Tet marks the beginning of both the lunar New Year and spring and usually falls in late January or early February. Other holidays celebrate national heroes and the birth of Buddha and Confucius.

Religion. The majority of the Vietnamese believe in some form of Buddhism, even if they do not practice it consistently. There are two religious sects, the Hoa Hao and the Cao Dai. Cao Dai is a mixture of

Confucianist, Christian, Buddhist, and Taoist beliefs, together with spiritualism. This sect claims twice as many followers as the Hoa Hao sect, which practices a form of Buddhism with a simplified ritual. Most of the Christians are Roman Catholics. Before the country was united, the government in the north discouraged the practice of religion except as prescribed by the Communist Party.

Population Distribution. The distribution of the population is extremely uneven. The mountainous areas are sparsely settled. Most of the people live in the deltas of the Red and Mekong rivers and on the fertile coastal plains. The war caused great destruction in the countryside, and hundreds of thousands of people fled from their villages to urban areas. After the war the government announced plans for resettlement of many people from Saigon into the countryside, as well as other shifts of population.

Language. The Vietnamese language is composed of single-syllable words that do not change. The same word, however, can have different meanings, which are expressed through different levels of pitch. Although Vietnamese was enriched with many Chinese literary, philosophical, and technical terms, it is basically unrelated to Chinese. Chinese ideographs, or characters, were originally used in written Vietnamese. But they came into disuse after Portuguese and French missionaries in the 17th century invented *quoc-ngu,* a system of writing Vietnamese in Roman letters. French, the official language during the colonial period, is still spoken by educated Vietnamese but is slowly being replaced by English as a second language.

Chief Cities. Vietnam's largest city, once the capital of the southern part of the country, is **Saigon** (sometimes called Ho Chi Minh City, after the Vietnamese Communist leader who died in 1969). It is the major port and the center of finance and industry in the south. The metropolitan area includes the twin city of **Cholon**, which handles much of the south's manufacturing. Saigon is a cosmopolitan city, with many European-style buildings and wide boulevards. Cholon, where many Chinese immigrants live, is more traditional in appearance. After Saigon came under French control in the middle of the 19th century, it grew rapidly. The population also increased considerably during the war, when people from the countryside sought refuge in the city.

Danang, the second largest city in southern Vietnam, has port facilities that rank next to Saigon's in that part of the country. During the war it had a large air force base and other military installations. **Hue**— situated about 50 miles (80 kilometers) from the 17th parallel, which once divided the northern and southern parts of Vietnam—is next in size in the south. Hue was the old imperial capital of Vietnam and the center of learning and traditional Vietnamese culture. In 1968 the city was temporarily occupied by the Vietcong (Communist guerrillas), and much of the city was destroyed during the fighting.

Hanoi, the former capital of the northern part of the country and the capital after the two parts were united, is Vietnam's second largest city. It is situated in the delta of the Red River. Under the Communists, it was rapidly industrialized until the mid-1960's, when many of its industries, as well as much of the population, were evacuated because of air bombing during the war. **Haiphong**, an industrial city that also serves as the port of Hanoi, was a major supply depot during the war years.

Parts of Saigon have a distinctly European look.

A crowded side street and food market in Saigon.

The latest news is eagerly read by these Vietnamese in Hanoi.

The Economy. The war shattered the economy of Vietnam. But up to that time the country was still predominantly agricultural. Rice was the most important crop. In normal times southern Vietnam had a substantial rice surplus. In the north, where more people were engaged in mining and manufacturing, the cultivated land produced hardly enough to feed the people. Northern Vietnam has extensive coal deposits. Mineral resources are generally rare in the south, but forests and sources of hydroelectric power are abundant in that part of the country. The northern section was much more advanced industrially. It had a complex of steelworks, modern shipyards, and a large tool factory. Local handicrafts played a large economic role in both parts of the country.

After the war the Communist government announced the nation's new five-year development plan. Under this plan vast areas of the south would be converted into the nation's rice bowl, and heavy industry would be concentrated in the north.

HISTORY

As a people occupying a given territory, the Vietnamese have a history older than that of most European nations. Chinese records show that a Chinese-Vietnamese state existed more than 2,000 years ago. It was called Nam Viet and was populated by Viet tribes who moved south when their kingdom below the Yangtze River was destroyed by the expanding Chinese empire. Nam Viet was founded in the 3rd century B.C. It extended from the vicinity of the present-day Chinese city of Canton down to the center of modern Vietnam. In 111 B.C. Nam Viet was conquered by the great Chinese Han dynasty, and thus the Viets fell under Chinese rule. Most Viets were absorbed by the Chinese in the course of a few centuries. But those who had settled in the Red River valley and south of it intermarried with the local inhabitants of Indonesian stock. They thus created a separate ethnic identity and nation.

During the more than 1,000 years of Chinese rule, the Vietnamese readily accepted the higher technology and many of the cultural values of the Chinese. But they remained firmly opposed to Chinese political domination, against which they staged many revolts. But real independence was gained only in 939, when the Chinese, after a long war, were driven out of the Red River valley. Many attempts were made by the Chinese to reconquer Vietnam. Only one, early in the 15th century, was temporarily successful. But after a long struggle, led by the national hero Le Loi, Vietnam regained its independence in 1427 and maintained it until its conquest by France in the 19th century.

During their 900 years of independence, the Vietnamese extended the southern border of their country to the Gulf of Siam. In this march to the south, the Kingdom of Champa was conquered, and the thinly populated and fertile Mekong Delta was taken from the declining Cambodian empire. During the 16th and early 17th century, Portuguese, Italian, and French missionaries had come to Vietnam and converted about 10 percent of the people to Catholicism. Persecution of the missionaries by the Vietnamese emperors was used by France as a pretext for invading Vietnam in 1858, and by 1883 the country had fallen under French domination. By the end of the 19th century the French Indochinese Union comprised Vietnam, Cambodia, and Laos.

The Indochina War. Economic exploitation and political suppression aroused strong national resistance against French rule. Between 1930 and 1945 the Communists became the strongest group in the anticolonial movement. In 1941 they founded the League for the Independence of Vietnam (Viet Minh) and brought many non-Communist nationalists under their leadership. When the Japanese, who occupied Vietnam during World War II, surrendered in 1945, Viet Minh troops seized power in Hanoi; and their leader, Ho Chi Minh, proclaimed the independence of Vietnam. In the South the French were able to suppress the national revolution. But their attempt to re-establish colonial rule over the North led to the outbreak of the Indochina War. The fighting lasted from 1946 to 1954. Finally, after being defeated by the Viet Minh in the important battle of Dienbienphu, the French agreed to a settlement of the war at a conference held in Geneva, Switzerland. Together with the French and Viet Minh were representatives of Great Britain, the Soviet Union, Communist China, the United States, Cambodia, Laos, and the State of Vietnam (South Vietnam). As a result of the conference, Vietnam was partitioned at the 17th parallel. This was a temporary dividing line to be abolished following national elections in 1956.

The Vietnam War. The Geneva Agreements left the Communist government of Ho Chi Minh in control of the northern half of the country. With United States support, an anti-Communist government developed in the South under Ngo Dinh Diem. Bao Dai, the last emperor under French rule, was deposed after a referendum in 1955. Diem proclaimed South Vietnam a republic and became its president. The nationwide elections were never held, and the Communists, deprived of what they considered certain victory, started a campaign of terror aimed at the overthrow of the Diem government. By 1960 the struggle between Diem and the Communists had developed into fierce civil war. Diem was supported by the United States, and the Communist guerrillas (known as the Vietcong) by North Vietnam. Dissatisfaction with Diem's dictatorial methods and his inability to defeat the Vietcong led to his overthrow and murder by South Vietnam army officers in 1963. The fall of Diem resulted in a period of political instability under a series of military governments. Only American intervention prevented the collapse of the South Vietnamese Government under pressure from Vietcong and North Vietnamese troops. With United States aid, the South Vietnamese Army tripled in strength, and a total of over 500,000 American troops were sent to Vietnam. South Korean forces and small numbers of Australian, New Zealand, and Thai troops also were fighting in South Vietnam. In 1965 the United States began the bombing of the North in response to increased attacks from Communist forces.

The sending of United States troops to Vietnam prevented a Communist take-over of the South. But the Communist Tet offensive, early in 1968, during which many southern cities, including Saigon, were attacked, proved that a military victory over the Vietcong and their North Vietnamese allies had not been achieved. At the same time, the war had caused enormous loss of life, both military and civilian, and great devastation to the land. The fighting also resulted in the dislocation of the people of South Vietnam, as millions of villagers, seeking safety, fled to the cities or were regrouped in camps. In 1968 peace talks between the United States and North Vietnam were begun in Paris. The talks were later extended to include representatives of the Vietcong and the South Vietnamese Government. In 1969 the United States announced a policy of gradual withdrawal of its forces from Vietnam. In 1972 the North Vietnamese launched a major offensive against the South, prompting the United States to renew the bombing of the North and to mine Haiphong harbor. Early in 1973 a ceasefire agreement was signed in Paris.

In spite of the ceasefire agreement, the long war continued. In 1975 North Vietnamese troops launched a new attack and drove deep into the South against crumbling resistance from the South Vietnamese. In the spring of 1975 the South Vietnamese Government surrendered.

GOVERNMENT

After the South surrendered, it was placed under a temporary government called the Provisional Revolutionary Government. This was the first step toward unifying the northern and southern parts of the country. Preparations then were made for elections to choose delegates from both parts to a new National Assembly, or legislature. After the elections, the delegates met in Hanoi. On July 2, 1976, they formally declared Vietnam to be a unified country named the Socialist Republic of Vietnam. Hanoi was selected as the capital, and the former North Vietnamese flag, anthem, and emblem were approved as national symbols. The delegates also elected a president, two vice-presidents, and a prime minister, as well as a committee to draft a constitution. The Vietnamese Workers' Party renamed itself the Vietnamese Communist Party. This is the governing party, and its leader is the country's top leader.

JOSEPH BUTTINGER, Author, *The Smaller Dragon*

MEKONG RIVER

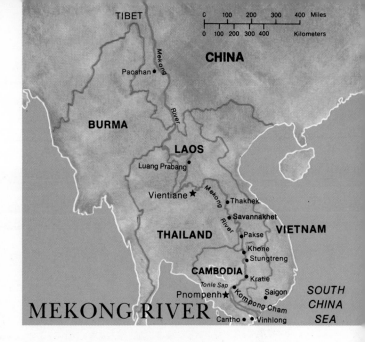

MEKONG RIVER

The Mekong, Asia's fifth longest river, is known by many names. In the Tibetan highlands, it is called Dza Chu. In China it is known as Lantsan Kiang. Mekong, its most common name, comes from the Thai expression *mae* (meaning "river") and Khong (the Thai name for the river). By whatever name, this great river has played an important role in the lives of Southeast Asia's people since earliest times. Millions of Thai, Lao, Cambodians, and Vietnamese depend on water from the Mekong to grow the rice that sustains them. In addition, the river traditionally has served as an important avenue of trade and communication.

The Mekong starts in the Tanglha Range, a barren mountain wasteland in China's Tsinghai province, near the Tibetan border. Flowing swiftly in a southeasterly direction, the Mekong first passes through wild mountain country and steep canyons. Then, as it enters the Indochina peninsula, the river moves at a slower pace through a region of lush green jungle and low, wooded hills. During one stretch, the Mekong forms the natural border between Laos to the east and Burma and Thailand to the west. Then it flows across Cambodia and the lower portion of Vietnam into the South China Sea.

The Mekong passes by the most important cities in Laos, including Vientiane, the river's main port, and Luang Prabang. It also flows by Pnompenh, Cambodia's chief city. The Mekong Delta, south of Saigon, Vietnam's largest city, has long been noted as one of the world's leading rice-producing areas. During the rainy monsoon season between June and October, the river becomes high and turbulent and flooding is common. The Mekong's principal flood reservoir is the Tonle Sap (Great Lake) in Cambodia's major rice-farming region.

Because of sandbars and dangerous rapids, most of the river cannot be navigated by large vessels. However, ships with a draft of up to 15 feet (4 meters) can sail up the river to a point some 350 miles (560 kilometers) from the China Sea. A plan to develop the hydroelectric power potential of the Mekong through the construction of several dams was delayed because of the Vietnam War.

Reviewed by Richard Butwell, The American University
Author, *Southeast Asia Today—and Tomorrow*

A Sikh policeman directs traffic in Kuala Lumpur.

MALAYSIA

Malaysia, in Southeast Asia, was founded in 1963 and is therefore one of the youngest countries in the world. Its people—a mixture of many races, nationalities, and cultures—have forged a national state out of territories that were formerly part of the British Empire. Malaysia consists of the 11 Malayan states and two former British colonies on the island of Borneo. Until 1965 it also included the island country of Singapore. The lands that make up Malaysia have been important in history because of their natural wealth and because of their favorable location for the conduct of trade between East Asia and lands farther west.

THE LAND

Malaysia is divided into two parts, separated by roughly 400 miles (640 kilometers) of the South China Sea. The western half of the country occupies most of the long Malay Peninsula. The eastern and larger half of Malaysia is composed of the two states of Sarawak and Sabah on the northern rim of the large island of Borneo. Malaysia occupies over 128,000 square miles (332,000 square kilometers). A long mountain range runs much of the length of the Malay Peninsula. To the east and west of these heights are low-lying plains that reach to the sea. A large part of these lowlands is covered by swamps and thick forests and is unsuitable for cultivation. The western lowlands of the peninsula are the most heavily populated areas. Here are located the principal cultivated lands and the main mining centers. Few people live along the peninsula's eastern coast, which is a generally undeveloped region of dense jungle and swampland. The peninsula's major rivers are the Pahang and the Perak.

The eastern part of the country, consisting of Sarawak and Sabah, is largely a land of coastal swamps, rain forests, and rugged mountain ranges. Only about 15 percent of the country's people live in this region. The Crocker range, Malaysia's tallest mountains, extends from Sabah into Sarawak.

Climate. Malaysia has an equatorial climate with generally high temperatures. Rainfall is heavy, particularly during the late autumn and early winter months, and the annual average is over 100 inches (250 centimeters). Although days are often hot and humid, nights are cool because of sea breezes. In the mountains, temperatures are cooler, and there is considerably less humidity.

Natural Resources. Malaysia is the world's largest producer of tin, and the mining and processing of tin give employment to many people. Bauxite, used to make aluminum, is another valuable mineral. A large part of the world's natural rubber supply—about 40 percent—is produced in Malaysia. It is estimated that half of the adult male members of the country's labor force are involved in the rubber industry.

Malaysia is also an important producer of copra (dried coconut meat) and timber. Among the valuable types of wood found in Malaysia's vast forests are teak, ebony, and sandalwood. Oil deposits have been developed in Sarawak, and cattle are raised in Sabah.

A worker taps a rubber tree. Rubber is Malaysia's leading plantation crop.

THE PEOPLE

Over 10,000,000 people live in Malaysia, which is a relatively small population for so large a country. However, the current birthrate is high —over 3 percent a year. Malaysia is a melting pot for people of various racial and ethnic backgrounds. More than four out of 10 of the country's inhabitants are Malays, who are closely related to the Filipino and Indonesian peoples. Over one third of the population is of Chinese descent, and there are also fairly large numbers of Indians, Ceylonese, and Pakistanis. Tribal peoples include the Dayaks (Dyaks), Kayans, and Melanaus of Sarawak, and the Dusuns, Muruts, and Bajaus of Sabah.

Language and Education.   The official language of the country is Malay. The written form of the language employs the Arabic alphabet. English is widely spoken in the major cities and towns. Chinese and a south Indian language called Tamil are also commonly spoken.

Free education is provided for all children between the ages of 6 and 12. Efforts are being made to reduce the illiteracy rate, which is over 50 percent for the country as a whole. The country has three universities, the most important one being the University of Malaya in Kuala Lumpur.

Religion.   Most Malays are followers of Islam, which was introduced into the region hundreds of years ago by Arab traders. People of Indian ancestry are mainly Hindus. Those of Chinese background follow the teachings of Confucius or Buddha. Some tribal peoples practice the ancient faiths of their ancestors, and there is a scattering of converts to Christianity.

Way of Life.   Modern buildings are commonly found in all of the cities and larger towns, but in the villages people ordinarily live in traditional Malay homes. Erected on small posts a few feet above the ground, these dwellings offer protection against floods and wild animals. The walls of matting and the thatched roofs can be constructed quickly and inexpensively, and the houses are cool and comfortable in the warm climate.

A river ferry taking a group of Kayans to the market.

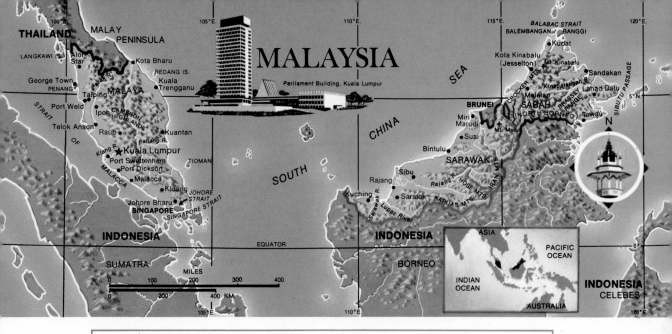

Parliament Building, Kuala Lumpur

FACTS AND FIGURES

MALAYSIA is the official name of the country. It was formerly called Federation of Malaysia.

CAPITAL: Kuala Lumpur.

LOCATION: Southeast Asia. **Latitude**—0° 52′ N to 7° 22′ N. **Longitude**—99° 38′ E to 119° 16′ E.

AREA: 128,430 sq. mi. (332,633 sq. km.).

PHYSICAL FEATURES: Highest point—Mt. Kinabalu (13,455 ft.; 4,100 m.). **Lowest point**—sea level. **Chief rivers**—Perak, Pahang, Sarawak, Rajang, Kinabatangan.

POPULATION: 10,700,000 (estimate).

LANGUAGES: Malay (official), English, Chinese, Tamil.

RELIGION: Muslim (official), Hindu, Buddhist, Confucian, Christian.

GOVERNMENT: Constitutional monarchy. **Head of state**—*yang di-pertuan agong* ("supreme head of state"). **Head of government**—prime minister. **Legislature**—parliament. **International co-operation**—United Nations, Commonwealth of Nations.

CHIEF CITIES: Kuala Lumpur, Kuching, Kota Kinabalu, Penang.

ECONOMY: Chief minerals—tin, iron ore, bauxite, oil, gold. **Chief agricultural products**—rubber, rice, palms, tea, coconuts, copra, hemp. **Industries and products**—tin mining, rubber goods and processing, forestry, handicrafts, fishing, livestock (oxen, buffalo, goats). **Chief exports**—rubber, timber, iron ore, tin, oil, copra, livestock, coconuts. **Chief imports**—foodstuffs (especially rice), machinery and transport equipment, chemicals, manufactured goods, petroleum products, beverages.

MONETARY UNIT: Malaysian dollar.

NATIONAL HOLIDAY: August 31, Freedom Day.

NATIONAL ANTHEM: *Negara-Ku* ("My Country").

Many styles of dress are worn. In the cities and towns, men often wear a loose shirt, trousers, and sandals. In the countryside, the sarong— a wraparound skirt reaching from waist to knees and sometimes lower— is usually worn. Many men also wear a brimless black velvet cap called a *songkok*. City women frequently wear Western-style clothes. Still, the most popular costume for women in cities and villages is the gaily colored sarong and blouse. Indian women favor saris and blouses, while Chinese women wear the pajamalike *sam-foo* or the *cheongsam,* a close-fitting dress with a slit skirt.

Malaysian food includes dishes made with rice, vegetables, fish, chicken, and meat. Many dishes are heavily spiced with curry sauces. Fruits, such as bananas, durians (a large, prickly fruit), and mangosteens (a reddish-brown fruit that combines the taste of peach and pineapple), are abundant. Since the majority of the people are Muslims, most Malaysians do not drink alcoholic beverages. Tea and coffee are the favorite drinks. People of Indian and Chinese ancestry eat foods that are adapted from the native cooking of their homelands.

Malaysians are fond of outdoor sports, particularly soccer, rugby, swimming, tennis, and cricket. There are large amusement parks in the principal cities. Among the different forms of entertainment are Chinese and Malay operas, puppet shows, the classical Malay dramatic play called the Menora, and traditional dancing known as *joget*. Celebrations take place on the main holidays. These include the Islamic holy day Hari Raya Puasa, the Chinese New Year, the Hindu Festival of Lights, and Freedom Day (August 31), the national holiday.

Cities.   Malaysia's capital, **Kuala Lumpur**, is the largest of the country's cities, with a population of over 500,000. It is situated on the Klang River in the heart of a tin- and rubber-producing region. The city is a blend of old Moorish-style architecture, introduced by the Arabs, and modern structures such as the imposing Parliament House complex, the National Museum, and the University of Malaya.

Sabah's capital, **Kota Kinabalu** (formerly Jesselton), is a major port and an important trading center. It has a population of about 25,000. Sarawak's leading city and capital is **Kuching** (population about 60,000), located on the Sarawak River. Livestock trading and fishing are its main industries. It is also a center for the export of timber and other products. Malaysia's major ports include Port Swettenham, Telok Anson (Port Anson), Port Weld, and Penang.

Economy.   More than half of all Malaysians earn their living from agriculture. Most farms are small and are worked by the farmer and his family, who consume much of what they raise. The most important crops raised on these smaller farms are rice, coconuts, fruit, and vegetables. Since the annual rice crop does not meet the needs of the population, large quantities must be imported. However, by using new farming methods, Malaysia hopes to be self-sufficient in rice by the mid-1970's. Rubber, oil palms, and coconut trees, which provide the chief cash crops, are grown mainly on large plantations.

Malaysians are skilled at various handicrafts, including basketry, silverware, the handweaving of shawls and sarongs, and the making of batik cloth. The batik process involves coating the fabric with wax, cutting designs out of the wax, and then dyeing the unwaxed portions of the fabric.

Women in traditional Malay dress dry small silver fish on mats.

GOVERNMENT

Malaysia is a constitutional monarchy. The *yang di-pertuan agong*—meaning "king" or "supreme head of state"—is elected for a term of 5 years. He is chosen from among the rulers of the nine original Malay states and elected by them. As head of state, the king appoints the prime minister and his cabinet. All of the ministers must be members of the Parliament, which consists of a senate and house of representatives. Thirty-two members of the Senate are appointed by the king. The other 26 are elected by the legislative assemblies of the 13 Malaysian states. All of the members of the House of Representatives are elected by popular vote.

HISTORY

The first groups of Malay people moved into the Malay Peninsula about 2000 B.C. For many centuries these people lived scattered along the coast in small villages. From the 8th to the end of the 13th century A.D., the ancient Indonesian Buddhist kingdom of Sriwidjaja dominated much of the peninsula. Later, for a shorter period, the Hindu kingdom of Madjapahit on Java gained control. The founding of the port of Malacca in the early 1400's aided the spread of Islam. The city grew into one of the principal trading centers in Southeast Asia, attracting Arab merchants who converted the inhabitants to the Muslim faith.

In 1511, the Portuguese captured Malacca, beginning more than 4 centuries of European colonial rule in this region. Malacca was taken by the Dutch in 1641, and in 1786 the British occupied the offshore island of Penang. British influence in the area expanded rapidly in the 19th century, and eventually all of the Malayan states came under British control. By the early 1900's, the British were also firmly entrenched in Sarawak and North Borneo (now Sabah).

Under British rule rubber plantations were established, mining was expanded, and railroads were built. In 1941–42, during World War II, the Japanese overran the Malay Peninsula. But at the end of the war, in 1945, the British regained control. Three years later they organized the nine Malay states into the Federation of Malaya, which became independent in 1957. At that time, Malacca and Penang became states in the federation. From 1948 to 1960, the federation was rocked by a Communist uprising that was finally put down with British military assistance.

In 1961, Malaya's prime minister, Abdul Rahman (later Malaysia's first prime minister) suggested organizing a Malaysian state by adding Singapore, Sarawak, Sabah, and Brunei to the Federation of Malaya. After 2 years of negotiations, the Federation of Malaysia was established on September 16, 1963. Brunei, a British-protected state in north Borneo, decided not to join; and Singapore left the federation in 1965.

Despite serious racial disturbances involving the Malay, Chinese, and Indian ethnic groups, Malaysia has flourished as one of the most promising of the Southeast Asian nations. International difficulties with Indonesia and the Philippines, which claim possession of parts of the North Borneo territory now included in Malaysia, have diminished. Thus, in the shaky world of Southeast Asia, Malaysia continues to be a beacon of progress and stability.

HYMAN KUBLIN, City University of New York; author, *The Rim of Asia*
Reviewed by PERMANENT MISSION OF MALAYSIA TO THE UNITED NATIONS

Raffles Place, named for the founder of modern Singapore.

SINGAPORE

A century and a half ago, Singapore was a quiet, unimportant little Southeast Asian island covered by jungle and swampland. But in 1819, a farsighted Englishman named Sir Stamford Raffles, acting on behalf of the British East India Company, leased Singapore from a Malay prince and established a trading station. The island was subsequently ceded to the British East India Company, which later turned it over to the British Crown. By the end of the 19th century, the trading post had grown into a thriving port city, and the island was a key outpost of the British Empire. Singapore remained under British rule until 1963, when it joined the Federation of Malaysia. In 1965 it left the federation to pursue an independent course as one of the smallest sovereign states in the world.

THE LAND

Singapore is located just off the southern shore of the Malay Peninsula, separated from the mainland by the narrow Johore Strait. The island is connected to the mainland by a causeway. South of the island is the Singapore Strait, which separates Singapore from several islands belonging to neighboring Indonesia. Singapore has an area of 224 square miles (581 square kilometers). It is 27 miles (43 kilometers) long and 14 miles

Singapore's crowded harbor is one of the busiest shipping centers in Asia.

(22 km.) wide. Most of the people live in the city of Singapore at the southern end of the island. The rest live in scattered villages. The central portion of Singapore is hilly, but the coastal areas are flat. The island's climate is hot, humid, and rainy.

Economy. From the days of Sir Stamford Raffles to the present, trade has been the lifeblood of Singapore, which is a major transshipment port. Goods from East Asian countries flow through Singapore on their way to ports in western Asia, Africa, and Europe. From Singapore's crowded docks and wharves sail ships carrying rubber, copra (dried coconut meat), timber, spices, and other products of the region. Other ships bring in manufactured goods and machinery from the industrial countries of the world. To accommodate incoming and outgoing cargoes, a great network of storage and handling facilities has been built over the years. Processing, packing, and handling these goods provides employment for tens of thousands of Singapore's residents. The port is also equipped to handle major ship repairs.

To provide jobs for a growing population, the government has recently encouraged the development of manufacturing and other industries. Rubber and tin from Malaysia are processed in Singapore's factories, and an industrial community has been established at Jurong, west of the city of Singapore. Singapore is a major commercial center, with numerous banks, insurance companies, and business firms. Singapore also serves as an important aviation and telecommunications center.

Farming does not play a large part in the country's economy. Although many farmers cultivate garden vegetables, much of the rice and vegetables consumed by Singapore's people must be imported.

THE PEOPLE

Singapore had few inhabitants when it was occupied by the British in the early 1800's. Since then its population has grown tremendously, and it now stands at over 2,000,000. Roughly three quarters of Singapore's people are of Chinese descent, mainly of immigrants who settled in Singapore during the past 150 years. The original Malay inhabitants—who now number about 300,000—have been steadily outnumbered by Chinese and other immigrants. There are also many people of Indian background, whose principal language is Tamil. Singapore has a small European population, which is mainly English-speaking.

Most of Singapore's Chinese residents cling to their ancient customs and religious beliefs. Their religion is a mixture of Confucianism, Taoism, and Buddhism. Nearly all of the Malay people are Muslims. Singapore's Indian population is mainly Hindu. Christians may be found among all of the ethnic groups.

Way of Life. Singapore combines both Eastern and Western ways in its pattern of living. The city of Singapore, where nearly eight out of 10 of the island's residents live, is a study in such contrasts. Tall office buildings and modern apartment houses line the broad avenues, along with buildings in the Victorian style popular a century ago. But along the side streets, the atmosphere becomes more Asian. Here are found food stalls and small shops run by Chinese, Indians, and Malays. Although many people wear Western clothing, one may also see Indian women in saris and others in the national dress of Malaya and China.

Singapore's harbor is one of the busiest in Asia. Thousands of ocean liners and cargo ships from all over the world put into the port each year. Mingling with them are the old-style wooden vessels of eastern Asia. Junks, sampans, and fishing vessels crowd the waterfront, in sharp contrast to the modern steamships. Both European and Asian cultures are represented in the city's libraries and museums. The city of Singapore also has a splendid botanical garden, and there are three large amusement parks, called the Great World, the Happy World, and the New World.

Education. All children receive 6 years of free primary education. Classes are taught in Chinese, Tamil, Malay, or English—the country's four official languages. Singapore has two universities, the University of Singapore and Nanyang University. There is also a technical institute and a small private college.

HISTORY

In the 13th and 14th centuries A.D., Singapore prospered as a center of trade. But with the founding of the port of Malacca (in what is now Malaysia) in the early 15th century, Singapore lost its importance. By the time Sir Stamford Raffles arrived in the 19th century, the island was virtually uninhabited. The trading post founded by Raffles in 1819 grew slowly but steadily. In 1826, Singapore was merged with Malacca and Penang into the British Straits Settlements, which became a crown colony in 1867.

FACTS AND FIGURES

REPUBLIC OF SINGAPORE is the official name of the country.

CAPITAL: Singapore.

LOCATION: Southeast Asia. **Latitude**—1° 15′ N to 1° 28′ N. **Longitude**—103° 40′ E to 104° E.

AREA: 224 sq. mi. (581 sq. km.).

PHYSICAL FEATURES: Highest point—Mount Timah (581 ft.; 177 m.). **Lowest point**—sea level.

POPULATION: 2,000,000 (estimate).

LANGUAGES: Chinese, Malay, Tamil, English.

RELIGIONS: Buddhist, Taoist, Confucian, Hindu, Muslim, Christian.

GOVERNMENT: Republic. **Head of state**—president. **Head of government**—prime minister. **Legislature**—parliament. **International co-operation**—Commonwealth of Nations, United Nations, Colombo Plan.

CHIEF CITY: Singapore.

ECONOMY: Chief agricultural products—fruits, vegetables, coconuts, tobacco, poultry. **Industries and products**—shipbuilding, petroleum refining, steel and metal products, electronics, chemicals, rubber products and processing, lumber processing. **Chief exports**—various manufactured goods, raw materials, fuels and lubricants. **Chief imports**—food, chemicals, machinery, manufactured goods.

MONETARY UNIT: Singapore dollar.

NATIONAL HOLIDAY: August 9, National Day.

NATIONAL ANTHEM: *Majulah Singapura* ("Forward Singapore").

SINGAPORE

In 1869 the Suez Canal was opened, enabling countries in Europe and Asia to trade more easily with one another. Because of its location at the crossroads of Southeast Asia, Singapore once again flourished. Singapore's importance was further heightened in the 20th century when the British constructed large naval and air bases on the island. However, despite its elaborate defenses, the island stronghold fell to the Japanese in 1942 and was occupied by them until the end of World War II (1945).

Singapore was granted internal self-government by the British in 1959. Four years later, in 1963, Singapore joined the Federation of Malaya and the former British colonies of Sarawak and Sabah on Borneo to form the new state of Malaysia. But serious differences developed between the Malays, who dominated the Malaysian Government, and Singapore's Chinese majority. As a result, Singapore left the federation in 1965 to become an independent country.

GOVERNMENT

The Republic of Singapore has a president as its head of state. The government is led by a prime minister, who presides over an 11-member cabinet. The prime minister is also the leader of the country's major political party, the People's Action Party. Members of the one-house legislature are elected by universal suffrage. In 1959, when self-rule began, a Singapore-born Chinese named Lee Kuan Yew took charge of the island's government. He became the country's first prime minister in 1965. Under Lee's direction, Singapore established a wide range of public welfare programs. It is now one of the most advanced as well as one of the most stable nations in Asia.

HYMAN KUBLIN, City University of New York; author, *The Rim of Asia*
Reviewed by T. T. B. KOH, Permanent Representative of Singapore to the United Nations

INDONESIA: AN INTRODUCTION

by Adam MALIK
Minister for Foreign Affairs, Republic of Indonesia

Introducing Indonesia to someone from another country is not an easy assignment. Such a task is made more difficult when one must write within a limited space. For after all, Indonesia is a very large nation that extends over an area of several thousand miles. Yet in most cases, mentioning Indonesia to a person from overseas is likely to stimulate only slight recollection of random facts learned at school or read in newspapers. Perhaps in some instances the individual may have had some firsthand experiences with the people of Indonesia.

But there are not many people who are thoroughly acquainted with Indonesia's history, culture, or even its geography. For example, how many people know about the large number of unusual plants and animals that are found in Indonesia? How many are familiar with the wide variety of subcultures among the many different ethnic groups of Indonesia? There are probably only a handful of people who have made an effort to learn about the hundreds of languages and dialects that are still used daily on the various Indonesian islands. So you can see that Indonesia is a very complex subject.

Of course, the easiest way to acquaint a foreigner with Indonesia would be to point out the territory of this vast country on a world map. Then it can easily be seen that geographically Indonesia consists of a chain of large and small islands, situated in the tropics below the equator, stretched between two oceans and two continents. It is this geographic setting that gives Indonesia the historical name of *Nusantara*. *Nusa* is the Indonesian word for "islands," and *antara* means "between." The name refers to the fact that Indonesia is situated between the Indian and Pacific oceans and the continents of Asia and Australia.

Another meaning of *Nusantara* is "archipelago" (a group of islands), and one often hears the phrase *Nusantara Indonesia*, which means "Indonesian archipelago." But although the term *Nusantara* may sound more poetic, it cannot take the place of the name "Indonesia." For "Indonesia" expresses more accurately the political and national identification of the Indonesian people. It is interesting to note that the term "Indonesia" was first used in 1850 by English scholars. Therefore, when Indonesian independence was proclaimed in 1945, the name was already nearly a century old.

Politically speaking, the meaning of "Indonesia" had become firmly established some years earlier, on October 28, 1928, at the Second Congress of Indonesian Youth, in Djakarta. On that day, the Sumpah Pemuda ("youth oath") was proclaimed, expressing the vow, "One Fatherland, One Nation, One Language: Indonesia." After that meeting, Indonesian nationalists began an intensified effort to formulate a single national identity that would unite all Indonesians regardless of ethnic and regional differences. The Day of the Youth Oath is generally accepted as marking the beginning of a national consciousness and a sense of unity among the Indonesian people. Subsequently, the use of the Indonesian language came to be regarded as a political expression against Dutch colonial rule (the Dutch language was then the official language) even more than as a means of self-expression and communication.

One other fact makes the eventful meeting on October 28, 1928, important in Indonesian history. During the course of that meeting, a young Indonesian named Wage Rudolf Supratman introduced one of his own musical compositions, *Indonesia Raya*. The song was warmly received by those who attended the congress, and today *Indonesia Raya* is the national anthem of the Republic of Indonesia. Because of these important events, the Day of the Youth Oath is now celebrated annually as a national holiday.

World War II, which brought Japanese occupation of the Indonesian islands, was an important factor in bringing about the independence of Indonesia. Efforts by Indonesian nationalists to end colonial rule came to a successful conclusion on August 17, 1945, when the Indonesian people declared their independence. However, it is wrong to view Indonesia's independence as a gift that came as a consequence of Japan's defeat in 1945. Actually, it was the result of bitter struggle and sacrifice on the part of the Indonesian people, who refused to accept the continuation of any form of colonial domination.

During Indonesia's relatively short history as an independent nation, there have been serious political difficulties on both domestic and international levels. These political events have hampered the country's efforts to develop itself to a point where conditions of backwardness could be eliminated or at least greatly reduced. Indonesia still lags far behind many other nations of the world in its activities to improve itself economically. Because of a lack of money and a shortage of trained technicians, Indonesia also has been unable to make adequate use of modern science and technology in order to fully exploit its natural wealth and potential resources.

The task of national improvement is not an easy one. The goal of modernizing Indonesia is made more difficult by the country's vast area. (If one were to include the territorial waters around the various Indonesian islands, Indonesia would have an area amounting to about 4,000,000 square miles, or 10,360,000 square kilometers.) In addition, there are problems created by the tremendously uneven distribution of the population. For example, more than half of Indonesia's inhabitants are concentrated on the island of Java. As a result, there are areas of Indonesia where the population density is more than 1,000 persons per square mile. By contrast, other sections of the country have less than 10 persons per square mile. Such problems of population distribution hold back development programs.

Indonesia is moving to confront these problems, and it is hopeful of solving them. The country is now in the midst of a 5-year program designed to improve all aspects of the economy—from food production to the creation of new industrial plants. Although the problems are quite serious, progress is being made. In conclusion something should be said about Indonesia's role in world affairs. As an independent and sovereign nation, Indonesia pursues its goals in international relations according to the philosophy expressed in the Preamble to the Indonesian Constitution. There it is stated that Indonesia's sincere wish is to contribute to efforts aimed at promoting world peace and international brotherhood.

Java, an island of rugged mountains, has many active and inactive volcanoes.

INDONESIA

Indonesia is an exotic land. On its many islands are found lush tropical forests, wild jungle animals, and scores of active volcanoes. In earlier times Indonesia was famous for its spices, and it was in order to find a shorter route to these "Spice Islands" that Christopher Columbus and Ferdinand Magellan set sail on their voyages of discovery. Beginning in the early 1600's, the Dutch gradually assumed control over the area, developing it into a colony called the Netherlands East Indies. Since 1949, however, the country has been an independent republic.

The Republic of Indonesia is the largest nation in Southeast Asia and the richest in terms of natural resources. With a population that may be as high as 120,000,000, Indonesia ranks among the six most populous nations of the world. Situated between the Pacific and Indian oceans, the country spans an area greater than the distance between the east and west coasts of the United States. Much of this surface is sea, however, and Indonesians refer to their country as Tanah Air Kita, which means "our land and water" and also "our fatherland."

THE LAND

Indonesia consists of thousands of islands, which cover an area of roughly 735,000 square miles (1,904,000 square kilometers). The islands stretch like a bridge of stepping-stones between the Asian mainland and the continent of Australia. Mountains rise steeply and often to great heights on many of the islands. The highest of these mountains, located on Irian Jaya (the Indonesian part of New Guinea), are permanently capped with snow. Many of Indonesia's mountains are active or inactive volcanoes. Probably the most famous Indonesian volcano is Krakatau (Krakatoa) on an island in the Sunda Strait, between Java and Sumatra. In 1883 Krakatau erupted in one of history's worst instances of volcanic activity.

Climate. Since Indonesia is located on the equator, its climate is tropical. Temperatures are usually high, but vary according to height above sea level. Most of the major cities are located along the coast or on the lowland plains. Here the weather is generally very warm and humid, and many townspeople go to the cooler hills when they can to escape the heat and humidity. Seasonal winds known as monsoons buffet the islands, resulting in two main seasons, a wet and a dry. Throughout most of Indonesia the rainfall is enough to make the land green and rich all year round. But in some of the smaller islands of the southwest, the dry season is long and severe. There the land is suitable mainly for grazing cattle and growing crops—such as corn—that require little moisture.

The Greater Sunda Islands. The four major islands of Indonesia are Java, Sumatra, Kalimantan (the Indonesian portion of Borneo), and Sulawesi (Celebes). Together they form what are known as the Greater Sunda Islands, with an area of about 493,000 square miles (1,277,000 sq. km.).

For nearly all of Indonesia's history **Java** has been the most important of these islands. Today over 60 percent of Indonesia's population, some 80,000,000 people, live on Java, and there is a serious problem of overpopulation. Fortunately the island is very fertile, particularly along the northern coastal plain. The richness of the soil is due in part to the ash from the island's numerous volcanoes. Farming is the primary means of support for the people, who cultivate rice, the staple food, in irrigated fields. During the dry season—and in those places where there is not enough water to grow rice—corn, cassava, and other dry-field crops are raised. There are many rubber, teak, and other types of plantations on Java. The principal cash crop is sugar, but this can no longer be sold abroad at a profit. In previous years Java was able to produce enough rice to supply all of its inhabitants. But Java's population has grown beyond the island's capacity to feed it, and rice must now be imported from overseas. Djakarta, the nation's capital, is located on Java along with several other important Indonesian cities and trading ports.

Sumatra is Indonesia's second most important island. A long chain of mountains, the Barisan Mountains, extends along the southwestern coast. The interior of the island is densely forested, and the eastern coastal region is covered by swampland. Sumatra has a great many large plantations, which produce rubber, tea, coffee, tobacco, and palm oil. The island is a major producer of petroleum. Tin is mined on the neighboring offshore islands of Bangka and Belitung (Billiton).

Borneo is the third largest island in the world, and most of it is

A street in Medan, Sumatra's largest city. Pedicabs and bikes are widely used.

Terraced rice fields may be seen throughout Batak country in northern Sumatra.

An outdoor market in Denpasar on Bali, one of the Lesser Sunda Islands.

Indonesian territory, called Kalimantan. The remaining portion of the island consists of Sarawak and Sabah (parts of Malaysia) and the oil-rich state of Brunei. Much of Kalimantan is mountainous terrain and nearly impenetrable jungle. Most of the population of about 4,000,000 lives along the coast. Rice and rubber are the principal agricultural products. Mineral resources include oil, tin, diamonds, gold, and silver. Forestry is a major economic activity, as it is on many of the islands.

Sulawesi, the last of the Greater Sunda Islands, consists of four mountainous peninsulas. Because of the rugged terrain, which divides islanders in one section from those in another, the population of about 7,000,000 consists mostly of small groups of people with different languages, customs, and religious beliefs. The island's chief products are rice, maize (corn), dried coconut meat (copra), spices, and various wood products, including rattan (palm). Makassar (Macassar), the island's principal city, was once a center of the international spice trade, which was so important in Indonesia's early history. Although spices are no longer the most important commodity for export, the city is still the trading headquarters for eastern Indonesia.

The Lesser Sunda Islands. Across the Banda and Flores seas from Sulawesi are the Lesser Sunda Islands, totaling about 28,000 square miles (73,000 sq. km.). Of this group, the most interesting is **Bali**. An island of rare beauty, Bali is noted for its ancient Hindu culture. Like Java, Bali is densely populated, and its people are mostly farmers who work tiny, irrigated rice fields. Among the other islands of this chain are Flores, Sumba, Lombok, and Timor, the largest.

The Moluccas (Maluku). There are hundreds of islands in this group, which were known historically as the Spice Islands. The principal

INDONESIA

FACTS AND FIGURES

REPUBLIC OF INDONESIA—Republik Indonesia—is the official name of the country.

CAPITAL: Djakarta.

LOCATION: Southeast Asia. **Latitude**—5° 54' N to 11° S. **Longitude**—95° 01' E to 141° 02' E.

AREA: 735,269 sq. mi. (1,904,347 sq. km.).

PHYSICAL FEATURES: Highest point—Mount Carstensz (Puntjak Djaja) (16,400 ft.; 4,999 m.). **Lowest point**—sea level. **Chief rivers**—Barito, Asahan, Kampar, Rokan, Hari, Musi, Solo, Brantas.

POPULATION: 120,000,000 (estimate).

LANGUAGE: Indonesian (Bahasa Indonesia); various regional languages and dialects.

RELIGION: Muslim, Christian, Hindu, Buddhist, Confucian.

GOVERNMENT: Republic. **Head of government**—president. **Legislature**—House of People's Representatives. **International co-operation**—United Nations.

CHIEF CITIES: Djakarta, Bandung, Surabaja, Semarang, Medan, Palembang, Makassar, Malang, Jogjakarta, Bandjermasin.

ECONOMY: Chief minerals—petroleum, tin, bauxite, iron ore, coal, manganese, gold, silver, diamonds, copper, nickel. **Chief agricultural products**—rice, maize, cassava, sweet potatoes, copra, sugar, tea, coffee, tobacco, palm oil, soybeans, groundnuts, spices (cloves, nutmeg, pepper). **Industries and products**—handicrafts (wood carving, silversmithing); forest products (teak, ebony, sandalwood, kapok, bamboo, rattan, cinchona); fishing, food processing, mineral refining, sugar-refining, shipbuilding, cement, textiles, tires, shoes, glass, paper. **Chief exports**—rubber, tobacco, palm oil, coffee, tin, copra, tea, petroleum, spices, quinine. **Chief imports**—rice, machinery and equipment, vehicles, textiles, iron and steel products, chemicals.

MONETARY UNIT: Rupiah.

NATIONAL HOLIDAY: August 17, Independence Day.

NATIONAL ANTHEM: *Indonesia Raya* ("Great Indonesia").

ones are Ceram, Buru, Ambon (Amboina), Ternate, Halmahera, and Tidore. From these islands came the rich spices—cloves, nutmeg, and mace—that were a magnet for traders from Europe and mainland Asia and that led to the colonization of Indonesia by the European sea powers. Copra is now the major crop raised in the Moluccas, while petroleum is the most important mineral resource.

Irian Jaya. The easternmost area of Indonesia is Irian Jaya, the

Indonesian part of New Guinea. (The other, or eastern, portion of New Guinea is the major part of Papua New Guinea.) Covered by dense forests, Irian Jaya has a population of some 800,000 Papuans, most of whom live at a Stone Age level of technology. In the interior are the Snow Mountains (Djajawidjaja Mountains), whose highest peak, Mount Carstensz (Puntjak Djaja), is 16,400 feet (4,999 meters) high. Most of the people on West Irian live along the coast. In fact, much of the interior of the island has only recently been explored. The Indonesian Government hopes to develop the area and its oil resources further.

Animals and Plants. The animals and plants of Indonesia are divided roughly into two groups by an imaginary line (called the Wallace Line after the naturalist who established it), which runs north–south between Kalimantan and Sulawesi. West of this line the plants and animals are much like those of Southeast Asia. There are dense tropical rain forests, vast numbers of palm and banana trees, and a great many varieties of wild flowers. Among the animals in this region are the orangutan (the word comes from the Indonesian *orang hutan,* or "man of the woods"), the tiger, the wild buffalo, and—very rarely now—the elephant and the rhinoceros. Another animal found here is the mouse deer (*kantjil*), a tiny creature whose legendary cleverness at outwitting larger enemies has made it the hero of Indonesian folktales.

East of the Wallace Line animal life is much more like that of Australia. Some of the more unusual creatures include the kangaroo and the brilliantly feathered bird of paradise. The reason for the sharp difference in natural life is probably the fact that the western islands of Indonesia, which lie in shallow waters, were once a part of mainland Asia. But the eastern islands, which are surrounded by deep seas, were always separated from the continent, and therefore did not develop the same type of wildlife.

Natural Resources. Indonesia has large deposits of several valuable minerals, including petroleum, tin, iron, bauxite, and coal, and smaller deposits of gold, silver, and diamonds. Forests are another major natural resource, and supply large amounts of teakwood, ebony, and sandalwood. Quinine, used in the treatment of malaria, is manufactured from the bark of the cinchona tree. Bamboo and rattan are used to make wicker chairs and other furniture. Since it is surrounded by the sea, Indonesia has an abundance of different kinds of fish, which form an important part of the Indonesian family's daily diet.

THE PEOPLE

Indonesia's population is a mixture of different ethnic groups. From island to island—and even within each island—different languages are spoken and different customs are observed. Most Indonesians are of Malay ethnic stock. They are usually slight in build and have brown skin and straight, black hair. But in the easternmost islands of Indonesia the people are bigger, darker in skin coloring, and curly-haired. These are the Papuans, the dominant people of Irian Jaya, whose language and culture are quite different from the majority of Indonesians. On a technological level the Papuans are behind most other Indonesians. This, and the difference in physical appearance and the fact that Irian Jaya has only been ruled by Indonesia since 1963, have combined to cause friction between the Papuans and other Indonesians.

Indonesian youngsters at school on Sumatra. Primary education is compulsory.

The major ethnic groups of Indonesia are found on the island of Java. They are the Javanese and the Sundanese. Together they make up over half of Indonesia's population. The way of life of these people shows traces of the old culture of the Indonesian kingdoms of past centuries, which were influenced by Indian religious (Hindu), political, and cultural ideas. The Sundanese are now orthodox Muslims, but many Javanese mixed Islam with their earlier Hindu and spirit-worshipping religions. As a result there is a sharp difference in Javanese society between strict Muslims, known as *santri,* and those people who practice a less orthodox Islam, who are known as *abangan.*

The most important of Indonesia's minor ethnic groups are the Batak and Minangkabau people of Sumatra. Both groups are noted for their interest in trade and are considered to be among the most aggressive and enterprising people of Indonesia. The Batak and Minangkabau have provided modern Indonesia with many of its political and economic leaders. In some areas there are groups of Christian Indonesians, whose ancestors were converted to Christianity by Portuguese and Dutch missionaries during the period of European colonization.

Language.   There are about 250 languages and dialects used on the islands of Indonesia. Each of these represents a group of people that considers itself culturally distinct. But there is a basic unity underlying this variety, since most of the languages belong to the Malayo-Polynesian group. The official national language is Indonesian (Bahasa Indonesia), a variety of Malay, which was previously the traditional trading language of the islands. It is generally familiar to Indonesians everywhere and is the language used in education and government.

Religion.   Islam is the major religion of Indonesia, having been introduced to the area some 500 years ago. Today nine out of 10 Indonesians are Muslims. The Muslim faith is particularly strong along the coasts, where it first took root. In the interior, particularly where the terrain is mountainous or heavily forested and thus hard to travel over, people still cling to older forms of religion, including spirit and ancestor worship. The Hindu faith is dominant on the island of Bali, and Confucianism and Buddhism are also practiced in Indonesia by people of Chinese descent. Roughly 5 percent of the people are Christians.

Way of Life

There is often a marked contrast between the way of life of the people living along the coasts and those in the interior, as well as between the lowland and the hill people. Historically the coastal areas, open to the sea and trade, have been centers of commerce and therefore more open to the introduction of new ideas, while the people of the inland plains have been more conservative in their way of life, which centers around farming. The mountain regions have been the most remote of all from government control and from new cultural and technological developments.

The great majority of Indonesians are farmers, farmworkers, or plan-

Balinese women perform traditional dances in front of a Hindu temple

Oxcarts are still an important form of transport in many parts of Indonesia.

tation laborers. Middle-class people are mainly employed in the large government bureaus. There is also a very small business and professional class, most of whom are Chinese. The Chinese, who number about 3,000,000, migrated to Indonesia in colonial days as laborers, and then worked their way up. The relative wealth of this minority has caused much jealousy among native Indonesians, and the Chinese frequently have been persecuted in the years since Indonesian independence.

Indonesians on the outer islands are generally better off economically than the people of Java. However, none are wealthy. Most farmers own very little land, either because fertile land is scarce or because they cannot farm large plots with the tools that are available. On Java many farmers do not own land and must earn their living by working on farms belonging to others. The problem of land ownership and reform was a controversial political issue in the early 1960's, and it continues to be a troublesome question.

On many of the Indonesian islands the people practice a crude form of farming known as shifting (or slash-and-burn) agriculture. A portion of the jungle is cleared by fire, and crops are planted. When the soil loses its strength after a few years, the farmers move to a new area. The worn soil, no longer protected by jungle foliage, is further eroded by heavy rains and loses its mineral content. As a result these areas become wasteland.

Indonesia's cultural diversity has resulted in many different styles of dress, food, and housing on the various islands. Nevertheless certain styles are commonly found in all regions. In the larger towns everywhere, most men wear Western-style clothes when they go out. At home, villagers and city people may wear a long wraparound skirt called a sarong. Many men, especially strict Muslims, wear a black velvet cap called a *pitji*. Western-style cotton dresses are favored by young city girls. But married women and country girls generally wear traditional garments— including a jacket, frequently of thin, flowered material (called a *kebaja*); a wraparound skirt (or *kain*) of heavy cotton; and a long scarf (*selendang*) draped over the shoulder. Most often the *kain* is decorated with beautiful colored designs, which are drawn or hand-stamped on the cloth as part of the batik dyeing process. The batik designs are one of Indonesia's most famous handicrafts and have often been imitated by Western textile designers. Orthodox Muslim women may wear a white scarf over their heads, especially on Friday, the Muslim sabbath. But they do not veil their faces, as is often the custom in Arab lands, and their social position is much higher than that of women in most Muslim countries.

An Indonesian woman weaver dressed in traditional flowered jacket and long skirt.

A typical Batak house in northern Sumatra. The Bataks are mainly rice farmers.

Houses also show certain common features. Traditional homes in Indonesia differ mainly according to whether they are designed for more than one family and whether they are built on the ground or raised above it. Some peoples, such as the Minangkabau and Batak of Sumatra, and the Dajak (Dyak) of Kalimantan, traditionally have organized themselves in clans or other large groups of relatives. To shelter all the members of the clan, they constructed huge longhouses. Often these were beautiful, elaborately carved wooden houses, which were built high above the ground on stilts. Many of the longhouses, especially in Sumatra, are now falling into disuse as the old ways of village life are changing. They are being replaced by modern, small-family units. Javanese houses, in contrast, have always been one-family buildings and are small, square, and peak-roofed. They are built directly on the ground and generally use the earth as a floor.

Most of Indonesia's farmers live in compact villages surrounded by their fields. Their yards are thickly planted with palms, fruit trees, and other plants, which are grown both for profit and to provide shade. In Java, poorer farmers live in small houses without windows and with walls made of woven bamboo matting. Those who are somewhat better off have wooden houses. A person who is well-to-do and modern will have a house built of whitewashed brick or concrete, and may even have glass panes in his windows. Most middle-class city homes are Western-style and are made of whitewashed stucco (a material made of cement and sand). Poorer city people live in shacks or village-type houses clustered about the edges of the town.

Rice is the basic food for Indonesians of all classes, although the poorest people and those who live in very dry areas must often live on maize (corn) or cassava (a plant whose roots provide a starchy food). Most Indonesians eat their rice together with various spices and sometimes a bit of dried fish or chicken. But when they celebrate a religious holiday or special event they will cook many small, elaborate dishes to go with the rice. These are served in a ceremonial dinner (*selamatan*), which is open to the entire neighborhood or community. *Rijsttafel,* a popular food in Holland, is derived from these special Indonesian dishes.

Cities. **Djakarta**, situated in northwestern Java, is Indonesia's capital and largest city (population over 3,000,000). A cosmopolitan city, Dja-

One of the main boulevards of Djakarta, Indonesia's capital and leading port.

karta is also the country's leading seaport and its most important center of trade and communications. During the time of Dutch rule, Djakarta was known as Batavia, and the Dutch influence is reflected in the network of canals that cut through the city. The city's broad avenues are lined with many modern buildings and crowded with cars, bicycles, and *betjaks* (bicycle-driven taxicabs). **Surabaja** and **Bandung**, Indonesia's second and third largest cities, are also located on Java. Surabaja is second only to Djakarta as a port city, and it is the center of Indonesia's limited manufacturing industries. Bandung, the most modern Indonesian city, is a popular resort town and is the home of the country's leading technical institute. Another important Javanese city is **Jogjakarta**, one of the old royal capitals and a center of fine arts and traditional Indonesian culture.

Sumatra's principal port cities include Belawan (the port of Medan), Padang, and Palembang. Located on the Musi River, Palembang is an important outlet for petroleum and rubber products, and its industries include shipyards and ironworks. Other leading Indonesian cities are Bandjermasin on Borneo, Makassar, the largest city on Sulawesi, and Ambon in the Moluccas.

Indonesian Art Forms. Indonesians are famous for their wood carving and other handicrafts. Traditional Indonesian homes often have hand-carved wooden figures at the entrance and various designs carved into pillars, roofs, porch ceilings, and walls. A number of beautifully constructed Buddhist and Hindu temples can be found on the island of Bali, in central Java, and elsewhere. An important part of Indonesian culture is the wayang (*wajang*), or shadow play, which dates back to the time of the Indonesian Hindu kingdoms. These plays are acted out by leather puppets, which cast shadows on a lighted screen. The puppet performers are accompanied by a narrator and an orchestra (gamelan), consisting of musicians playing traditional Indonesian instruments.

HISTORY

Prehistoric man is said to have lived on the islands of Indonesia more than 500,000 years ago. In 1891 Dutch scientists working on Java discovered human bones hundreds of thousands of years old. These were the remains of the famous Java Man, who many scientists believe was one of the earliest human beings. Today's Indonesians are descended from Malay people who began immigrating to the islands 4,000 to 5,000 years ago.

Even in ancient times the Indonesian islands were important in international trade, for the islands lay along the sea route that connected China, India, Persia, and the Roman Empire. Ships from many lands docked at Indonesian ports to take on cargoes of spices, resins, and precious woods, which were the principal items of this trade. The Indonesians themselves were skilled seafarers and merchants, and their trading ships traveled as far as the shores of India and China. Although the Indonesian islands were never under a single government before the period of Dutch colonial rule, they were tied to each other by a long history of contact and trade.

The many traders who visited the islands introduced foreign ideas and customs to the Indonesians. Indian culture, including the belief in royal power, became the most important influence by the early centuries A.D. The classical Indian idea of an all-powerful king appealed to ambitious Indonesian rulers, who until then had been little more than tribal chiefs. Both the Hindu and Buddhist faiths of India were adopted by the Indonesian ruling class. Indian ideas also filtered down to the rest of the people. Hinduism mixed with, or sometimes replaced, the traditional worship of spirits (known as animism).

Several Hindu and Buddhist Indonesian kingdoms came to power on the larger islands. The two greatest of these kingdoms were Sriwidjaja (from the 7th through the 13th century A.D.) in southern Sumatra and Madjapahit (from the 13th to the 16th century) in eastern Java. Both kingdoms were trading powers, and they controlled much of the commerce of the islands. During the 13th century the famous traveler Marco Polo became the first known European to visit the Indies. Shortly after Marco Polo's visit, the powerful Mongol emperor of China, Kublai Khan,

Statue of Buddha at Borobudur, an 8th-century A.D. Buddhist shrine on Java.

attempted to conquer Java, but his army was defeated. Between the 12th and 15th centuries the religion of Islam was brought to Indonesia by traders from India and the Malayan coast. Islam gained its first foothold in the coastal cities, whose princes were trying to break away from the powerful Hindu kingdoms on Sumatra and Java. Eventually Islam spread to nearly all of the Indonesian islands and became the area's major religion.

The year 1511 marked the beginning of European involvement in Indonesia. It was then that the Portuguese captured Malacca, which gave them control of the narrow strait between Malaya and Sumatra. For many years Europeans had sought to gain direct access to the Indies and its spices, and the Portuguese were followed by the Spanish, Dutch, and British. During the 16th century the great European sea powers competed for control of the Indies, particularly the Spice Islands (the Moluccas). Early in the 17th century the Dutch East India Company set up trading posts in the Spice Islands and on Java. The Dutch established their main base at the present-day city of Djakarta. Gradually the Dutch extended their influence to the rest of the islands. By the end of the 17th century Holland was the major European power in what became known as the Dutch, or Netherlands, East Indies. Dutch rule spread slowly through the islands. On the larger islands, particularly Sumatra and Java, many large plantations were developed by the Dutch and other non-Indonesians. On some of the islands, such as Java, the Dutch ruled directly. But in the outlying islands the Dutch merely backed the local ruler or tribal chief, and the people there had little contact with European social, political, and economic ideas. As a result Indonesia did not develop as a unified nation, but rather as a group of islands with varied customs and life styles. These regional differences have continued to trouble the young nation in the years since independence.

The Dutch governed with a firm hand while exploiting Indonesia's natural riches. Resentment against Dutch rule led to the rise of an Indonesian independence movement, which began in 1908 with the founding of the Budi Utomo ("noble endeavor"), an association of Javanese intellectuals. Other nationalist groups were formed during the next 2 decades, including the Indonesian Nationalist Party (PNI) in 1927. The PNI was led by Sukarno (many Indonesian people have only one name), who later became the first president of the Republic of Indonesia. In order to pacify the nationalist groups, the Dutch established a Volksraad ("people's council") in 1918 to give Indonesians some voice in the government. But nationalist feeling continued and an unsuccessful Communist-led rebellion took place in 1926–27. Many of the leaders of the independence movement, including Sukarno, were later jailed or sent into exile.

During World War II the Japanese occupied Indonesia (1942–45). In order to win support in their war against the Western Allies, the Japanese released the imprisoned leaders and encouraged the nationalist movement. Toward the end of the war the Japanese allowed the nationalist leaders Sukarno and Mohammed Hatta to establish an Indonesian-run government. On August 17, 1945, a few days after Japan's surrender to the Allies, the nationalist leaders declared independence for Indonesia. However, the Dutch did not recognize this new government, and 4 years of fighting and lengthy negotiations followed. Finally, in December of

1949, the Dutch recognized the independence of what was first called the United States of Indonesia. (The name was changed to the Republic of Indonesia in 1950.) The Dutch, however, retained control of western New Guinea (now Irian Jaya) until 1962. The Indonesians administered the area until 1969, when it was formally incorporated into the republic.

GOVERNMENT

The Indonesian Constitution of 1945 provided for a strong presidential system of government. This was changed in 1950 to a parliamentary form of government, with a president, vice-president, premier, cabinet, and one-house legislature. The parliamentary system proved to be unworkable in Indonesia because of differences within the many political parties and among the various islands and ethnic groups. Serious economic and social problems remained unsolved, and a new form of government, "guided democracy," was devised by President Sukarno and other political leaders. The Constitution of 1945 was restored, which gave almost unlimited power to Sukarno, who became both head of government and head of state. In addition, the People's Consultative Assembly was created, with the power to choose the president and to formulate national policy. The Consultative Assembly consists of the members of parliament and representatives of various national groups and organizations, including the military, labor and religious groups, and the professions. The parliament is called the House of People's Representatives.

Under this system President Sukarno became the central political figure, gradually assuming many of the powers of a traditional Indonesian king. The political parties and the press were placed under strict control. To promote Indonesian unity Sukarno encouraged anti-foreign campaigns against the Dutch in Irian Jaya (1949–63) and the new state of Malaysia (1963–66). Meanwhile the economic situation worsened. Popular sympathies began to center around the Army on the right and the Communist Party on the left. In 1965 a group of army officers attempted a coup against the country's military leaders, whom they accused of plotting Sukarno's downfall. The military high command blamed the Communists for the coup, and bloody civil strife erupted. Perhaps as many as 500,000 people, including Communists and non-Communists, were killed. In the aftermath of the coup President Sukarno's power was broken. He was replaced in 1967 by General Suharto, who had put down the coup and had taken charge of the government in 1965.

Indonesia's government under Suharto has followed a conservative path. Civilian governmental institutions have been maintained, but the Army remains the dominant political force in the country. As its first goal the government has aimed at stabilizing and improving the economy, a necessary and difficult task. To this end Suharto launched a new economic development plan. In foreign affairs Indonesia has adopted a policy of neutralism. But in contrast to President Sukarno, who sought closer ties with the People's Republic of China, Suharto favors expanding relations with the United States and other Western nations. Foreign businesses are being encouraged, since the Suharto government considers international investment and aid vital to Indonesia's economic recovery and development.

RUTH McVEY, School of Oriental and African Studies, University of London
Editor, *Indonesia, Its People, Its Society, Its Culture*

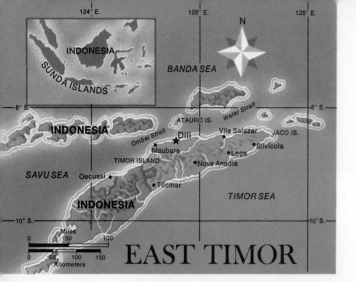

Some of the most beautiful vegetation in the world grows in East Timor, a tiny area in the Lesser Sunda islands of Southeast Asia. The territory lies in the path of the Asiatic, or northwestern, and the Australian, or southeastern, monsoons. This monsoon climate has produced the magnificent plant life for which Timor is noted.

The Land.   The island of Timor is located approximately 525 miles (845 kilometers) northwest of Australia. The western portion of the island has long been a province of Indonesia. East Timor, which was once an overseas province of Portugal, comprises the eastern section plus the small district of Oecussi Ambeno. Its total land area of about 7,300 square miles (19,000 square kilometers) is largely mountainous, with many beautiful lakes and forests. The largest town and major port in East Timor is Dili, the capital.

The People.   The vast majority of the population of about 600,000 is of mixed Malay and Papuan ancestry. There is a Chinese community of several thousand. Primary schools and educational institutions run by Roman Catholic missionaries were established during the period of Portuguese rule. Despite Western influence, the Timorese retain many ancient customs. Their religion is a blend of Catholicism and local ritual, and each village has its sacred house, presided over by a priest who invokes the blessing of ancestral spirits. The economy is chiefly agricultural, with corn, rice, and coffee the main crops. Livestock raising and fishing are also important.

History.   The Portuguese came to the eastern part of the island of Timor in 1512. They were followed about 100 years later by the Dutch, who established a trading base farther west, and for three centuries the two countries divided the island between them. When the Dutch relinquished their colonies in Southeast Asia in 1949, the western section of Timor became part of the newly independent Republic of Indonesia. In 1975 fighting erupted in East Timor when independence groups clashed with those favoring union with Indonesia, and the area became a center of dispute between the governments of Indonesia and Portugal. Independence forces announced the formation of a government, but the move was followed shortly by the intervention of Indonesian troops. In 1976 the Indonesian Government declared East Timor a part of Indonesia, ending 400 years of Portuguese rule.

Reviewed by LEÃO LOURENÇO, Permanent Mission of Portugal to the United Nations

BRUNEI

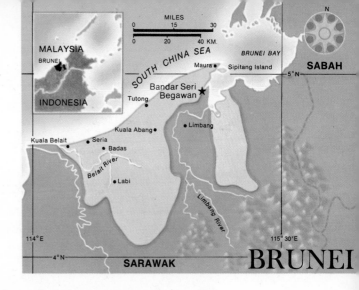

On the lush green coast of northwestern Borneo is the tiny oil-rich state of Brunei. Its neighbors on Borneo are parts of the independent nations of Malaysia and Indonesia. But Brunei is a protected state of Great Britain, ruled by a young sultan, Hassanal Bolkiah.

The Land and the People.   Brunei's area of 2,226 square miles (5,765 square kilometers) is made up of two regions. The western section is mostly low and swampy, while the eastern section is hilly. Brunei lies close to the equator, and the climate is hot and humid with considerable rainfall. A little over half of Brunei's 116,000 people are Malays and about a quarter are Chinese. The rest are largely Dayak and members of other groups native to Borneo. The official language is Malay, but Chinese and English are also used. Islam is the official religion.

For centuries in Bandar Seri Begawan, the capital, people have lived in a settlement of wooden houses built on stilts over the tidal flats of the Brunei River. But the government is encouraging families to leave the "water town" by offering them small landholdings and loans for building houses. The simplest village houses are one-room rectangular structures of wood, with a porch in front where women do much of the housework. Rice and fish are favorite foods.

Economy.   Oil is Brunei's main export and source of income, and farming and fishing are the chief occupations. People grow rubber for export, and rice, sago, pepper, coconuts, and fruit. The government is attempting to increase fish catches by encouraging modern fishing methods. A development program includes the building of a large airport and a deepwater port at Muara.

History. Historians know little of Brunei's early history. Islam reached the country in the 15th century, and by the early 16th century Brunei had become a powerful independent state, controlling all of northwestern Borneo and a number of small islands. After 1600 Brunei began to decline, and by the late 19th century it included only its present area. By terms of a treaty signed in 1888 Brunei came under the protection of Great Britain. In 1959 Brunei had its first written constitution. Great Britain remained responsible for the state's defense and foreign relations, and the sultan for internal affairs. When the Malaysian federation was formed in 1963, Brunei remained outside it. Brunei has made it clear that it favors remaining under British protection.

Reviewed by HYMAN KUBLIN, The City University of New York

PHILIPPINES: AN INTRODUCTION

by Carlos P. ROMULO
Secretary of Foreign Affairs, Republic of the Philippines

The Philippines today is one of the most progressive and also one of the fastest developing countries of Asia. It is republican in government, democratic in its institutions, and modern in its outlook and aims. As a result, the nation is well on the way to becoming an important factor in Asian and world affairs. The Filipinos seek to fashion their proper role in the family of man by combining their traditional Asian traits with the influences that have come from other parts of Asia, from Europe, and from the Americas.

Filipinos take much pride in having been among the first people in Asia to begin an independence movement. In fact, the Philippines was the first Southeast Asian country to revolt against a colonial ruler—Spain in 1896. While under the United States, the Filipinos continued their independence campaign—encouraged by the Americans themselves—until independence was gained in 1946. But the long years under colonial rule and the difficult struggle toward nationhood left many problems still unresolved. Much of the energies of the present and the next generations will be spent on efforts to solve these problems.

The saga of the Philippine people began many thousands of years ago when early man first migrated to these islands. Gradually the Philippines was peopled by small bands and by occasional migrations in larger groups, first over land bridges during the Ice Age and later across the sea by boat and raft. The peoples who came 2,000 years ago and later, bringing stone and metal cultures, were the ancestors of the present Malayo-Polynesian peoples—who now inhabit the Pacific islands, Southeast Asia, and the island country of Madagascar, which lies in the Indian Ocean off the southeastern coast of Africa.

During the centuries that these early Filipino peoples lived in comparative isolation on these islands, they developed a distinct culture of their own. This culture eventually became quite uniform throughout the islands, as there was constant mingling among the various island peoples. While there was much sharing, cultural differences also developed, which account for the great variety of languages and customs that distinguish one group of Filipinos from another. These early Filipinos also maintained active contacts through trade with the rest of Asia. In fact, Filipino society of that time can be called a trading culture, in which products of the sea and the forest were exported in exchange for the manufactured goods of China, Siam (Thailand), Khmer (Cambodia), and probably India. To a large extent the Philippines today is still a trading society. Nevertheless, in common with other developing countries of the world, the Philippines hopes to become an industrial, manufacturing society.

Relations with the outside world became more active in the 13th and 14th centuries, when the Islamic religion began to spread through the islands, first in Mindanao and Sulu and later in the central and northern islands. The new religion, brought by Arab and Indian traders, replaced native belief in spirits. In the 16th century, the first European

(Spanish) expeditions arrived. Ferdinand Magellan planted the flag of Spain on Philippine soil in 1521, and later, in 1565, Miguel López de Legazpe completed the conquest of Luzon and the Visayas.

As a result of these different influences, the Filipinos were divided into several groups. One group consisted of those who were conquered, converted to Catholicism, and administered by Spanish civil and church authorities. These people lived mainly in settled communities—villages called barrios—and were wet-rice farmers. Then there were those who were never conquered and retained their native beliefs. Most of the people in this group lived in the mountains and hinterlands of the major islands. A third group of people, who also were never conquered, lived under four sultanates and followed the Muslim faith. They made their living as fishermen and traders. Among Christian Filipinos social and economic classes developed, similar to those in pre-industrial Europe.

On top of the original ethnic differences came new ones. An important group arose known as the mestizo, or half-breed. The first mestizos were a mixture of Filipino and Spanish or Filipino and Chinese, and they had important roles in society. Later other minority groups, including Indians, Europeans, and Americans, mixed with the native population.

Spanish rule had a profound effect on the Filipino people. The Catholic religion jarred them loose from their original cultural moorings, leading to a present-day crisis of identity. Economic and status differences between different Filipino groups resulted in a problem of unity. A similar situation developed between the Christianized Filipinos and various cultural minority groups. On the other hand, Spanish rule had the effect of creating a sense of nationalism, based on a consciousness of kind and the common cause of freedom.

The period of American rule intensified the Filipinos' sense of nationhood. While contact with Spanish civilization merely gave the people a glimpse of modern life, the American experience presented them with the entire range of ideas and tools needed to create a modern society. While Spanish influence came mainly from the Catholic Church, that of the United States came by way of secular education and the great liberalizing influence of the American democratic political tradition. Thus the Filipino experienced a flowering of his modern political institutions under American teaching. Educational opportunities were opened to the Filipino masses; and this, in turn, opened new channels for citizen participation in national affairs. However, less attention was paid to economic development, as compared to the development of the political and legal aspects of society.

In the wake of independence, Filipinos faced their future with some uncertainty but also with courage and confidence. Independence and changing conditions in Asia have made Filipinos aware of two things: first, a need for more self-reliance in economic and other matters; and second, the importance of closer identification with other Asian nations, through economic co-operation and cultural exchange. Today Filipinos are embarked on a continuing program to correct the economic and social imbalances that still exist; to unify the nation; and to intensify efforts aimed at modernization and development. Unless these problems are solved, the Philippines will not be able to fulfill its hopes for a free, democratic, and prosperous society. Fortunately, the foundations for such a society have been laid.

The Ifugao rice terraces rise on mountain slopes.

PHILIPPINES

On July 4, 1946, the Stars and Stripes was lowered from its flagstaff in Manila, and the Republic of the Philippines was proclaimed an independent country. Nearly 50 years of American rule, coming after more than 300 years of Spanish rule, were at an end. The Philippines was the first Asian nation to rise as an independent state after World War II and is today one of the stablest republics in Asia.

THE LAND

The Republic of the Philippines is a slender network of islands in the Pacific Ocean, northeast of Borneo and south of Taiwan. The total land area of 115,830 square miles (300,000 square kilometers) is divided among 7,100 islands. The largest are Luzon in the north and Mindanao in the south. Fewer than 500 have an area of 1 square mile (2.59 sq. km.) or more. The whole coastline of the archipelago, with its myriad bays and gulfs, is longer than that of the United States. The ocean reaches one of its greatest depths, over 34,000 feet (10,400 meters), in the Mindanao (or Philippine) Trench on the eastern margin of the Philippines.

Mountains, which are part of a belt of volcanoes circling the Pacific, ridge the islands, and there are countless rivers and lakes in the valleys between the ranges. Highest of the peaks is Mount Apo, 9,690 feet (2,954 m.), a dormant volcano on Mindanao. The Philippines also lie in an earthquake belt.

Throughout the islands the rainy seasons vary, but there is an average rainfall of over 80 inches (200 centimeters). Temperatures average between 75 and 85 degrees Fahrenheit (24–30 degrees Celsius). In the summer and fall, typhoons, called *bagyós,* cause property damage.

Economy

The Filipinos are basically an agricultural people. At one time about 80 percent earned their living by farming. Today the figure is closer to 65 percent. The soil was never lush, and poor farming methods and heavy rainfall have caused erosion of more than half the land. Most farms are small, but in the plains of central Luzon there are large estates held by individual owners and worked by tenant farmers. A government land-reform program is aimed at ending this system.

The greatest crop is rice, the people's staple food. The central plain of Luzon is considered the rice granary of the Philippines. But in modern times the country has not produced enough rice to meet its needs and has had to import supplies. The crop of 1968 indicated, however, that the nation may now be able to produce all the rice it needs. The secret of the record crop is a new "miracle rice" developed at the International Rice Research Institute at Los Baños, near Manila. IR-8 (International Rice) withstands heavy winds and yields several times the normal crop.

For about a quarter of the people, corn replaces rice as a staple food. The major corn-growing areas are the central Visayan Islands, the Cagayan Valley of Luzon, and Mindanao.

More coconuts are grown in the Philippines than in any other coun-

A coconut plantation. The Philippines leads in world coconut production.

try, and the nation leads the world in the export of copra (dried coconut meat). Coconut oil, used in making soap and in other industries, is also a major export. Coconuts are grown almost everywhere on the islands, but eastern and southern Luzon are the main coconut-growing regions.

Sugar, grown mostly on the island of Negros, ranks among the principal exports. Abaca (Manila hemp), tobacco, and pineapples are other important cash crops.

Natural Resources

More than half the land is covered with forests, and the timber industry is an important source of income. No fewer than 3,000 species of trees are found in Philippine forests. Most famous of these is the narra (*Pterocarpus indicus*), called Philippine mahogany. Rubber plantations have been started in Mindanao, and pulp and paper production is a possible future industry.

No other archipelago in the world contains such plant wealth as the Philippines. Botanists say there are 10,000 species of ferns and flowering plants in the country, including nearly 1,000 different orchids. The country also teems with an abundance of animal life. Of special interest are three Philippine animals—the mouse deer, among the smallest deer known; the tarsier, a small animal related to the monkey, with big owl-like eyes; and the tamarau, an animal that looks like a dwarf buffalo and is found only in the islands. The ponderous, powerful carabao (water buffalo) is the favorite animal of the Filipinos. It draws the plow in the rice paddies, pulls *carretons* (bull carts), and drags logs; and its meat and milk are the farmers' food.

Lakes, rivers, and surrounding seas abound with more than 2,000 kinds of fish. Some of the largest and smallest fish in the world are found in Manila Bay. The mammoth whale shark (*Rhincodon typus*) is more than 50 feet (15 m.) long, while the tiny Philippine goby (*Pandaka pygmaea*) is 11.5 millimeters long—less than ½ inch.

A great wealth of minerals is found in the Philippines, and mining is a leading industry. Copper, gold, iron ore, manganese, chromite, and silver are mined, and copper smelting has been introduced.

PLACES OF INTEREST

Everywhere one looks there are beathtaking views: towering mountains; mist-capped volcanoes; rolling valleys carpeted with coconut groves or fields of rice, sugarcane, or tobacco; rivers bordered by bamboo and flowering plants; foaming waterfalls; and sparkling lakes.

A visitor to Luzon can discover a number of places of interest on a drive south from Manila. He will pass Las Piñas, with its unique century-old organ made of bamboo, and the quaint lakeshore town of Calamba, birthplace of José Rizal (1861–96), a physician, man of letters, and valiant nationalist, who is the hero of the Philippines. On the same drive the visitor can see Los Baños, with its hot springs, and Pagsanjan Falls, 200 feet (60 m.) high. Mount Mayon, the world's only volcano with a perfect cone, rises in the southeastern part of the island.

About 180 miles (290 kilometers) north of Manila is the mountain resort of Baguio, its air scented with strawberries and pine. Not far away are the famous Ifugao rice terraces, built on the slopes of the high sierras 2,000 years ago and rising like a giant's staircase into the clouds.

FACTS AND FIGURES

REPUBLIC OF THE PHILIPPINES is the official name of the country.

CAPITAL: Manila.

LOCATION: Southeast Asia. **Latitude**—4° 38′ N to 19° 35′ N. **Longitude**—116° 57′ E to 126° 36′ E.

AREA: 115,830 sq. mi. (300,000 sq. km.).

PHYSICAL FEATURES: Highest point—Mount Apo (9,690 ft.; 2,954 m.). **Lowest point**—sea level. **Chief rivers**—Cagayan, Pampanga, and Agno on Luzon; Agusan on Mindanao. **Major lakes**—Laguna de Bay and Lake Taal on Luzon, Lake Lanao and Lake Mainit on Mindanao.

POPULATION: 37,000,000 (estimate).

LANGUAGE: Tagalog (national), English, Spanish, local languages.

RELIGION: Roman Catholic, Philippine Independent Church (Aglipayan), Iglesia ni Kristo ("church of Christ"), Protestant, Muslim.

GOVERNMENT: Republic. **Head of state**—president. **Head of government**—prime minister. **Legislature**—National Assembly. **International co-operation**—United Nations, Association of Southeast Asian Nations (ASEAN), Asian Pacific Council (ASPAC).

CHIEF CITIES: Manila, Quezon City, Cebu, Davao, Iloilo, Bacolod.

ECONOMY: Chief minerals—copper, gold, iron, chromite, manganese. **Chief agricultural products**—rice, corn, coconuts, sugar, abaca, tobacco, pineapples. **Industries and products**—construction, fishing, forestry, mining; cigar and cigarette making; processing of sugar, rice, coconut oil. **Chief exports**—copra, sugar, timber, abaca, coconut oil, copper, chromite, manganese. **Chief imports**—machinery, mineral fuels and lubricants, transport equipment, cereals, dairy products.

MONETARY UNIT: Philippine peso.

NATIONAL HOLIDAYS: June 12, Independence Day; July 4, Philippine-American Friendship Day.

NATIONAL ANTHEM: "Philippine National Anthem," beginning *Tierra adorada* ("Land of the morning").

Cities

The oldest city in the Philippines is the port of Cebu, founded in 1565, on the east coast of Cebu island. Davao on Mindanao is the hemp metropolis of the south. Bacolod on Negros is a major sugar center, and Iloilo on Panay is an important port. Quezon City, the country's second largest city and the capital from 1948 to 1976, is just 10 miles (16 km.) northeast of Manila. It has a population of over 500,000.

Manila. The Philippines' capital and largest city is Manila, which is also one of the great cities of Asia. A gay metropolis of about 1,500,000 people, it is the chief port and the commercial and cultural center of the islands. In 1975 Manila and many nearby communities, including Quezon City, were joined into a greater Manila, known as Metro Manila, which has about 5,000,000 people. Manila was once ruled by the Muslim warrior Rajah Soliman. In 1571 the Spaniard Miguel López de Legazpe captured it and rebuilt it as a Christian city.

The Pasig River flows through Manila, the largest city in the Philippines.

Manila Bay is said to be the finest harbor in the Far East. At its entrance is the rocky island-fortress of Corregidor. Opposite this island is the peninsula of Bataan. The sunsets over Manila Bay are among the most beautiful in the world. At twilight the Luneta esplanade along the shore of the bay is thronged with people who come to watch the flaming sunsets. Near the Luneta is Intramuros, a section of Manila that was once a walled medieval Spanish city. It was almost totally destroyed in World War II, and little of it is left today but the walls, 20 feet (6 m.) thick, and the Church of Saint Augustine, the oldest church in the islands. A beautiful capitol, built after the war, stands a stone's throw from the walls. Across the Pasig River, which runs through Manila, is the Escolta, the city's famous business thoroughfare. On the north bank of the Pasig is palatial Malacañang, the official residence of the president.

Manila is a city of 12 universities, the oldest of which is the University of Santo Tomás, founded in 1611. The University of the East, founded in 1947, has an enrollment of 65,000, making it the largest university in Asia. A 40-acre (16 hectares) tract of land extending into Manila Bay is the site of the Cultural Center of the Philippines, which opened in the summer of 1969.

THE PEOPLE

The Philippine population of over 37,000,000 is concentrated in a few areas. The islands of Luzon, Negros, Panay, and Cebu, for example, are crowded, while the large island of Mindanao is thinly settled.

A fishing village. Fish is a major source of food in the Philippines.

The majority of Filipinos are descendants of people of Malay stock who migrated to the islands, beginning thousands of years ago, from Malaysia and Indonesia and from the mainland of Asia. In modern times groups of Chinese, Spaniards, and Americans moved to the Philippines, and many married Filipinos. The children of these mixed marriages are known as mestizos. There is also a small group of people, known as Negritos (Pygmy blacks), who live in remote areas and have been in the islands longer than any other group.

The Philippines is the only predominantly Christian nation in the Far East. More than 90 percent of the people are Christians, mostly Roman Catholics. Other Filipinos belong to Protestant churches or to one of two churches found only in the republic: the Philippine Independent, or Aglipayan, Church, founded in 1902; and the Iglesia ni Kristo, or "church of Christ," founded in 1914. About 4 percent of the people are Muslims. They live in southern Mindanao and the Sulu Archipelago. There are also a small number of Buddhists and some groups who have their own ancient forms of worship.

Family ties are strong in the Philippines, and the word "family" has a broad meaning. Children are close not only to their parents and brothers and sisters, but to grandparents, uncles, aunts, and cousins as well.

Women are held in the highest regard. Politically, socially, and economically, they are the equal of men. They were the first among Asian women to have the right to vote, to be elected to public office, and to be educated for the professions.

The contrast between life in the cities and life in small villages is marked. The cities are bustling, expanding centers of business and industry. Families live in apartment houses and modern homes in the city itself, or in the suburbs. But three quarters of the people live outside the cities and suburbs. The smallest farming village, or barrio, consists of a few houses grouped together near the fields. The *población,* or town, has schools, shops, a town hall, a church, and a market.

A typical one- or two-room Filipino village house has walls of nipa or coconut palm leaves and bamboo, a thatched roof, and a bamboo floor. The windows are square openings in the walls. Because of the dampness, houses are commonly raised on poles above the ground. The space under the houses can be used for storage or to shelter animals. In the Sulu Archipelago, people build houses on stilts in the water and keep their outrigger canoes under their homes.

Fish and rice are the basis of the diet, but in spite of the fish in the waters all around them, Filipinos depend on some imports. Ocean fishing has not yet developed enough to supply the country's needs. Besides rice, people eat corn and a number of root crops (sweet potatoes, Irish potatoes, manioc, or cassava, and taro). The root vegetables are boiled and served as they are, or with meat, fish, or other vegetables. Tropical fruits include mangoes, bananas, papayas, jackfruits, and star apples.

In the cities people dress in fashionable Western clothes, and even in small villages the basic style of dress is Western. Women wear skirts and blouses or dresses, and men wear loose shirts and trousers. But for fiestas (religious festivals), both men and women wear traditional costumes. The Filipino farmer, who goes barefoot the rest of the year, puts on shoes and his best *barong tagalog,* the shirt that is the national costume for men. It is generally made of silky piña cloth, a fabric woven of pineapple fiber. The shirt has a starched collar, embroidered front panel, and long cuffed sleeves. It is worn outside the trousers.

The gala dress for women is the *balintawak,* a two-piece dress with a *camisa* (top) made of hemp fiber dyed in gorgeous colors. Ruffled fan-like pieces fit over the shoulders like butterfly wings. The *saya* (skirt) is made of cotton and covers the ankles. For special occasions, well-to-do Filipino women may wear another national costume, the *mestiza terno.* It has ruffled wing pieces like the *balintawak* and a skirt that fans out gracefully and touches the floor.

Three main languages, English, Spanish, and Tagalog, are spoken in the Philippines, and more than 80 dialects are in use as well. The Filipinos take great pride in being the only people in Asia who speak English universally. Educated Filipinos also know Spanish. Tagalog, which was adopted as the national language in 1937, must be taught in all schools. It is now spoken throughout the archipelago.

Public education became firmly established under the Americans, and the main language of instruction is English. Boys and girls must all go to school until they finish fourth grade. After that, they may attend intermediate school for 2 years and high school for 4 years.

In school programs and in public concerts Filipinos sing the folk songs and dance the folk dances of old Malaysia. Sweetest of the ancestral songs is the *kundiman,* with which young men serenade their sweethearts. Of the numerous folk dances the *tinikling* ("jumping bird") is the most popular.

Bayanihan dancers do the traditional "tinikling" in view of Malacañang Palace.

The Filipinos are at their happiest during a fiesta. Every barrio, town, and city has its own patron saint, whose feast day is marked with a celebration. Scarcely a day passes without a fiesta somewhere in the Philippines. During a town fiesta the streets echo with the blare of brass bands or with gay tunes of *musikong buho* (musicians playing bamboo instruments). But when the Angelus sounds, a solemn procession of people carrying lighted candles winds around the town. Trailing them is a carriage bearing the image of the patron saint, accompanied by the town priest, prominent citizens, and the brass bands. After the procession come the fireworks and the *moro-moro,* a drama re-enacting the old wars between the Christians and Muslims (Moros).

HISTORY

The oldest skeletal relic of man so far unearthed in the Philippines is a fossilized human skull, 22,000 years old. This relic supports the theory that as early as 25,000 years ago the Negritos began their migration to the Philippines from the Asian mainland. They walked across the land bridges that then linked Asia and the fringing archipelagoes. From 5000 to 1500 B.C. the Indonesians came in outrigger canoes in two migratory waves. They were followed by the sea-roving Malays, who arrived in various waves of migration from 200 B.C. to A.D. 1500.

The Spanish period in Philippine history began when Ferdinand Magellan, a Portuguese-born navigator in Spain's service, reached the Philippines on March 16, 1521. Magellan was killed while invading Mactan, an island near Cebu, in April, 1521. In the years that followed, Spain dispatched a series of expeditions to the East. It was Ruy López de Villalobos, the leader of an expedition in 1542–43, who named the islands Las Filipinas (the Philippines) in honor of Prince Philip of Asturias, who later became Philip II of Spain. Miguel López de Legazpe, who led the last expedition (1564–65), established a settlement in Cebu and laid the foundation of Spanish sovereignty in the Philippines.

The Rise of Filipino Nationalism and the Philippine Revolution

Before the Spanish period, the inhabitants of the Philippines were divided into tribes, each with an independent ruler. Although it was never its intention to foster Filipino nationalism, Spain nonetheless contributed to the welding of the diverse and warring tribes into one nation. Spain gave the multi-island country one name (the Philippines), a centralized government under one sovereign (the Spanish monarch), a common religion (Christianity), and a common culture and language (Spanish). Regarding Spain as a common enemy also brought the different Filipino tribes into closer union.

The execution of three priests on February 17, 1872, set fire to Filipino nationalism. The men died in defense of Filipino rights and were acclaimed as martyrs. Foremost among the leaders who rallied to carry on the fight was José Rizal, whose two novels, *Noli Me Tangere* (*Touch Me Not*) and *El Filibusterismo* (*Treason*), paved the way for the revolution against Spain.

Inspired by Rizal's ideas, Andres Bonifacio, a hero of the common people, founded a secret revolutionary society. In August, 1896, Filipinos in all parts of the archipelago rose to fight under the red banner of this society. General Emilio Aguinaldo of Cavite later became their leader.

Spain rushed troops and generals to Manila, and the Spanish Governor-General ordered the execution of Rizal in Manila on December 30, 1896. The revolution raged even more furiously. The Spanish authorities persuaded General Aguinaldo to leave the country and go to Hong Kong in exchange for a huge sum of Spanish gold and a promise of reforms. Meanwhile, the Spanish-American War broke out in April, 1898.

The American Years

On May 1, 1898, the United States Asiatic Squadron under the command of Commodore George Dewey destroyed the Spanish fleet in Manila Bay. On May 19 General Aguinaldo returned from Hong Kong and resumed the revolution. He proclaimed the independence of the Philippines on June 12.

As a result of United States victories in the Philippines and Cuba, Spain sued for peace. By the terms of the Treaty of Paris, signed on December 10, 1898, Spain gave up its sovereignty in Cuba and ceded the Philippines, Guam, and Puerto Rico to the United States. The Filipinos spurned the Paris treaty, however, and refused to accept American rule. On January 23, 1899, they established the First Philippine Republic, with General Aguinaldo as president. United States soldiers opposed the Filipino patriots and in March, 1901, captured President Aguinaldo. In April, 1902, the Filipinos surrendered.

After 3 years of military rule (1898–1901), civil government was established. William Howard Taft, who later became the 27th president of the United States, served as the first civil governor. (In 1904 the title was changed to Governor-General.) In 1916 the United States Congress enacted a law creating an all-Filipino legislature of two houses. The preamble to the law promised independence to the Filipinos as soon as they were prepared to have a stable government.

Senate President Manuel L. Quezon and House Speaker Sergio Osmeña were the leaders of a peaceful crusade for Philippine independence. From 1919 to 1934, 12 missions went to the United States to remind the government of its promise to grant independence to the Philippines. In 1934 the United States passed the Tydings-McDuffie Act, known as the Philippine Independence Act, which provided that the islands would become independent on July 4, 1946, after a 10-year transition period as a commonwealth. The Commonwealth of the Philippines was established on November 15, 1935, with Quezon as president and Osmeña as vice-president.

When war broke out between Japan and the United States in December, 1941, Filipinos, under the command of General Douglas MacArthur, fought side by side with American troops in Bataan and Corregidor. Manila, although declared an open city, was occupied by the Japanese on January 2, 1942. For 3 dark years the Filipino people suffered under the Japanese occupation, while Filipino guerrillas carried on the fight against the enemy. In October, 1944, General MacArthur returned from Australia, to which he had escaped, and crushed the Japanese in an island-to-island campaign. By July 4, 1945, the Philippines was liberated.

The Republic of the Philippines

On July 4, 1946, American rule ended, and the Philippines became an independent republic with Manuel A. Roxas as president and Elpidio Quirino as vice-president. In the years since independence the Philippines has had to face a number of serious problems. One of these is the Filipino love of politics. Politics permeates the armed forces, diplomatic service, public works projects, and government hospitals. Politicians have become so powerful that they can place their friends, no matter how poorly qualified, in good jobs and can even protect dishonest men.

The Huks are a second problem. They were formerly guerrilla fighters of the Hukbong Bayan Laban sa Hapon ("people's army against Japan") and are now peasant Communists who want to overthrow the republic. Ramón Magsaysay, who later became president, suppressed the Huks in the early 1950's. But in 1968, during the administration of President Ferdinand E. Marcos, the Huks again became active.

The upsurge of crime is another grave problem. In addition, the islands face the threat of overpopulation. The present population is about 37,160,000, but the annual rate of increase, over 3 percent, is one of the highest in the world.

Finally, a territorial conflict exists between the Philippines and Malaysia over the state of Sabah in northern Borneo. The Philippines claims Sabah, although it is now a part of Malaysia. What happened is that in 1946 the territory was ceded to Great Britain, and when the Federation of Malaysia was formed in 1963, Sabah became part of it. The Philippines continues to press its claim but Malaysia rejects it.

Government. Under its 1935 Constitution the Philippines had a presidential form of government patterned on that of the United States. In 1972 President Ferdinand E. Marcos, in an effort to end corruption and crime as well as radical left-wing attempts to overthrow his government, declared martial law. A new constitution was adopted giving the Philippines a parliamentary form of government. Marcos continued to head the country under the new constitution.

GREGORIO F. ZAIDE, Far Eastern University, Manila
Author, *Political and Cultural History of the Philippines*

Tibet, the Roof of the World.

TIBET

An autonomous region of Communist China, Tibet is a place of vast distances and great heights. It has been called the Roof of the World and the Land of Snow. For centuries it was a land of mystery, isolated from the world by the great mountains surrounding it.

According to Tibetan mythology, a monkey was sent by Avalokitesvara (the Compassionate), the patron god of Tibet, to fulfill himself by meditation in the Land of Snow. Seduced and threatened by a mountain ogress (a mythical monster or demon) disguised as a beautiful woman, he decided to sacrifice his own salvation by marrying her, so that she would not have to marry one of her own kind and breed more ogres and ogresses who would prey on every living creature in the Land of Snow. So, as the myth goes, the Tibetan people are the offspring of a saintly monkey and a beautiful demon.

THE LAND

Tibet is an extremely high plateau, averaging about 15,000 feet (4,500 meters) in elevation, surrounded by the world's highest mountains. The Kunlun mountains are in the north and the Himalayas in the south. To the west, between the Kunlun and the Himalayas, is the Karakoram mountain system. In the east three great rivers, the Yangtze, the Mekong, and the Salween, cut deep parallel ravines through the mountains, which are relatively low there. Tibet is the source of other great Asian rivers—the Brahmaputra (called Tsangpo in Tibet), the Indus, and the Sutlej.

The Tibetan plateau comprises an area of approximately 500,000 square miles (1,300,000 square kilometers) and is made up of three land regions: the Tsangpo Valley, Chang Tang, and Kham. The Tsangpo Valley is the economic, cultural, religious, and political center of Tibet. All three of Tibet's largest cities—Lhasa, Shigatse, and Gyangtse—are located in this region. A windswept highland, Chang Tang, or "northern plain," is the highest of Tibet's regions. Chang Tang is barren and cold, with many salt lakes. The eastern part of Tibet is known as Kham. Kham gets more rain than the other regions, and has large forests, abundant grazing land, and rich mineral deposits. Although deep ravines formed by the great rivers and their tributaries make the building of roads difficult, Kham is heavily populated. Many people live on the high mountain slopes.

Climate.   Much of Tibet is dry and cold most of the year. With the exception of the Kham region and parts of the Tsangpo Valley, the country rarely gets more than 10 inches (25 centimeters) of rainfall per year. Harsh winds, blizzards, and snowstorms are common. Temperatures in the capital, Lhasa, are moderate. However, summer temperatures of 90 degrees Fahrenheit (32 degrees Celsius) and winter temperatures of –40 degrees F. (–40 degrees C.) have been recorded in other parts of Tibet. Climate varies with elevation and the position of the mountains. Even in the same region, temperatures may vary from place to place.

Animal Life.   Wild animals, such as the giant panda, the musk deer, and the kiang (Tibetan wild ass), roam the country. Domesticated animals include yaks (a shaggy, long-haired variety of ox), sheep, goats, ponies, mules, and the Tibetan mastiff (a large dog). The most important animal is the yak. It is the main beast of burden and a source of food and shelter. Yak hair is woven into tents. Sheep and goats are also raised, for food and for their wool and hides.

Economy.   Tibet is an agricultural country. The most common grains and vegetables grown include wheat, buckwheat, barley, peas, and beans. Recently, millet, maize, and some 42 kinds of vegetables have been brought to Tibet for experimentation. Most of them are reported to have grown well in this high elevation. Until recently a small number of Tibetans were engaged in metal-working, weaving, pottery-making, and wood carving. Since the late 1950's coal mining has become the most important industry. Tibet exports wool, yak hair, hides, and musk and imports tea, cotton goods, and household wares. Mineral deposits include salt, borax, gold, iron, coal, and oil.

Traditionally, transportation was the most difficult problem in Tibet. Before 1952, wheeled vehicles were never used, and animals were the only means of transportation. Since then, several thousand miles of highways have been completed, and a railroad and air service links Tibet with other parts of China.

THE PEOPLE

The plateau is the homeland of over 1,000,000 Tibetans, although more live in the surrounding regions. The Tibetans belong to the great Mongoloid stock, which includes peoples native to East and Southeast Asia. However, many ethnologists—scientists who study the racial divisions of mankind—claim to have found two major types of Tibetans: one is tall and long-headed and lives in the north and east, especially in Kham; the other is short and round-headed, with high cheekbones and a flat nose. The spoken language of Tibet belongs to the Tibeto-Burmese family. The Tibetan alphabet is based on Sanskrit, an ancient Indian language.

Perhaps the most unusual thing about the Tibetan people is their religion, Lamaism, the Tibetan version of Buddhism. About 3,000 lamaseries, or monasteries, served as centers of learning and trade as well as of religion. The head of the Lamaist church, the Dalai Lama, who is held to be the reincarnation of Avalokitesvara, is the supreme political and spiritual ruler of Tibet. The present Dalai Lama (the 14th) has been in exile in India since 1959 as a result of an unsuccessful rebellion against the Chinese Communists.

Before the 1959 revolt there were some 100 noble families in Tibet. Most of them owned large estates and engaged in trade. They shared political power in the state with high-ranking lamas. Most Tibetans were peasant farmers who did not have land of their own but worked as tenants or as hired laborers on land owned by the government, monasteries, high lamas, or noblemen. After 1959, all large landowners were purged by the Communist authorities under the so-called democratic reform campaign. Their land and other properties have been confiscated and distributed to poor peasants.

There are several thousand nomads in Tibet. Dressed in bulky sheepskin clothing, the nomads roam about the northern plateau of Tibet with their flocks of yaks, sheep, horses, and camels. During the coldest months they move southward to the Tsangpo River valley to get such necessities as tea and barley. At the first sign of spring they leave for the north again. The nomads live in rectangular tents made of yak hair.

In Tibetan marriages monogamy (one husband and one wife) is the usual practice. But polyandry (one wife with more than one husband) is popular, especially the type in which several brothers share one wife. When a child is born in such a marriage the most prosperous and socially important husband will be considered the father and the rest uncles.

The basic food of Tibetans is parched barley (*tsamba*). Yak meat, mutton, butter, and cheese are eaten by those who can afford them. The most common beverages in Tibet are tea and beer. Tibetan tea is churned with butter and salt. Many Tibetans drink 50 cups of tea a day.

Major Cities.   The capital of Tibet, **Lhasa** (population about 170,000), lies in a sheltered valley near the Tsangpo River in southern Tibet. Lhasa is Tibet's main trading city and the center of Lamaism. Northwest of the city is the Potala, a gold-roofed, many-windowed palace, the traditional seat of the Dalai Lama. Tibet's second largest city, **Shigatse**, is in the southeastern part of the country. The famous lamasery Tashi Lumpo, headed by the Panch'en Lama (who ranks second to the Dalai Lama), is situated nearby. **Gyangtse**, the third largest city, is known for its handloomed woolen cloth and its carpets.

FACTS AND FIGURES

CAPITAL: Lhasa.

LOCATION: Central Asia (southwestern China). **Latitude**—27° 20′ N to 36° 30′ N. **Longitude**—78° 24′ E to 98° 57′ E.

AREA: Approximately 500,000 sq. mi. (1,300,000 sq. km.).

PHYSICAL FEATURES: Highest point—Mt. Everest, on the Tibet-Nepal border (29,028 ft.; 8,484 m.). **Chief rivers**—Yangtze, Mekong, Salween, Brahmaputra (Tsangpo), Indus, Sutlej. **Major lakes**—Nam Tso, Manasarowar, Rakas.

POPULATION: 1,000,000 (estimate).

LANGUAGE: Tibetan.

RELIGION: Lamaist Buddhist.

GOVERNMENT: Autonomous region of China.

CHIEF CITIES: Lhasa, Shigatse, Gyangtse.

ECONOMY: Chief minerals—coal, salt, borax, gold, iron, oil. **Chief agricultural products**—wheat and other grains, vegetables, milk and milk products. **Industries and products**—mining, animal husbandry (especially yaks and sheep), handicrafts, meat, yak hair and hides, wool, musk. **Chief exports**—wool, yak hair, hides, musk. **Chief imports**—tea, cotton textiles, household goods.

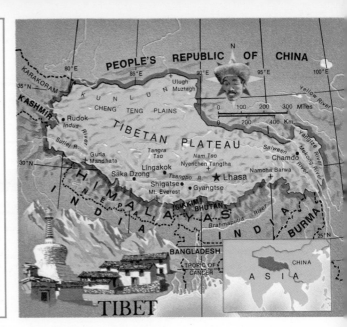

TIBET

HISTORY AND GOVERNMENT

Around the year A.D. 620, Songtsan Gampo, a tribal chieftain of Tibet, established a kingdom in the Tsangpo River valley. Soon he became so great a threat to the T'ang dynasty of China as to force the emperor to send him a Chinese princess, Wen Chen, to be his wife. By the 8th century, Tibet was a military power in Central Asia.

Because of internal strife, Tibet's power declined during the next 3 centuries. Although there was internal weakness, the high mountains protected Tibet from invaders until it was finally conquered in the 13th century by the Mongols, who overran all of Central Asia.

In 1270 Kublai Khan, the Mongol conqueror who became the emperor of China, adopted Lamaism as the state religion of his empire. He appointed a lama as priest-king of Tibet. For the next several centuries Tibet was governed by its lamas under the protection of China.

Toward the end of the 19th century, China itself became weak and was unable to rule Tibet effectively. In 1904 a British expedition made its way to Lhasa. This led to Tibet's granting trading rights to Great Britain. In 1913–14 a conference was held at Simla in northern India to discuss relations between Great Britain, Tibet, and China. One of the results of this meeting was the McMahon Line (named for Sir Arthur Henry McMahon, the British delegate to the conference), which marked the boundary between India and Tibet. The Chinese delegate, however, refused to sign the treaty, and the disputed McMahon Line has led to border clashes between India and China in recent years.

Until 1950 Tibet was ruled by the Dalai Lama as an independent country. In October, 1950, the Chinese Communist Army entered Tibet. By an agreement reached the following year Tibet became a self-governing region of China. In reality, however, the Dalai Lama had no real power. In an effort to free themselves from Chinese Communist oppression, the Tibetans revolted in 1959. But the attempt was quickly crushed. The Dalai Lama, who was implicated in the revolt, escaped to India. The Tibetan Government was dissolved and actual control of Tibet was placed in the hands of a representative of the Peking Government.

TAO CHENG, Trenton State College

A Mongol herdsman on the vast Mongolian plateau.

MONGOLIA

The name Mongolia is commonly used to refer either to the Mongolian People's Republic or to the vast plateau in Central Asia of which the Mongolian People's Republic is a part. The plateau is the traditional home of the Mongol people, although many places are now inhabited by peoples other than Mongols and Mongol nomads roam over areas outside the plateau. In the 13th century the various Mongol tribes, united under Genghis Khan, swept down from their plateau homeland to carve out a great empire that eventually stretched from China to eastern Europe. But the empire of the Mongols did not last, and today the plateau is divided among three nations: the Mongolian People's Republic, China, and the Soviet Union.

THE LAND

The Mongolian plateau is made up of three general regions. The northwest is a massive mountain complex dotted with river basins. The central region consists of the Gobi, a great desert. The southeast contains many mountain chains plus an extensive zone of lowlands. The Mongolian People's Republic (Outer Mongolia) comprises a little less than half the plateau, with an area of 604,248 square miles (1,565,000 square kilometers). It is surrounded by Chinese territory to the south and east and Soviet territory to the north.

Outer Mongolia has an over-all elevation of over 5,000 feet (1,500 meters). The climate is marked by generally limited rainfall and extremes in temperature. Even in the same place temperatures vary greatly from one season to another. In Ulan Bator, the capital, the average tempera-

A nomad family. Their white tents are called yurts.

ture in January is 17 degrees below zero Fahrenheit (−27 degrees Celsius), while in July it is 64 degrees F. (18 degrees C.).

Vegetation, Animal Life, and Mineral Resources. With the exception of the Gobi, most of Mongolia is prairie and grassland, which provides the Mongols, who are chiefly stockbreeders, with excellent pastures. Mountains are generally treeless, except in the northwest.

The most important domesticated animals are camels, horses, cattle, yaks, sheep, and goats. Mongolian camels are the two-humped, or Bactrian, variety. The horses are small but compact and strong, with abundant tail and mane. Camels and horses are the traditional and most important forms of transportation, although rail and air service now links Mongolia with the Soviet Union and China. All domestic animals must be hardy to endure the harsh environment. Wild animals include deer, gazelles, and antelope, the brown bear and snowshoe hare, wolves, badgers, wolverines, and lynx, and such valuable furbearing species as sable, ermine, marten, and fox. Minerals found in Mongolia include coal, iron, oil, copper, lead, silver, tungsten, and gold.

THE PEOPLE

Of the approximately 3,000,000 Mongols in the general region, a little over one third live in the Mongolian People's Republic. Most of the rest live in the Chinese Inner Mongolian Autonomous Region (Inner Mongolia), where they make up about 20 percent of the population. The balance, some several hundred thousand, live in the Buryat Autonomous Republic of the Soviet Union. Non-Mongols who live in different parts of the plateau include various Turkic peoples (Khotons, Kazakhs, Tuvinians, Uighurs), Tungus, Chinese, and Russians.

The Mongols belong to one of the major racial divisions of mankind, the Mongoloid, of which they are the classic type. They have round heads, oval or round faces, high cheekbones, and flat noses. Their eyes are not deeply set and have a slanted appearance. The color of their skin ranges from near-white to brown with a yellow tinge, which has given rise to the term "yellow race." Their hair is dark and coarse on the head, but so light and thin on the body as to give an appearance of hairlessness. They are generally stocky in build. The Mongols are divided into numerous tribes with different dialects, customs, and dress. The major tribal group in the Mongolian People's Republic is the Khalkhas in the east, who number about 700,000. Smaller groups are the Oirats in the west, and the Buryats, who are not native to the country but emigrants from the Soviet Union. Of the Turkic groups, the Khotons are farmers while the Kazakhs and Tuvinians are stockbreeders like the Mongols. There is also a significant likeness in appearance between the Kazakhs, Tuvinians, and Mongols. The Tungus are hunters and nomads. Most live in Siberia and only small groups are found in Mongolia.

The Mongol Way of Life. Traditionally the Mongol was a nomad, who traveled with his herds of sheep, goats, horses, camels, and cattle from season to season seeking pasture. His home, the yurt, a circular tent made of a wooden frame covered with felt, was easily portable and suited to the nomadic life. His herds provided most of his food—meat, butter, cheese, and *airag,* or fermented mare's milk. In recent years many of the Mongolian nomads have settled down in various parts of the plateau and have given up their traditional wandering life. This is partly because of the development of industries and partly due to the deliberate policy of the government.

The most popular sport among the Mongols is wrestling, and the matches are often held with great fanfare. Archery is also popular, as is horse-racing. The Mongols have always been renowned for their horsemanship, and riding is still a necessary skill, which is acquired early by the children. They usually become expert riders by the age of 5.

Religion and Language. Three great religions have held sway in the Mongolian plateau—shamanism, Islam, and Lamaism. Their influence on the life of the common people is still felt, even though the official policy of the Communist government is to discourage religion.

Shamanism, an ancient animistic faith, was the main religion of Mongolia before Lamaism. It had a total of 99 gods, who were highly personified in the myths and legends of the people. Islam is the religion of two of the Turkic peoples, the Kazakhs and the Khotons. Lamaism, the Tibetan form of Buddhism, was introduced into Mongolia in the 16th century and became the state religion. For more than 3 centuries the Jebtsun Damba Khutukhtu ("living Buddha") was both the spiritual and secular ruler of the Mongols. The last Living Buddha died in 1924, and no successor has been permitted by the government since.

The Mongolian language is made up of various dialects. The most important is the one spoken by the Khalkhas, and it is understood by all other Mongol tribal groups. It is the official language of the Mongolian People's Republic and is now written in the Russian (Cyrillic) alphabet. The influence of the Soviet Union in Mongolia is strong. The Russian language is taught in schools and is used as a medium of communication between the two countries on higher governmental levels.

Economy. Livestock is still the foundation of the Mongolian economy. It is also the basis of much of the country's industry. Coal mining is

MONGOLIA

FACTS AND FIGURES

MONGOLIAN PEOPLE'S REPUBLIC—Bügd Nairam-dakh Mongol Ard Uls—is the official name of the country.

CAPITAL: Ulan Bator.

LOCATION: Central Asia. **Latitude**—41° 36′ N to 52° 09′ N. **Longitude**—87° 47′ E to 119° 54′ E.

AREA: 604,248 sq. mi. (1,565,000 sq. km.).

PHYSICAL FEATURES: Highest point—Tabun Bogdo in the Altai Mountains (15,266 ft.; 4,653 m.). **Lowest point**—Lake Khukhu-Nur (1,745 ft.; 532 m. above sea level). **Chief rivers**—Selenga, Kerulen, Onon, Tes, Kobdo, Dzabkhan. **Major lakes**—Ubsa Nor, Khara Usu, Airik Nor, Kirgis Nor.

POPULATION: 1,250,000 (estimate).

LANGUAGE: Khalkha Mongolian (official).

RELIGION: Lamaist, Muslim.

GOVERNMENT: Communist republic. **Head of state**—chairman of the presidium of the Great People's Khural. **Head of government**—premier. **Legislature**—Great People's Khural. **International co-operation**—United Nations, Council for Mutual Economic Assistance (COMECON).

CHIEF CITIES: Ulan Bator, Choibalsan, Uliassutai, Tsetserlik, Kobdo, Sukhe Bator.

ECONOMY: Chief minerals—coal, petroleum, alabaster, fluorspar, gold, tungsten, copper, lead, salt, silver, uranium. **Chief agricultural products**—wheat, oats, barley, millet, vegetables, potatoes, hay, livestock (sheep, goats, horses, cattle, camels, pigs). **Industries and products**—processed foods (milk, butter, cheese), leather goods, textiles, hides and skins, soap, bricks, meat packing. **Chief exports**—cattle, wool, hides, butter, meat products. **Chief imports**—consumer goods, machinery.

MONETARY UNIT: Tughrik.

NATIONAL HOLIDAY: July 11, People's Revolution Day.

NATIONAL ANTHEM: *Bügd Nairamdakh mongol ulsyn suld duulal* ("Anthem of Our Country, the Free Revolutionary Land").

the most important mineral industry. Agricultural products are mainly grains, millet, oats, and barley. Economic aid is provided by the Soviet Union, with which Mongolia does most of its trading.

Cities. Ulan Bator, the capital and largest city of the Mongolian People's Republic, has a population of over 200,000. It is the economic center of the country and the seat of government. With Russian aid, Ulan Bator has developed into a modern city with broad avenues, a university, museums, hotels, housing developments, and an airport. The city also has some ancient places of interest, including the Gandun Monastery, which once housed thousands of Buddhist monks. Other Mongolian cities include Choibalsan, Uliassutai, Tsetserlik, and Sukhe Bator.

HISTORY AND GOVERNMENT

Many kingdoms and peoples rose and declined in the Mongolian plateau. But it was not until Genghis Khan united the various warring

Ulan Bator, the capital of the Mongolian People's Republic.

tribes in the 13th century that the Mongols created their great empire. After the death of Genghis Khan in 1227, portions of the empire were administered by his sons and grandsons under the leadership of one son elected as khan, or ruler, of all the Mongols. By 1260 the empire had been divided into several khanates, and its unity had been destroyed. The chief khanate, comprising China and Mongolia, was ruled from Peking by Kublai Khan, a grandson of Genghis Khan. It was known as the Yüan dynasty in China and lasted until 1368. Later Mongol leaders aspired to restore the old glory of Genghis Khan's empire, but their aspirations always ended in petty struggles for power among themselves.

After the Manchus conquered China in 1644, they gradually extended their power to Mongolia. By the middle of the 18th century they had made the entire Mongolian plateau their subject state.

The Chinese revolution of 1911 and the abdication of the Chinese emperor gave the Outer Mongolians a chance to assert their independence briefly. But the Chinese re-asserted their influence until, in 1921, Soviet troops with some Mongolian units marched into Urga (now Ulan Bator). In 1924 Outer Mongolia declared itself the Mongolian People's Republic. Although remaining a part of China, it was in fact a self-governing satellite of the Soviet Union. In 1946, after a plebiscite, the Mongolian People's Republic received its independence from China.

Government. The chief organ of the state is the Great People's Khural, a national assembly elected every 3 years by universal suffrage. Between the annual meetings of the Khural, its affairs are handled by a seven-member presidium elected by the Khural. The chairman of the presidium is the titular head of state. Executive responsibilities are handled by a council of ministers, headed by a premier. The premier is also the first secretary, or head, of the country's only political party—the People's Revolutionary Party, a Communist party.

TAO CHENG, Trenton State College

CHINA: AN INTRODUCTION

by Pearl S. BUCK
Author, The Good Earth

What is China like? This is the question I am asked more often, I believe, than any other question. It is a good question, for today China is one of the three greatest powers, and yet very few Westerners know what China is like or what its people are like. Actually, China has always been one of the great powers, both because of its vast size and because of its people. Of these two I put first its people, for other nations have had great size, and yet they have died. The reason China has survived is that its remarkable people have been able to build a workable and therefore indestructible civilization. This civilization has not had a fixed structure. The Chinese, while not a changeable people, are nevertheless people who are able to change when they see the time has come for change. They are basically a practical people. They do not cling to a custom, or a tradition, or even a religion just because "it has always been that way." When they see that something no longer works, they change it.

Even today we see this adaptability. Many people thought Communism could not happen in China because it was so different from the traditional ways of the Chinese. But when early in the 20th century it became apparent that there had to be a change of some sort, since the old ways were no longer working, the change came. I doubt that China would have turned to Communism if Sun Yat-sen, the founder of the Chinese Republic, had not died when he did. For Sun himself said that he did not think that Communism would suit his people. Nevertheless, he accepted help from the Soviet Union at the time because he could not get help from any other country. What might have happened if Sun had not died too young in 1925, no one can tell.

The second question I am asked most often about China today is whether Communism will last. My only answer is that if it works it will last. If it does not work, it will not last. Since the Chinese are, as I said, a very practical people, they will probably keep its achievements and reject its failures. Communism will certainly leave some results in China, but no one can tell what the final form of government will be. It will, however, be practical.

Another quality of the Chinese people that has had much to do with their long-continuing life is their remarkable freedom from the narrowing influences of prejudice. The Chinese, while they are intensely proud people—as they have every right to be, considering the magnificence and age of their culture—are, on the whole, more free from prejudice than any other people, in my opinion. For example, there has never been any persecution of Jews in China. From time to time Jewish people have gone to China to live, either as businessmen or as refugees from persecution elsewhere. No one prevented their coming and staying. Indeed, when twice their synagogue was destroyed by storms and floods, the Chinese Government rebuilt it for them as a courtesy. Similarly, when China was invaded by the Mongols and later by the Manchus, the Chinese accepted the conquerors calmly. So well structured was the Chinese Government, so superior their civilization, that these foreign conquerors allowed the Chinese to go on administering the government as they had always done. The result was that soon the Chinese absorbed their conquerors. Five hundred years before the birth of Jesus Christ, the Chinese philosopher and teacher Confucius said, "All under Heaven are one family." Without losing their pride as Chinese, the Chinese people have accepted other peoples with tolerance and respect, if respect was deserved.

A third element in Chinese life has been their instinct for and belief in democracy. One may wonder at my use of this word, so basic in Western thought, but there were very sound democratic roots in the Chinese way of life and thinking, too. Take, for example, the form of government of old China. Chinese history, up to the creation of the republic in 1912, may be divided into 24 dynasties. A dynasty is the length of time during which one family retains the imperial power. Twenty-four times the Chinese began all over again, so to speak, with a new imperial family. Two of those times were when the Mongols and Manchus took power by force. But the other times the new ruling family began with a man from among the Chinese people.

How was this man discovered? When a ruling family became weak and corrupt, as they all did sooner or later, affairs in the nation began to go badly. Even prices went up until the people could not live comfortably. There is an old Chinese proverb that says, "When the price of rice is higher than the people can pay, Heaven ordains a change of rulers." In other words, it was time for a new dynasty. When this was apparent, there rose from among the people young ambitious men, each with an army to support him. These young men and their supporting armies fought each other until at last one was the victor. He then was acknowledged as the first emperor of the new dynasty. During the struggle the people waited, staying as far as possible from the actual battlefields, until the best man proved himself. Sometimes, indeed usually, this man was of very humble origin, the son of a peasant, or the leader of a robber gang, or a renegade priest. But he had shown in victory the necessary qualities of daring and courage and especially of leadership. Although I daresay he made a rough-and-ready emperor, even in his fine imperial robes, and although his manners were probably not the best, he brought fresh life and energy to the government. And there were plenty of teachers for his children and advisers for him.

How did the people manage to live as usual while this struggle was going on? That they were able to do so was also due to the structure of their society. I shall mention only two of the most important points of this structure. One was the Chinese civil service. The civil service in any government is composed of the people who are the workers, and whose jobs continue no matter what political party or dynasty is in power. The Chinese had an excellent system of civil service, so good, in fact, that the British patterned theirs after the Chinese. And since the United States patterned its civil service after the British, to that extent Americans are indebted to the Chinese. But it was the way the civil service in China was chosen that was unique. No other people at that time had discovered so practical a way to discover its best brains. The Chinese found their most able and brilliant men through the imperial examinations, which were given once every 3 years. Any man could try out for them, provided he had the preparation. The subjects for the examinations were very difficult. They covered the whole range of the complex Chinese civilization, and only men with the best minds could possibly learn enough to pass them. But these men could come from anywhere in society, and in this was the democratic element. The son of a village barber, for example, might be bright enough to be noticed by the other villagers. Even though the villagers might not be able to read or write, if they thought a boy worth educating for the imperial examinations, they pooled their money and sent him to school. If he passed the examinations he was given a job in the government and so he brought honor to the entire village. In this way the Chinese imperial government was administered by able men. Of course there must have been some corruption—men have their weaknesses everywhere—but the method was good, for it meant that the government was continually being renewed and invigorated from the entire people, and not just from wealthy or intellectual groups. With such a civil service, the periods of change of dynasty did not disturb the people greatly. The work of the government went on as usual.

Another stabilizing element in Chinese society was the family. Instead of having an elaborate police system, the Chinese simply made every family responsible for each member of that family. The family meant all one's relations, not just one's parents. It meant uncles, aunts, grandparents, and cousins to the remotest degree. A boy (or a girl) had very little chance to misbehave when all his relatives were watching him. And the whole family was watching, because they had to pay the price of any crime, or even any mischief, that a young person committed. Responsibility went so far that if a son became so unmanageable that even the family could not control his evil ways, the family had to put him to death. I myself knew of this happening, but only once. A father shot his son, and the whole family, especially the father, mourned for the lost child. But this universal family control and responsibility resulted in an orderly, well-behaved society, even during times of change.

In our own times there has been a great change in China, a change new, I daresay, in the whole of Chinese history. The end of the Manchu dynasty, the last imperial dynasty, came about following a revolution by young Chinese, who felt that the old form of government was outmoded. The new form of government they chose was a republic, based on that of the United States. They chose this kind of government because many of their leaders, including Sun Yat-sen himself, had been to American missionary schools in order to learn more about the West. In 1912 they established the Chinese Republic and for 10 years they tried to make it work and could not. The republican form of government was new to the Chinese people and they did not understand how to make it work. The Soviet Union offered to help Sun Yat-sen and, in despair, he accepted their help. When the Communists came into China, however, they knew that they could never succeed unless they destroyed the entire family system. This meant a profound change, for first of all Confucianism, with its emphasis on family relationships and loyalties, had to be destroyed. Then there had to be a physical separation of the members of the family. Older people were separated from the younger, and children were taken from their parents and placed in institutions where they were taught from the Communist point of view. With the victory of the Chinese Communists in 1949, this process was accelerated. There are young Chinese today who know nothing of their own history, except from the Communist viewpoint, and who know nothing of the Western nations except those that are Communist. For Westerners this is a real danger, for these young people know nothing that is good about us. They have been taught that we, and particularly the United States, are their enemies. Of course we are not their enemies, however we may disagree with their present government.

And yet I hope, and I believe with reason, that the old virtues and graces of the Chinese people are not lost, and that a civilization that is more than 4,000 years old cannot be destroyed in less than 50 years. Indeed, there are a few signs that this may be true. We have heard of the great "cultural revolution" in recent years in China, when thousands of young people, known as Red Guards, took matters into their own hands and endeavored to hasten the completion of Communism in their country. This revolution has failed, or has been quieted, for its extremes shocked the Chinese themselves. A few of the young Red Guards—perhaps 100 or so out of thousands—fled the Chinese mainland. They tell us of their disappointment in finding out that what they were taught was not always true. And yet they have not been happy in non-Communist countries either, for they feel that it is immoral, too, for some people to be richer than others.

We do not know what will be the result of this generation of young Chinese, who have known only Communism. We do not know what will happen when, freed from political restrictions, they discover again the past glories of their own civilization, for they have heard only of its faults and failures.

Nations, like people, have different ages. China is the most ancient on earth. Centuries ago the Chinese understood the folly and waste of great wars and consequently they rejected the creation of certain weapons. Long ago, for example, a Chinese scientist understood the principles of rocketry, which the West only fairly recently began to know and use. The scientist was not allowed, however, to continue his work to include the making of rockets, for it was considered inhumane to produce weapons that would inevitably involve innocent people who had nothing to do with war, the people whom we call civilians. The Chinese understood very well that there are in every nation, and perhaps always will be, those men who will create wars of some sort. But as a Chinese emperor once said, "Let them fight each other and only with swords." In the same way, though the Chinese invented gunpowder, its use was limited to fireworks.

The Chinese are people of naturally scientific minds, and some of the finest scientists in the world today are Chinese. But in old China, humanism and morality were in control even of the scientific mind. What will new China be like? Faced with nations too young yet to know the wisdom of the rejection of great wars, in self-protection the new China may have to meet the present world as it is. Thus, until all nations can achieve what old China achieved in wisdom and morality, no nation will be safe.

The Great Wall, perhaps the best-known symbol of old China.

CHINA

The study of things Chinese, which is called sinology, has a long history in Western scholarship. In recent decades, however, sinology has developed into two separate but complementary fields—traditional and modern. The first deals with the study of China from ancient times to the coming of the West—roughly in the middle part of the 19th century— and stresses the continuing traditions of old China. The second deals with modern China and emphasizes the changes the country has undergone since the 19th century. Whether traditional or modern, the study of Chinese civilization and culture is an important task, for it is concerned with a people who make up almost one fourth of the human race and who represent the oldest continuous civilization in the world.

A nation with a recorded history of over 4,000 years, China has made a variety of important contributions to the world. Such commonly known inventions as gunpowder, paper, and printing (and perhaps the magnetic compass) originated in China. In addition, the Chinese over many centuries developed a unique form of government based on Confucianism; a highly productive agricultural economy; a society emphasizing harmony and order; an artistic and literary heritage of great accomplishments; and a high level of sophistication in science and technology. These and many other aspects of traditional Chinese culture have made fascinating subjects for study by scholars in various fields.

China in the traditional period has often been described in such terms as *continuity, stability, changelessness,* or *historical stagnation.* In sharp contrast, China in modern times has been changing at a very fast pace in every aspect of its national life. Since 1949, China under Communism has undergone so many radical changes that people outside China find it difficult to follow the events, let alone to understand the meaning of the changes. But whatever the difficulties, it is vital that efforts be made to understand contemporary China, because what happens there affects not only 850,000,000 Chinese, but the entire world.

THE LAND

China is a vast country. It has an area (not counting Taiwan) of almost 3,692,000 square miles (9,561,000 square kilometers), which makes it slightly larger than the United States. Like the United States, China has a continental type of climate and vegetation and a wide variety of geographical features. Within China's borders there are tall mountains, great rivers, fertile plains, forbidding deserts, and different temperature zones. In latitude, too, China and the United States are generally comparable. For example, Peking, the capital of China, is only 1 degree farther north than the American capital, Washington, D.C. But unlike the United States, which extends from the Atlantic to the Pacific and is a sea power, China has always been primarily a land power.

Although large in area, much of China is covered with mountains, hills, and plateaus. Only about 12 percent of the land is plains. Both China's mountains and its great rivers extend from west to east, dividing the country into three geographically distinct regions. In the west lie the lofty Tibetan highlands. To the north is the Sinkiang-Mongolia region.

FACTS AND FIGURES

PEOPLE'S REPUBLIC OF CHINA—Chung Hua Jen Min Kung Ho Kuo—is the official name for mainland China.

CAPITAL: Peking.

LOCATION: East Asia. **Latitude**—21° 09′ N to 53° 54′ N. **Longitude**—73° 37′ E to 135° 05′ E.

AREA: 3,691,512 sq. mi. (9,561,000 sq. km.).

PHYSICAL FEATURES: Highest point—Mount Everest on China-Nepal border (29,028 ft.; 8,848 m.). **Lowest point**—Turfan depression (below sea level). **Chief rivers**—Yangtze, Yellow (Hwang Ho), Si Kiang, Amur. **Major lakes**—Tungting, Poyang, Tai, Hungtze.

POPULATION: 850,000,000 (estimate).

LANGUAGE: Chinese (various dialects, including Mandarin, Cantonese, Shanghai) is spoken by the great majority of the people.

RELIGION: Confucian, Buddhist, Taoist, Muslim, Christian; ancestor-worship also has been observed by most Chinese. The Chinese Government, however, generally discourages organized religion.

GOVERNMENT: Communist republic. **Head of government**—actual leader of the country is the chairman of the Central Committee of the Chinese Communist Party. **Legislature**—National People's Congress. **International co-operation**—United Nations.

CHIEF CITIES: Shanghai, Peking, Tientsin, Shenyang, Canton, Wuhan, Chungking, Nanking, Lüta, Sian, Harbin, Tsingtao, Chengtu, Taiyüan.

ECONOMY: Chief minerals—coal, iron, petroleum, tin, manganese, salt, sulfur, antimony, tungsten. **Chief agricultural products**—wheat, rice, soybeans, sorghum, maize (corn), cotton, millet, barley, tea, sugarcane, tobacco, jute, groundnuts (peanuts). **Industries and products**—iron and steel, textiles, chemical fertilizer, cement, sugar, cotton, timber, paper. **Chief exports**—cotton, tea, silk, pork products, tobacco, tung oil, tungsten. **Chief imports**—machinery and industrial equipment.

MONETARY UNIT: Yuan.

NATIONAL HOLIDAYS: January or February, New Year's Day; October, National Day; May 1, May Day.

NATIONAL ANTHEM: *I yung chun chin hsing ch'u* ("The March of the Volunteers") reportedly has been replaced by "The East is Red" as the official song of the government.

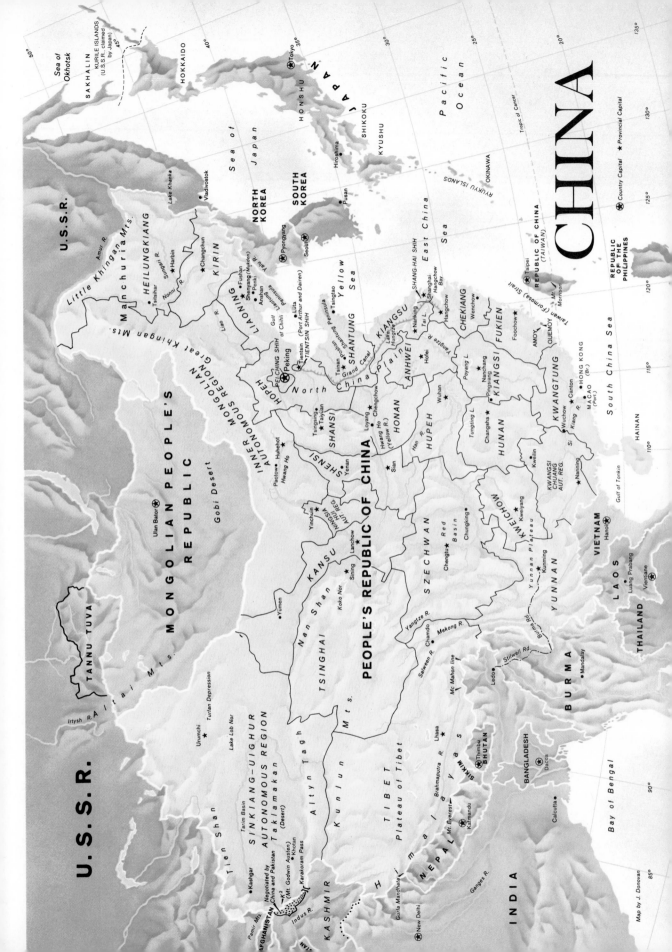

Large parts of these two regions consist of mountains, deserts (including the great Gobi in Mongolia), or high plateaus suitable only for pastureland. And although these regions make up about half of China's area, they contain only 5 percent of its people. (Separate articles on TIBET and MONGOLIA appear in this volume.) The third major region is Eastern China, where 95 percent of the population is concentrated. In contrast to the west and north, where one can often travel for days without seeing a house or person, in Eastern China almost every bit of good land is crowded with farmers.

Of the three regions, Eastern China is by far the most important. Not only is nearly all the densely populated fertile farmland located there, but all the important industrial centers as well. Eastern China in turn is composed of three subregions—Northeast (Manchuria), North, and South. These subregions themselves are divided by mountain ranges and differ significantly in vegetation and climate. Three of China's major rivers, the Yellow River (Hwang Ho), the Yangtze—China's longest river —and the Si Kiang, flow through the farmlands of the North and South and serve as important waterways. (There are separate articles on the YELLOW RIVER and the YANGTZE RIVER in this volume.)

The shape of China's coastline is determined to a great extent by its hill systems. North of Hangchow Bay, with the exception of the Liaotung and Shantung peninsulas, the coast has low, sandy beaches. South of Hangchow Bay, the coast is generally rocky.

Climate

In a country so large and with such a diversity of physical features, it is only natural that the climate varies a great deal, too. In winter cold, dry air flows down from the plateaus of the northwest. In summer warm, wet air flows inland from the southeast. The winter winds last much longer than the summer winds, but during the summer destructive winds such as typhoons, which carry large quantities of rain, often cause great damage to coastal areas. Temperatures can vary widely from region to region. Winter temperatures in northern Manchuria may reach 17 degrees below zero Fahrenheit (–27 degrees Celsius) as compared with 68 degrees F. (20 degrees C.) in Hainan in the South. The difference is less marked in the summer, which is generally hot and humid.

Rainfall shows even greater regional contrasts than temperatures; it is the major climatic condition that marks the difference between North China and South China. Rainfall is heaviest in the Southeast, ranging from 60 to 80 inches (150–200 centimeters) annually. It steadily decreases toward the north and northwest to less than 10 inches (25 cm.) outside the Great Wall. In North China rainfall occurs almost exclusively during the summer months. In the South there is some rain every month, although most falls in the summer.

The growing season for crops varies with the weather. The season is shortest (less than 140 days) in the far Northeast. It is longest along the tropical South China coast, where crops can be grown all year long.

Animal and Plant Life

China has a wealth of animal life. Among the larger mammals are different species of deer and bear, the giant panda, wild sheep, kiang (wild ass), wild boar, tigers, leopards, wolves, and monkeys. Com-

One of the more fertile areas of China.

A less hospitable, mountainous region near the Great Wall.

mercially valuable fur-bearing animals, such as sable, marten, mink, and fox, are found in the North and Northeast. Rodents and various kinds of birds abound. Snakes (a few poisonous), alligators, and tortoises are also found. Freshwater fish, which form an important part of the Chinese diet, include perch, shad, sturgeon, bass, and pike. Pigs, sheep, and goats are the most important domestic animals raised for food. Donkeys, oxen, camels, occasionally horses, and, in the South, water buffalo are the chief work animals.

Plant life is equally varied. Trees used mainly for lumber are the evergreens, oak, elm, and birch. There is a wide range of fruit trees and shrubs and flowering plants. Plants grown for food include various vegetables and grains. The principal food grains are wheat, millet, barley, sorghum, and maize (corn), grown in the North, and rice, which is grown in the South. China also produces large quantities of soybeans and tea, as well as cotton, jute, groundnuts (peanuts), bamboo, tobacco, and sugarcane.

Mineral Resources

China has substantial mineral resources, although less than such countries as the United States and the Soviet Union. Its most important mineral is coal, of which China is one of the world's leading producers. There are smaller deposits of iron ore, much of it mined in Manchuria, but these are sufficient to support China's iron and steel industry. Other important minerals that China has in fairly considerable amounts are petroleum, tin, manganese, salt, and sulfur. China is also the world's largest producer of antimony and tungsten, which are often used in alloys to harden metals.

THE PEOPLE

The population of China, in the absence of official government figures, is estimated to be about 850,000,000. This vast number of people is distributed very unevenly. Since agriculture is the most important occupation, the availability of fertile land and climatic conditions determine the density of population from region to region.

Generally the most heavily populated and most intensely cultivated regions are in the eastern part of China: Manchuria, the North China plain, the middle and lower Yangtze River valleys, and the southeast. The eastern coastal province of Kiangsu, for example, has over 1,000 people per square mile. Population density decreases sharply toward the northwestern and western borders of China. Inner Mongolia has less than 30 persons per square mile, and Tibet, the most sparsely populated region, less than three.

The great majority of China's people still live in the countryside, but there is a growing trend toward urbanization. There are about 17 cities with populations of more than 1,000,000, and of China's total population, over 14 percent now live in cities and towns.

Population Problems

Although China has vast territory, only about 11 percent of its land is suitable for cultivation. China's population is increasing at the rate of 15,000,000 a year. This, together with its already immense numbers and increasing urbanization, poses a serious problem. For although China

Rice, a chief food crop, requires careful cultivation and plenty of water.

An irrigation project. It will help to provide food for China's many millions.

This young Chinese is selling candied crab apples.

can feed its teeming millions, a high standard of living will be difficult to achieve.

Rapid industrialization and a more diversified economy may relieve the population pressure for a time, and migration to less densely crowded regions has helped. Earlier in the 20th century millions of Chinese moved from the heavily populated provinces of Hopeh and Shantung to the Northeast. During the 19th century many Chinese emigrated to Southeast Asia. More recently, the Communist government has encouraged the migration of people from the coastal areas to the border regions of Sinkiang, Tibet, and Inner Mongolia.

But China's Northeast, after absorbing millions of people, is almost saturated. Migration to outside areas, such as Southeast Asia, is no longer practical. And there is little possibility for the southwest and northwest regions of China to support a much larger population. Therefore, the only real answer to the population problem seems to be birth control— making the birthrate equal to or near the death rate. The Chinese Government undertook a large-scale birth control program prior to 1958. However, it is doubtful that the rural Chinese, who traditionally have valued large families, were deeply involved.

An outdoor food stand.

The Diversity of the Chinese People

The overwhelming majority of the people of China are known as Han-Chinese. Han is the name of the first long-lasting imperial dynasty, or ruling family, which lasted from the beginning of the 3rd century B.C. to early in the 3rd century A.D. The Han-Chinese have a dominant position, and they occupy the most productive regions of China. But China has always been a multi-national state. In addition to the Han majority, there are many different minority peoples, representing about 6 percent of the entire population. These minorities number from almost 8,000,000 to no more than a few thousand. Though numerically weak, they are scattered over more than half the territory of China, mostly in regions that are strategically important. The larger nationalities, in accordance with Communist government policy, have been organized into autonomous (self-governing) regions. At present there are five such regions, for the Mongols, the Hui (Chinese Muslims), the Uighurs, the Chuang, and the Tibetans. Smaller minority groups, such as the Koreans, the Miao, and others, have been organized into smaller autonomous districts.

The Chuang. With a population of almost 8,000,000, the Chuang are the largest minority group in China. They are concentrated mainly in south central China, in the province of Kwangsi; others live in Yunnan and Kwangtung. The Kwangsi-Chuang Autonomous Region was established in 1957. The Chuang are a settled agricultural people, like the Han. They have had long and extensive contacts with the Han and are culturally more assimilated with the Han-Chinese than the other ethnic groups. However, their spoken language and some of their customs are somewhat different.

The Uighurs. The Uighurs, numbering about 5,000,000, live in the Sinkiang-Uighur Autonomous Region, which was established in 1955. They speak a language related to Turkish and practice the religion of Islam. They are primarily farmers.

The Hui. These are descendants of Chinese who adopted Islam. There are approximately 4,000,000 Hui. Most live in northwest China, where the Ningsia-Hui Autonomous Region was created in 1958. Other Hui communities are found in Kansu, Sinkiang, Tsinghai, Hopeh, and Yunnan provinces.

The Tibetans. There are about 3,000,000 Tibetans in China. They are distributed over the entire Tibetan highland area, including the autonomous region of Tibet and the provinces of Szechwan, Tsinghai, and Kansu. A great majority of the Tibetans are engaged in farming. Some are herdsmen. They are followers of the Lamaist version of Buddhism, which originated in Tibet.

The Mongols. The Mongols, who number about 1,600,000, live in Inner Mongolia (the Inner Mongolian Autonomous Region), Northeast, and Northwest China. The typical Mongol is either a nomad, who travels with his herds of sheep, goats, cattle, and horses, or, where soil conditions permit, a settled farmer. The settled Mongols outnumber the nomads. The Mongols are also followers of Lamaism.

Problems of Minority Peoples. The presence of large numbers of ethnically and culturally different minority groups has presented China with certain problems. Striving to become a modern nation-state and a world power, China under Communism has redoubled its efforts to achieve greater national unity. This has required the steady integration of the various minorities into the national fold. Particularly since the Great Cultural Revolution of 1966, there has been a relentless struggle against what some Chinese leaders call "regional nationalism." Severe measures have been adopted against all political and social groups that are identified with the interests of the national minorities. In recent years, many thousands of young men and women in China have been forcibly removed to the border regions, where most of the minorities live. There are two main reasons for this: to suppress regional nationalism and to ease the tensions caused by the large concentrations of young people in China proper.

Chinese Customs and Traditions

Religion. Traditionally, the Chinese believed that the world was peopled not only by human beings but also by supernatural forces that accounted for happenings that were obviously not man-made. In religious belief, the Chinese have been polytheistic (worshiping many gods) rather than monotheistic (worshiping a single god). There were gods of all sorts—harvest gods, river gods, town gods, kitchen gods,

gods of disease, gods of war, and many others. But the Chinese did not recognize one awesome, supreme divinity as Jews, Christians, and Muslims do. Their relationship with the spiritual forces, the gods, and the ancestors was rather businesslike. They paid homage to the spirits, offered sacrifices, and sometimes engaged in fasts and meditations. However, the main purpose was to achieve harmony between men and the "other world," largely by appeasing the gods and spirits.

Ancestor-worship was the oldest and most widespread of Chinese religious practices. Except for those who believed in Islam, Lamaism, and more recently Christianity, all Chinese households observed ancestor-worship, regardless of social class or geographical location.

Most homes had a small altar or a shelf containing wooden tablets inscribed with the names, titles, and birth and death dates of the deceased members of the family. Usually on the 1st and 15th of each month of the lunar calendar (which is based on the phases of the moon), and on other dates of festivals, such as the lunar New Year, ceremonies were performed. These consisted of offerings of food and wine, incense burning, and occasionally the burning of mock-silver ingots. These offerings were made to the ancestors who, according to ancient beliefs, watched over their descendants as guardians and benefactors. The cult of ancestor-worship developed gradually over the centuries, and represented the original form of Chinese religious belief and practice.

Taoism was originally a philosophical system derived from the teachings of Lao-tzu (or Lao-tse), who lived during the 6th century B.C., and Chuang-tzu, who lived in the 4th century B.C. It emphasized harmony between man and nature and advocated passive behavior. Over the centuries, this system of philosophy became a religion and, under the influence of Buddhism, developed gods of its own, with temples and a priesthood.

Taoism separates human nature into its material and spiritual aspects. Although the freeing of the spirit (or soul) is the ultimate goal, the Taoists are also concerned with the investigation of the physical world. It is this concern that drove many Taoists to the pseudoscience (or false science) of alchemy, searching for an elixir that would bring immortality.

Taoism was very influential in China, but declined with the growth of Buddhism and the official support given to Confucianism, which used ancestor-worship as a form of religious expression.

Confucianism is not really a religion as much as a moral and social philosophy. It is based on the teachings of Confucius (K'ung Fu-tzu), who lived from 551 to 479 B.C. Confucius emphasized the importance of ethical relations and the dignity of man. The two chief Confucian concepts are *jen* and *i*. *Jen* was simply defined as love of man, or the principle of human association; and *i* was merely what is appropriate, or in modern words, one's obligations to one's fellow beings. In essence, Confucianism is Chinese humanism.

According to the Confucian school of thought, the promotion of human welfare must begin with the cultivation of individuals through education. It then proceeds through the regulation of the family and national life to the peace of the world and, ultimately, to the creation of an ideal commonwealth. According to Confucians, human nature is best manifested in love between parents and children. For this reason, emphasis is placed on the teaching of respect between parents and children, both in the schools and in society. When a person is respectful to his parents, he can be expected to be loyal to his ruler, kind to his brothers, and faithful to his friends.

Buddhism came to China from India around the beginning of the Christian Era. It became the predominant religion and was widely practiced. Although many Confucian scholars deplored the influence of Buddhism, they were unable to stop it from spreading. Perhaps the main reason for this was because from the late Han dynasty (2nd century A.D.) until the late 6th century there was little peace and unity in China. Many people sought refuge in Buddhism. In Buddhist monasteries one did not need to worry about the world's insolvable problems, but only to read the scriptures, perform the ritual, and meditate.

The Buddhism that flourished in China was the Mahayana variety, which taught that everybody could win salvation through faith and sincerity. Since the Confucians ignored the question of life after death, Buddhism offered an outlet for those concerned with the "other world." Popular Buddhism had many gods or spirits, but in China the image of

A statue of Buddha in bronze covered with gold, from the T'ang dynasty. The Metropolitan Museum of Art, New York. Rogers Fund, 1943.

Kuan Yin. The stone statue dates from the 6th century A.D.
Museum of Fine Arts, Boston. Francis Bartlett Donation.

the Goddess of Mercy (Kuan Yin) was the one found in most households and temples.

Islam, in the form of a Muslim community, existed in China as early as the 8th century, as a result of China's contact with the Arab world. However, the Islamic faith penetrated China slowly. Today it is estimated that there are more than 10,000,000 Muslims scattered throughout China. In most places, they tend to segregate themselves from the rest of the community and try to preserve their own social customs, their religious ritual, and their dietary laws. Where they live in the midst of non-Muslims, there have been clashes and communal riots.

Christianity was introduced into China as early as the 7th century, but it was not until the arrival of the Jesuits in the 16th century that this religion began to win important converts. Many Jesuits, such as Matteo Ricci (1552–1610), became favorites at the imperial court and were given official posts. Disagreements with other Catholic missionaries, however, and Pope Benedict XIV's decision in 1742 to forbid Chinese Catholics to worship their ancestors led to the suspension of the Jesuit mission. The modern phase of Christian missionary work began at the time of China's defeat by the European powers in the mid-19th century.

Since the Communists are materialists and atheists, they have discouraged virtually all organized religion. Christians have been persecuted most severely because of their former ties with foreign missions. Buddhist monasteries have been stripped of their temples, income, and priesthood. Only Islam is permitted to exist as an organized religion for political reasons, since China wishes to maintain good relations with the governments of Muslim countries. It is expected that the Communist government will continue its policy of doing away with organized religion as a social institution.

The Family. China for many centuries was a family-centered society. In traditional China the father had absolute authority over the family. Men had higher status than women. Elder brothers had authority over younger brothers until they became heads of their own families. The continuation of the family was considered the most important task of its members. Allegiance to one's family, including distant cousins, was considered more binding than allegiance to the state. In modern times, however, particularly under the Communist regime, this traditional pattern of the family has been drastically changed.

Marriage. The belief in the perpetuation of life through the family explains in part the importance that the Chinese attached to marriage. In traditional times, marriages were arranged by the family. Usually the bride and bridegroom did not see each other until the wedding ceremony.

The man's family generally gave money or presents to the woman's family. The money was used to provide clothing and adornment for the bride on her wedding day. The calendar was carefully studied and the most favorable date chosen for the wedding. The ceremony was as long and as magnificent as the finances of the families would permit.

After the establishment of the Chinese republic in 1912, marriage ceremonies became less elaborate. A young couple today may invite their friends to dinner and at the end announce that they are husband and wife. Or a man and woman may merely go to court and ask a judge to declare them married.

This trend has continued and is encouraged by the Communist government. The Marriage Law of 1950 states that marriage should be based on the consent of the man and woman concerned; that bigamy (marrying again while still legally married to another), concubinage (keeping a woman without marriage), demands for money or gifts as payment for marriage, and interference with the remarriage of widows are forbidden. In the family, equality between husband and wife is recognized. Although it took some time to implement this law, marriage customs in China today have become comparable to those of the Western countries.

Birth and Death.   In an agricultural society with an ancestor cult, the male child was always more treasured than the female, because a boy could not only add to the family labor force but also carry the family name into the future. This explains why in traditional China the birth of a son was regarded as a happy event and was celebrated with enthusiasm, usually when the child was 1 month old and thereafter on his birthdays. Well-to-do families, however, often treated the birth of a girl with equal enthusiasm.

Preparation for death and burial was a very serious matter in China. Elaborate rites, often combining Confucian and Buddhist beliefs, were performed before and after death. Confucius said that one should pay to the dead the kind of respect they would enjoy if they were still alive. Coffins were made of extremely hard wood. The well-to-do often selected their own coffins and places of burial before their death. They or their descendants would choose a location thought to bring the greatest fortune to future generations.

When the time came for the last farewell and burial, the eldest son made the responses during the ceremony. Relatives and friends, carrying papier-mâché figures of men, women, and various kinds of household articles, formed the funeral procession. After the burial, a banquet was prepared for the mourners. The grandeur and duration of the mourning period varied according to the age and rank of the deceased.

Death ceremonies have been much simplified in the 20th century, largely as a result of the process of modernization, in which most of the old beliefs and practices have been rejected.

Festivals and Holidays.   Over the past several thousand years many festivals and holidays have developed in China. The most important ones are New Year, Tuan-Wu, and Chung-Ch'iu.

The Chinese **New Year** comes on the first day of the first month of the lunar calendar (somewhere between late January and late February). This is the most important festival of all, as important as Christmas is for Christians. About a week before New Year's Day the whole house must be cleaned. Women spend days in the kitchen preparing special food for the occasion. On the 23rd day of the 12th moon, the kitchen god goes to heaven. The family then gathers in the kitchen and prepares a feast for the god. Sweets are offered to the kitchen god so that he will say sweet things about the family in heaven.

On New Year's Eve, everybody stays up all night, except small children. It is often the only time that all members of the family are together during the year. Incense is burned and lanterns are lighted. Homage is paid to gods and ancestors, and a feast is set for them. A big family dinner is prepared for the night, usually very late. The children are awakened by the sound of numerous firecrackers. Everybody wears his best clothes. On New Year's Day, the gods and ancestors are greeted again and their blessing is asked. For the next few days (sometimes for as long as two weeks), relatives and friends visit one another and exchange gifts. New Year is without question the happiest time of the year.

Tuan-Wu falls on the 5th day of the 5th moon. The origin of this festival is unclear, but it is universally observed. Coming in early summer, it is the occasion for housecleaning and disease prevention. Certain medicinal branches and leaves are hung on the doors to protect the house from evil spirits. This holiday is also known as the Dragon Boat

Festival. All day long boats of all sorts race up and down the rivers or other bodies of water. In the evening lamps are set afloat on the water. Spectators gather along the river banks, shouting and applauding. Rice dumplings wrapped in leaves are served during this festival period. Legend has it that this type of food was originally designed to be dropped in water for the purpose of commemorating the death by drowning of an ancient wise man.

Chung-Ch'iu, or Mid-Autumn, occurs on the 15th day of the 8th moon. The Chinese are great lovers of the moon. When the 8th moon is full, marking the end of summer, they gather in their courtyards, offer feasts to the moon, and enjoy cakes known as moon cakes. Friends and relatives also exchange gifts, while scholars write poems during the night. Like the New Year and Tuan-Wu festivals, the Mid-Autumn Festival used to be the occasion on which debts were collected.

There are numerous other festivals and holidays. The 15th day of the 1st moon is called the Feast of Lanterns. In the spring, during the holiday known as the Day of Pure Brightness, the family goes out to tend the graves of the dead. In late autumn on the 9th day of the 9th moon comes the Festival of Climbing the Heights. Children are particularly delighted with these holidays because they bring such outdoor fun as kite flying and picnicking.

The Chinese Language

The Chinese written language is unique and differs from most of the languages of the world in that it has no alphabet based on sound. Instead, it is made up of about 40,000 ideographs, or characters. Most have one syllable when pronounced, so the Chinese language is called monosyllabic. But though a complete Chinese dictionary may list 40,000 or more characters, no more than 5,000 or so are in common use, and nowadays a person is considered literate if he knows about 2,000 of the characters.

While written Chinese is the same throughout China, spoken Chinese has a number of different dialects (such as Mandarin, which is the most widely spoken dialect, and Cantonese), and a Chinese who speaks one dialect may not understand a Chinese speaking another. One problem that most non-Chinese find when learning to speak the language is that although many characters have the same sound, they are pronounced in different tones, and a difference in tone may make a difference in meaning.

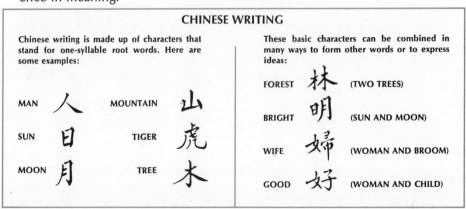

CHINESE WRITING

Chinese writing is made up of characters that stand for one-syllable root words. Here are some examples:

MAN	人	MOUNTAIN	山
SUN	日	TIGER	虎
MOON	月	TREE	木

These basic characters can be combined in many ways to form other words or to express ideas:

FOREST	林	(TWO TREES)
BRIGHT	明	(SUN AND MOON)
WIFE	婦	(WOMAN AND BROOM)
GOOD	好	(WOMAN AND CHILD)

This stone lion stands guard outside the Forbidden City in Peking.

CHINA'S PAST

In the long history of China there are several important themes. One has to do with the continuous expansion of the Chinese people from their original home in the North China plain. In this process, the Chinese managed to establish themselves as the dominant civilization in East Asia. At one time or another, China's neighbors, such as Korea, Annam (now part of Vietnam), Burma, and the Ryukyu Islands, were under some degree of Chinese political control and paid tribute to the Chinese empire. Because of geographical isolation, the Chinese had relatively little contact with other centers of civilization until modern times. And because of China's dominant position in East Asia—due to its vast territory, large population, and cultural accomplishments—the Chinese gradually came to regard themselves as the center of the civilized world. All other peoples were considered barbarians, who could be civilized only through exposure to Chinese cultural influence.

Ever since China became a unified empire in the 3rd century B.C., there has been a long succession of imperial dynasties, or ruling houses. Broadly speaking, these dynasties have had two variations. There have been several long-lasting dynasties, notably the Han (early 3rd century B.C.–A.D. 220), the T'ang (618–906), the Sung (960–1279), the Ming

"One of the Kings of Hell." A painting from the late Sung dynasty.
The Metropolitan Museum of Art, New York. Rogers Fund, 1929.

(1368–1644), and the last imperial dynasty, the Ch'ing, or Manchu (1644–1912). Between these dynasties there have been a number of short-lived ones, some lasting no more than 20 years or even less, as was the case of the Ch'in, which preceded the Han.

The other variation is between the so-called native and conquest dynasties. Although most of the dynasties were founded and ruled by the Han-Chinese, at times one or another of the peoples on China's northern and northwestern borders succeeded in defeating the Chinese Army and establishing themselves as rulers of the Chinese empire. Since these dynasties came as a result of military conquest, they have been known as conquest dynasties. The best-known of these are the Yüan, founded by Kublai Khan and his Mongol horsemen in 1260; and the Ch'ing, founded by the Manchus (who came from Manchuria).

Dynastic Cycle.   Whether long or short, native or alien, the Chinese dynasties over the centuries have risen and fallen in a pattern that has been described as the "dynastic cycle." A new dynasty usually came into existence after a period of disunity and civil war resulting from the collapse of the previous dynasty. Once established, the new dynasty went through a period of good government and economic prosperity, followed by a gradual decline in power because of population pressure, military adventures, or other factors, and finally by a period of disintegration and collapse. In Chinese political philosophy, the loss of dynastic rule meant the loss of the Mandate of Heaven. Theoretically, the emperor was the absolute ruler because he represented the will of heaven; in fact, emperors were known as Sons of Heaven.

The relatively changeless character of traditional China was to a large extent due to Confucianism, which reigned supreme for more than 2,000 years. Because of the Confucian emphasis on ethics, the concept and force of law tended to remain weak in traditional China. Instead, education was always stressed, for it was considered the most effective way of moral persuasion. Consequently, education based upon the teachings of Confucius became the chief method by which officials in government and leaders in society were selected. The system created for such selection was the civil service examination, which was open to all educated people at regular intervals. There were several kinds of degrees, awarded at the end of examinations, on four different levels. These were roughly comparable to modern school degrees, ranging from the high school diploma to the doctorate. The examinations were very competitive, and only a small portion of the candidates were allowed to pass. Those who did pass were eligible for government appointment. When not in active government service, they became community leaders.

Classes in Old China.   Old China had four classes—scholars, farmers, artisans (craftsmen), and merchants, in that order. Since agriculture was the basis of economic life, farmers were honored. Merchants were given a low position because they lived off other people's labor. Actually, there were only two classes—the scholars were the ruling class, and the others were the ruled. But people did move about in the social order, and it was not uncommon for a farmer's son to receive a classical Confucian education and to become a government official through success in the examinations. Women were not permitted to compete in the examinations, although some of them, especially those from upper-class families, were educated at home.

Like all other civilizations, traditional China had its weaknesses. But unlike other ancient civilizations, China has somehow survived for a longer period of time. One major reason for its long life has been the Confucian ideology. Although not without its shortcomings, especially from a modern point of view, Confucianism, under traditional conditions, proved workable for the Chinese and even for some of their neighbors. It was fundamentally humanistic and rational, and therefore relatively easy to accept. This partially explains why the Chinese never had religious wars and why Confucianism succeeded in meeting the challenge of other systems of thought until modern times.

A supporting factor has been the unity of the Chinese written language, which brought the teachings of Confucius to all parts of China. The Chinese speak many different dialects, and even today a man from Peking in the North speaking to a man from Canton in the South may have difficulty in being understood. But they will understand each other's writing. This is because the Chinese written language has been uniform since the 3rd century B.C.

In addition, the long tradition of political centralization helped to insure China's cultural continuity. In spite of periods of disunity and civil war, China has largely been a centralized empire. The imperial government had a great deal of control over all parts of the country, and even the local magistrates were appointed by the emperor. The armed forces, taxation, the civil service, law courts, education, weights and measures, the maintenance of postal roads—these were all controlled and directed by the central government, although there were regional variations.

CHINA IN MODERN TIMES

The modern phase of China's history began not so much because of great changes in China itself but because of the coming of the West. In the early 19th century China was still very much an isolated empire in East Asia. There had been earlier contacts with the West. But such contacts did not bring about any important changes, and the Chinese remained in a state of ignorance about the Western world and highly reluctant to make adjustments to changing world conditions. The relationship between tradition-bound China and the dynamic West in the early 19th century has been likened to a fast-moving object coming in the direction of an immovable object. A collision was unavoidable.

The first collision came in the form of the first Anglo-Chinese War, often called the Opium War of 1839, when Great Britain used force to open China to trade. China's defeat in that war resulted in the signing of the first of a series of what the Chinese have called "unequal treaties." Every time China was defeated at the hands of one or a combination of the Western powers, it lost some of its territories or some of its rights as an independent state. By the end of the 19th century, China had lost all its so-called "fences"—areas that used to be China's protectorates. Korea and Taiwan were lost to Japan; large areas in North and Northeast China were lost to Russia; Annam went to France, and Burma to Great Britain. Moreover, the Western powers and Japan penetrated into all parts of China, claiming their "spheres of influence." In less than 50 years, the once great empire of China was reduced to the status of a semi-colony, on the verge of being swallowed up by the imperialist powers.

In the second half of the 19th century, China's ability to cope with the challenge from overseas was sharply reduced because of internal troubles. Population pressure, heavy taxation, and government corruption threatened the old system. In almost all parts of the empire rebellions broke out, plunging the country into war and destruction. The most devastating of all was the Taiping Rebellion, which started in 1851 and was not put down until 1864. The Taiping rebels fought bitterly for the overthrow of the Manchus. They overran many of the most productive regions of China and caused millions of casualties and untold losses in property. At the same time, the Nien, another rebel group, were active in North China, while the Muslims revolted in the southwest and later in the northwest.

Foreign aggression and domestic disturbance combined to drain the strength of the decaying empire. Although the Manchu dynasty managed to survive the rebellions, it was hard-pressed in the area of foreign relations. In the face of the aggressive West, with its superior military power, intense nationalism, and modern techniques in organization and communication, the Chinese were powerless. It was under these circumstances that the rulers of China began to shift from a policy of blind resistance to the West to one of learning from the West. The first efforts at modernization were in the area of national defense. Arsenals, shipyards, and railways were built with the help of foreigners. New schools were opened to train technical personnel, including foreign language experts. These costly projects of modernization proved completely useless when, in 1895, China was defeated by Japan in the Sino-Japanese War. Thereafter, attempts were made to change China in the political and economic fields as well, for the defeat of 1895 made it very clear that China could not be saved merely by trying to build a modern army and navy when the political and economic systems of the country remained backward.

Creation of the Republic. In the meantime, such new and refreshing ideas as democracy, nationalism, and individualism, brought to China by the Western powers, attracted increasingly large numbers of Chinese intellectuals. Some advocated reforms; others worked for a revolution. While the reformers met with little success, the revolutionaries, under Dr. Sun Yat-sen (1866–1925), overthrew the Manchu dynasty and established a republic in 1912.

The new republic was beset with many problems. With no real military power, Sun Yat-sen was soon forced to give up his position as provisional president of the republic in favor of Yuan Shih-k'ai, who controlled the army in the North. Yuan proceeded to undermine the new parliamentary system of government and attempted to restore a monarchy by proclaiming himself emperor. Following Yuan's death in 1916, the period of so-called warlordism began. Military men in various parts of China fought among themselves while the central government in Peking changed hands according to the fortunes of war. The revolutionaries under Sun Yat-sen organized themselves into the Nationalist Party, known as the Kuomintang, and found a base in South China. After Sun's death in 1925, the leadership of the Nationalist Party gradually went to Chiang Kai-shek. In 1926 Chiang led his army on a northern expedition against the warlords. He defeated most of them and finally established a new national government in Nanking.

China under Chiang Kai-shek suffered from dissension within the country and threats from abroad. Within China, the Chinese Communist Party, founded in 1921, challenged the central authority and staged armed rebellions. The threat from abroad was a militaristic Japan determined to expand its empire by conquering China. In 1931 the Japanese occupied Manchuria. Five major military campaigns in the early 1930's finally dislodged the Communists from their bases and drove them to the remote northwest, but the threat from Japan remained serious. In spite of these difficulties, the Nationalist government under Chiang pushed forward a modernization plan and made some impressive gains. Fearing a unified and strong China, the Japanese attacked in July, 1937, thus beginning 8 long years of a war of resistance. Threatened by a common enemy, the Nationalists and the Communists formed a United Front.

When World War II ended in 1945, the Communist army had grown in numbers and strength, while the Nationalist government was sorely taxed with many problems. Military morale steadily sank, inflation went out of control, and the government's ability to rally popular support suffered because of inefficiency and corruption. In spite of efforts by the United States to bring peace between the two fighting parties, civil war raged all over China and in the end Chiang and his government were forced to retreat to the island of Taiwan (Formosa). On October 1, 1949, the Communists under Mao Tse-tung proclaimed the founding of the People's Republic of China in the ancient city of Peking.

CHINA TODAY

Most of the territory of China is now ruled by the People's Republic of China. The Republic of China, under the leaders of the Nationalist Party, retains control only over the large island of Taiwan and some of the smaller islands in the Taiwan (Formosa) Strait and off the East China coast. Great Britain and Portugal possess the colonies of Hong Kong and Macao respectively. (Separate articles on TAIWAN, HONG KONG, and MACAO appear in this volume.)

The People's Republic of China is divided into 21 provinces, three municipalities (Peking, Shanghai, and Tientsin), and the five autonomous regions where the large groups of minority peoples live. These three main divisions are subdivided into smaller administrative units such as special districts, counties, and people's communes.

Chinese Society Under Communism

One of the major aspects of Communist thought is the idea of class struggle. Even before the Communists came to power, Mao Tse-tung had already divided the people of China into several social classes. Workers and peasants (the proletariat) are to be the new masters of the society. They are supported by the Army, officially called the People's Liberation Army, which is drawn from the workers and peasants. The landlords in the countryside and the capitalists in big cities are considered to be enemies of the proletariat, so they have to be either eliminated or reformed. The land reform movements in the early 1950's did away with the landlords; at the same time private property was confiscated or put under state control. The middle classes (the bourgeoisie) were screened according to their attitude toward the new regime, but all were required to go through the so-called thought reform. This is a form

The faces of young China. A kindergarten class.

The traditional grace of China. An old artist at work.

of intensive indoctrination, designed to make intellectuals and others accept the Maoist doctrine of socialism.

As early as 1950, the Communist regime adopted the Marriage Law, which was intended to change the position of women in Chinese society. Under this law, women are given equal status and rights as men. Women with sufficient reason can sue their husbands for divorce, and many have done so. Education for women has also been encouraged. Many women are active in politics, in social organizations, and in all kinds of production work. Both in cities and in the countryside nurseries have been set up for the care of small children. Usually these nurseries employ elderly women, so that the younger ones may work or go to school.

All forms of religion are considered by the Communists to be superstition. Believers in Christianity, Islam, Buddhism, and other religions are supposed to be permitted by the 1954 Constitution to practice their faiths. But social and political pressures are often so heavy as to keep most of them away from their places of worship. Other practices of a religious nature, such as ancestor-worship and funeral processions, are frowned upon by the government and are therefore rarely observed.

The traditional Chinese large family pattern has given way under Communism to small (nuclear) families. Nor is the family the cornerstone of Chinese society any more. People are now organized into many different kinds of units. The Communist Party itself has about 20,000,000 members; for youth there is the Communist Youth League; and for children there is a national organization called the Young Pioneers. Industrial workers and craftsmen have their unions, and peasants are orga-

Produce from a farm co-operative.

nized into production brigades in the rural communes. The most important fact in modern Chinese social life is that everything is under the control of the Communist Party.

Life in the Cities

In old China, most of the larger cities were mainly administrative centers where the government carried on its business. Beginning in the 1840's, as more and more ports were opened for trade by the Western powers, the modern type of commercial cities began to appear and greatly changed the style of city life. One of the major characteristics of modern China's cities was the evidence of strong foreign influence. Many of the cities contained within themselves foreign settlements, known as concessions, where the foreigners ran their own affairs without interference from the Chinese Government.

Today, after more than 20 years of Communist rule, life has changed greatly in Chinese cities. The changes have come about because of the fundamental beliefs and principles held by Communist leaders. In the first place, the Communists believe that the cities have been centers of corruption, bourgeois decadence, and exploitation. As such, the cities had to be completely rebuilt, not so much physically as spiritually. Secondly, although urbanization is inevitable in the process of modernization, the Communists are determined to reduce, if not completely eliminate, the differences between city and rural life. Thirdly, as government policy, over the years millions of city people have been resettled in the countryside in order to relieve some of the population pressures.

Strollers in Wan Shien Tsen garden in the city of Soochow.

A folk dance troupe performs at the Peking opera house.

In one important way city life in China today differs very much from life in most other cities in the world. And that is the absence of glitter and excitement. The elimination of what the government considers imperialist and bourgeois influences has given Chinese cities an unmistakable air of austerity. Since practically all commercial activities are controlled by the state, even business sections of the cities lack the usual hustle and bustle so common in other cities. There are few neon lights, traffic is sparse, and night life is almost nonexistent. There are no bars, nightclubs, gambling casinos, or dance halls, and theaters show only party-approved movies, operas, and other theatrical productions. All forms of entertainment are used for political purposes, the commonest themes being the virtue of socialism and the greatness of Chairman Mao Tse-tung. The elimination of private wealth has been accompanied by an equally successful elimination of criminal activities. Compared with cities elsewhere, the crime rate in Chinese cities is very low. Such social problems as alcoholism, drug addiction, gambling, prostitution, and juvenile delinquency are considerably less serious than in other cities. This is largely because of the intensity of Communist education and the severity of party watchfulness.

Compared with the countryside, the level of material well-being is in many respects higher in the cities. This is why many Chinese still prefer to live in the cities, in spite of the government's effort to encourage moving to the rural areas.

The clothing of the great majority of Chinese men now consists of a tunic, or jacket, which is buttoned to the neck, and rather wide trousers.

Women of the younger generation are similarly dressed, and foreign visitors often find it difficult to distinguish between the sexes. Use of cosmetics is looked down upon as a sign of decadence, although women on occasions do wear colorful clothes. Blue and gray are the dominant colors in clothes, which in most cases are made of fairly coarse cotton material. Cloth shoes are usually worn by older people; younger ones wear canvas and rubber tennis shoes, or sneakers. Clothes made of woolen or synthetic materials and leather footwear are considered luxury items. Cloth and other materials are sometimes rationed and hard to get.

Food has also been rationed from time to time. Although there has been no widespread famine, at times there have been food shortages. The supply of meat and poultry has for years been inadequate, but the government has tried to insure fair distribution of food grains and vegetables. Queuing (lining up) for food is a common scene in the cities, often beginning very early in the morning. In restaurants some of the traditional Chinese delicacies are still served, but most city-dwellers have to be content with what the housewife can produce in the kitchen, which is usually rather meager.

Because of the large population, most cities in China have experienced an acute housing shortage. In addition to the repair of old buildings, new buildings have been built for the working people, especially industrial workers. But even in the new buildings, space is hard to come by. A family of five usually has no more than two rooms and shares a community kitchen and bath with several other families. However, cities in China have improved greatly in sanitation, so much so that even in the old slum areas there is the feeling of cleanliness. Rents are very low, generally representing less than 10 percent of the family income. Since most

Two elderly men practice Tai Chi Ch'uan, a form of unarmed self-defense.

enterprises are owned and operated by the state, the majority of working people in cities are protected by health and other forms of insurance.

There are virtually no privately owned motor vehicles, although bicycles are a common sight. Public transportation exists in the forms of buses and electric trolleys. Goods are moved more often by pushcarts than by trucks. For exercise, many Chinese do calisthenics, and the younger people take part in group sports, such as basketball, soccer, and volleyball. Table tennis is extremely popular; in fact, for several years the Chinese team won championships in international tournaments.

By American or Western European standards, the Chinese people, whether in cities or in the country, lead a very hard life. The Communists have so far failed to lift their country to a level of economic development where the people can enjoy at least the basic comforts of modern life. But Mao Tse-tung and his followers have their own ideas about life, society, and man's position in society. At this point, they are more concerned with re-educating the Chinese people, to make them better organized, disciplined, and mobilized for the building of a new China. Therefore, the future of China will depend to a great extent on the Chinese people's acceptance or rejection of Mao's ideas.

Some Major Cities

Peking. Peking, the capital and second largest city of China, with a population of about 7,000,000, lies in the North China plain about 35 miles (56 kilometers) south of the Great Wall. Though it has been known by different names, Peking has existed since ancient times and, except

Early morning in Peking. These cyclists are on their way to work.

for brief periods, it has been the capital of China for centuries. The Mongol conqueror Kublai Khan made the city his capital in the 13th century, calling it Khanbalik (Cambaluc). As Peking, meaning "northern capital," it was the capital first of the Chinese empire and then of the Chinese republic from the 15th century until 1928. In 1928, when Nanking ("southern capital") became the capital of the Republic of China, Peking was renamed Peiping ("northern peace"). The name Peking was restored when the Communists made it the capital of the People's Republic of China in 1949.

Today Peking is China's center of higher learning and culture, as well as the seat of government. It has many universities, and it is the home of the Chinese Academy of Science and the National Historical Museum. Peking Library is the largest in the country. Although Peking was not an important industrial city before 1949, it now produces iron and steel, tools, and textiles.

In the heart of Peking are two old walled towns. The northern town, or Inner City, contains the Forbidden City, where the Chinese emperors once lived. The imperial palaces are now a national museum. Surrounding the Forbidden City is the old Imperial City, where members of the emperors' courts and officials lived. Today this section is the seat of government of the People's Republic. Political meetings and parades are often held in Tien An Men, Peking's main square.

In the southern walled town, called the Outer City, are two famous buildings, the Temple of Agriculture and the blue-roofed Temple of Heaven, which is now a public park. Outside the two walled towns new

Children play near the Forbidden City, once the home of the Chinese emperors.

districts have been built, with homes, factories, universities, and various public buildings.

Shanghai.   China's largest city, major port, and great industrial center is Shanghai. With a population of about 10,000,000, it is one of the world's largest cities. Shanghai lies just off the east coast of China on the Whangpoo River, near where the Yangtze flows into the East China Sea.

Although it was settled about 1,000 years ago, Shanghai remained relatively unimportant until the middle of the 19th century. The city's rapid development came after it was opened to trade with the West following the Opium wars. Its location made it a natural outlet for the entire Yangtze Valley, and goods passed through Shanghai to the interior of China. Shanghai was then known as the adventurer's paradise, because it offered all things to all people. It was the center of commerce, foreign trade, banking, industry, and shipping, as well as education and publishing. At the same time, Shanghai was notorious for all kinds of crime, from gambling and prostitution to underworld struggles for control of the narcotics trade. This aspect of Shanghai's past is only a dim memory today.

The oldest and most heavily populated part of the city is the southern section. Its maze of narrow streets was once surrounded by a high wall, but this has been torn down and the narrow streets have been replaced by wide ones. To the north is the area of the former foreign settlements. It was here that Westerners settled and established their legations, banks, and businesses.

Shanghai's famous Nanking Road, with its department stores and restaurants, is located here as is the Bund, a broad avenue lined with buildings along the Whangpoo River. Farther north is the City Center. This is the newest part of Shanghai. It was built in the 1930's and rebuilt after World War II. The administrative offices of the municipality of Shanghai are situated here.

Shanghai, China's major port.

Vegetables are unloaded for sale in a village outside Shanghai.

A main street in Shanghai.

Tientsin.   Together with Peking and Shanghai, Tientsin is one of the three municipalities controlled by the central government. A port and industrial city with a population of about 4,000,000, it is located on the Hai River, not far from Peking. Tientsin's rapid development came after it was opened to Western trade in the middle of the 19th century. Rebuilt extensively at the beginning of the 20th century, Tientsin is a modern city. Its chief industries are cotton and wool textiles and iron and steel products.

Shenyang.   Shenyang (formerly known as Mukden) is the chief city of Manchuria, the heavily industrialized Northeastern region of China, and the capital of Liaoning province. It produces machinery, machine tools, trucks, automobiles, tractors, and locomotives and is also a transportation center with an extensive railroad network. The population is about 4,000,000. Long a center of a Chinese colony in Manchuria, the city was taken by the Manchus and made their capital before their conquest of China. From 1931 to 1945 it was occupied by the Japanese. The city gave its name to the Mukden Incident of 1931. Japan, claiming the destruction of Japanese property by Chinese, used this "incident" as a pretext for the invasion and conquest of Manchuria. The old section of the city contains the former imperial palace. The new section, built largely during the Japanese occupation, has wide avenues and tall buildings.

Canton.   A port and manufacturing city, lying near the mouth of the Pearl (Canton) River, Canton is South China's largest city, with a population of about 3,000,000, and the capital of Kwangtung province. Iron and steel, machinery, cotton and silk textiles, cement, paper, and refined sugar are its chief industrial products. It is also famous for such handicrafts as lacquerware and carved jade and ivory. Canton was one of the first Chinese cities to have contact with the West, beginning with the arrival of Portuguese traders early in the 16th century. After the Opium wars, it was one of five ports opened to foreign trade. The modernizing of the ancient walled city was begun in the 1920's. The walls were torn down and broad streets and parks were built.

Wuhan.   Wuhan is an industrial and commercial city, the leading city of central China, and the capital of Hupeh province. It is made up of the three towns of Hankow, Hanyang, and Wuchang, with a combined population of over 2,000,000. The tri-city is situated at the juncture of the Han and Yangtze rivers. Bridges over the two rivers link the towns. Hankow, the largest of the three towns, was a small fishing village until 1858, when it was opened to foreign trade. Wuchang is the oldest of the towns and the administrative center of the city. Wuhan's industrial output includes iron and steel, machine tools, agricultural machinery, and textiles.

Other important Chinese cities include Chungking, the largest city of Southwestern China, and the provisional capital of China during the Japanese occupation; Nanking, which became the capital of the Republic of China in 1928, and is now the capital of Kiangsu province; Lüta, the former Port Arthur and Dairen; Sian, the chief city of Northwestern China, and capital of Shensi province; Harbin, an important city in Manchuria, and the capital of Heilungkiang province; Tsingtao, in Shantung province; Chengtu, the capital of Szechwan province; and Taiyüan, the capital of Shansi province.

The Economy of China

It is very difficult for a backward and agricultural society like China to be transformed into a modern industrialized state. The Communists, however, are determined to advance China to a stage of high economic development in the shortest possible time. After the land reform of 1950–51, the government launched its first Five Year Plan in 1953. With massive assistance from the Soviet Union, China managed to lay a foundation for industrial growth, especially in the Northeast and the lower Yangtze Valley area. In 1958 the government claimed that the average annual rate of growth for industry reached 28 percent, and for agriculture nearly 10 percent. In that same year, a nationwide campaign known as the Great Leap Forward was launched, accompanied by the organization of people's communes in the rural areas. Because of poor planning, bad climatic conditions, and lack of co-ordination, the Great Leap campaign proved to be a disaster, and China's economic development suffered serious setbacks. Since the early 1960's, agriculture has been stressed more than industry. This is partly because of the withdrawal of Russian help, resulting from China's dispute with the Soviet Union over the question of how to build Communism.

Until mid-1966, when the Great Proletarian Cultural Revolution seized the nation, China was recovering from the economic losses suffered during the Great Leap Forward. The rate of development slowed down, compared with the First Five Year Plan period, but there were indications that the overall economic situation was improving. The Cultural Revolution, however, has once again caused trouble for the national economy.

A steel plant. Industrial development is one of China's chief economic goals.

A textile mill. Cotton textiles are one of the country's important manufactures.

Rice threshing. Agriculture is still dominant in the Chinese economy.

Aside from the economic ups and downs during the past 20 years or so, there are some fundamental problems that make economic development difficult for China. Although industrialization is pushed with great enthusiasm, China today is still a land of farmers. Agriculture remains the dominant form of economic life, accounting for almost half of the national income and employing about 80 percent of the labor force. The low productivity of agriculture, together with the pressure of a rising population, means that very little surplus is available for investment in industry. It also means that China's ability to buy industrial goods from abroad is limited. Moreover, with population increase comes a larger working force. When neither agriculture nor industry is sufficiently developed to absorb the additional manpower, the problem of unemployment and underemployment becomes serious.

The question is often asked, Have the Chinese people enjoyed any substantial improvement in their standard of living since the Communists came to power? In earlier years there were reports of food rationing, malnutrition, and housing shortages. The drabness of clothing was also remarked upon by many foreign observers. On the other hand, a number of travelers who have gone to China since President Nixon's visit in 1972 have brought back more positive accounts, of a people seemingly well-fed and decently housed. Undoubtedly China under Communism has made great strides in certain economic areas. At the same time, by Western standards it still has a long way to go, both in economic development and in the material well-being of its people.

GOVERNMENT

According to China's Constitution, all power of the state belongs to the people, who exercise power through the National People's Congress and the local people's congresses. The National People's Congress is a unicameral (single-house) legislature. Deputies are elected for terms of 5 years by the people's congresses of the provinces and municipalities. Special quotas of deputies are also allotted to national minorities, the armed forces, and overseas Chinese.

The National People's Congress has formal authority over all important matters affecting the nation, including finances, economic planning, declarations of war, and treaties of peace. It also has the power to elect and remove the highest government officials, amend the constitution, and pass laws. During the long intervals between the yearly sessions of the National People's Congress, its Standing Committee exercises power.

Under the old (1954) constitution the functions of the head of state were performed by the chairman of the People's Republic of China, who was elected by the National People's Congress. The chairman of the Republic represented China in its foreign relations and commanded the armed forces. He carried out laws and decrees and appointed and removed high state officials. In 1975 a new constitution was adopted which abolished the office of chairman of the Republic. The ceremonial duties of the head of state will now be carried out by the chairman of the Standing Committee of the National People's Congress.

The State Council is the highest organ of state administration. It is composed of a premier, several vice-premiers, the ministers of the different departments, the chairmen of various commissions, and a secretary-general. The State Council is an administrative rather than a policy-making body. It serves as a "brain trust" formulating proposals to the National People's Congress or its Standing Committee.

At the local levels of government are the people's congresses and people's councils. Deputies to the congresses at each level are elected by the congress at the next lower level. Deputies to the basic administrative units are elected directly by the people. The basic units were formerly the *hsiangs* (villages) and towns, but these have been replaced by people's communes.

The people's courts and the people's procuratorates are the judicial organs of the People's Republic of China. The people's courts administer justice independently but they are responsible to the people's congresses at corresponding levels. The president of the Supreme People's Court and the chief procurator of the Supreme People's Procuratorate are elected to and may be removed from their posts by the National People's Congress. A new, simpler constitution is reported to have been drafted in 1970, giving Mao Tse-tung supreme authority in national affairs, including command of the Army and the designation of head of state.

The Chinese Communist Party

It is difficult to understand the operation of the Chinese political system without realizing the key role of the Chinese Communist Party (CCP). After its founding in 1921, the CCP struggled long and hard against the Nationalists to bring about a revolution and the creation of the People's Republic. Since 1949 it has been the guiding force behind the political development of China. The 1969 Constitution of the CCP states that all organs of leadership must accept the party's authority. The

Wall posters are seen almost everywhere in China. This one is in Shanghai.

A political rally in Peking. The portraits are of Mao Tse-tung and Ho Chi Minh.

Some children stop to admire prints and posters for sale.

CCP stands alongside every organized unit of state and society. Leaders of the government organizations are generally also members of the party.

The National Party Congress is theoretically the highest source of authority within the party. There have been 10 party congresses since the establishment of the CCP. The most recent one was held in 1973. The party congress adopts long-term policies and approves changes in party leadership. The members of the party congress transmit the new policies to the local governments and social organizations, and help insure that they are carried out.

The Central Committee represents the top leadership of the party. But the Politburo and its Standing Committee exercise the authority of the Central Committee when it is not in session. Because of its size and infrequent meetings, the Central Committee is not a day-to-day decision-making body. Major decisions are made in the Politburo and the Standing Committee, although the Central Committee is supposed to approve major policy decisions. Party organization parallels the administrative divisions of the country. Its units extend to all political, military, economic, and social organizations.

THE CULTURAL REVOLUTION

In the years of Communist rule, one of the most important events in China was the Great Proletarian Cultural Revolution, which lasted from the middle part of 1966 to early 1969. This revolution was closely identified with Mao Tse-tung, because it was Mao who decided to purge some of the top leaders of the party and the government for ideological reasons. The Cultural Revolution was, first of all, the result of a major split between the Maoist faction, or group, which was radical, and the faction of Liu Shao-chi (the former chairman of the Republic), which was considered "revisionist." In order for Mao and his followers to get rid of those in power, they used both the Army and the youth, who were organized into the now famous Red Guards. The basic quarrel between the two factions had to do with national policies affecting all aspects of Chinese life. Mao believed that the revisionists were betraying the cause of Communism by following a practical road, using material incentives to increase economic production, trying to compromise with the Soviet Union, and encouraging the training of a new educated elite. In the struggle, Mao Tse-tung and his followers succeeded in destroying their opponents, who for decades had been their close comrades. The conflict, however, caused great damage to government administration, the economy, social order, and education.

LATER EVENTS

Since the end of the Cultural Revolution, a number of events have taken place that affect both China and the rest of the world. In 1971 the People's Republic entered the United Nations, taking the seat formerly held by the Republic of China (Taiwan). American presidents have twice visited China, helping to improve relations between the two countries. The most dramatic event in recent years was the death, in 1976, of China's longtime leader, Mao Tse-tung. What influence this will have on China remains to be seen. Its results, however, will be of great importance not only to the Chinese but to people everywhere.

C. T. HU, Columbia University; co-author, *China: Its People, Its Society, Its Culture*

YELLOW RIVER

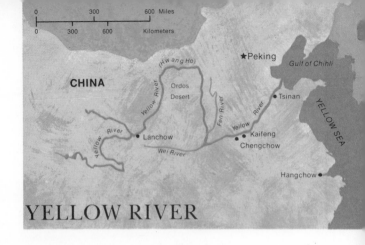

YELLOW RIVER

The Yellow River (in Chinese, Hwang Ho) has been called China's Sorrow. For this great river, which was the cradle of China's civilization, has often threatened to destroy that civilization by floods. The unpredictable waters of the vast river have destroyed millions of lives and areas of farmland, and for the people who live along the river's banks, life is a constant gamble. For this reason, few large cities have grown up on the river's banks—Lanchow, Kaifeng, Chengchow, and Tsinan are the most important. And although the river is the second longest in China (after the Yangtze), its waters defy navigation in all but a few places.

From its source in the Tibetan highlands of Tsinghai Province, the Yellow River begins its 2,900-mile (4,700 kilometers) journey eastward across North China to its mouth in the Gulf of Chihli (or Po Hai), an arm of the Yellow Sea. The course of the river has changed many times in its history. At one time it flowed into the East China Sea, the mouth also of the Yangtze River.

One of the few places where the Yellow River is navigable for small vessels is the Ordos loop in Inner Mongolia. This stretch of the river encircles the Ordos Desert. Beyond the Ordos loop the river is joined by its main tributaries, the Wei and Fen rivers. From here the river passes through a series of gorges that form the entranceway to the great plain of North China. This fertile plain of the Yellow River covers parts of five provinces and merges with the Yangtze River plain to the south. Here where the Yangtze and Yellow River valleys converge was the center of the early imperial dynasties of China. In this area, too, the yellow-colored deposits of silt (tiny particles of rock) that give the river its color and name begin to accumulate. The flooding of the river is caused mainly by the continual buildup of silt, which elevates the water level above the surrounding plain. Although dikes have been built, they are often unable to hold back the floodwaters.

Because of the floods, the vast agricultural potential of the Yellow River delta has never been realized. The Chinese embanked and deepened the river for the first time about 220 B.C., and efforts to check its raging waters have continued since that time. The Grand Canal, constructed in ancient times to connect the Yellow and Yangtze rivers, now extends from Peking to Hangchow, a distance of some 1,200 miles (1,900 km.), but much of it is choked with silt. The canal is now being rebuilt as part of a long-term plan to control the river and to make it work for, not against, the people of China.

Reviewed by C. T. HU, Columbia University
Co-author, *China: Its People, Its Society, Its Culture*

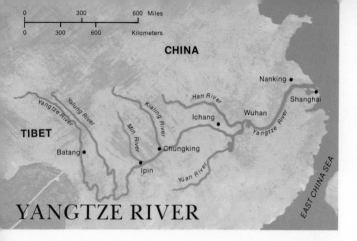

YANGTZE RIVER

YANGTZE RIVER

To the Chinese it is known as Chang Kiang ("long river"), and it is the wellspring of much of China's history and life. To the rest of the world it is better-known as the Yangtze—the longest river in Asia. Originating high in Tsinghai Province in Tibet, the Yangtze alternately rages and flows 3,430 miles (5,520 kilometers) through eight provinces, past some of the world's most breathtaking scenery, to its mouth in the East China Sea. Together with its tributaries, the Min, Han, Yalung, and Kialing rivers, the Yangtze carries millions of tons of freight a year, and its delta provides the Chinese with their chief rice-growing region. Along its banks lie some of China's major cities—Chungking, Wuhan, and Nanking—whose commercial importance is due largely to this mighty river. And the area drained by the river is one of the most densely populated in China.

From its source, the Yangtze rushes first east and then south through Tsinghai and Yunnan provinces. Turning sharply eastward again, it flows snakelike across Yunnan and into the fertile Red Basin of Szechwan Province. At the city of Batang, some 600 miles (960 km.) from its source, the river begins a series of spectacular falls that lower its level from 16,000 feet (4,800 meters) above sea level to 630 feet (190 m.) by the time it reaches Chungking. Called by various names in its upper course, the Yangtze proper begins at Ipin, south of Chungking, where it merges with the Min River. Here the river first becomes navigable for small boats. About 250 miles (400 km.) from Chungking lies the gateway to the magnificent triple gorges of the Yangtze. Their names—Bellows Gorge, Witches' Gorge, and the Gorge of the Western Grave—evoke both their beauty and the terror with which they are regarded by Chinese sailors who must navigate the tortuous twists and turns.

At Ichang the gorges end, and the river takes a more leisurely course on its journey to the sea. Beyond Ichang the Yangtze is choked with junks, sampans, and other small rivercraft, which form virtual floating cities. At this point the river also becomes navigable for oceangoing ships. Here the lower course of the Yangtze begins.

The triple city of Wuhan, one of China's greatest industrial centers, lies on the Yangtze's lower course, at its junction with the Han River. At Wuhan is the Yangtze Bridge, which links north and south China. This double-deck steel bridge, built in 1957, is considered by many Chinese to be one of their great modern engineering feats. Past Wuhan, the river courses northwest until, near Shanghai, it finally empties into the East China Sea.

Reviewed by C. T. HU, Columbia University
Co-author, *China: Its People, Its Society, Its Culture*

HONG KONG

HONG KONG

Hong Kong, a British crown colony on the southeastern coast of China, is a dazzling burst of light and color, people and activity. Hong Kong's deep, sheltered harbor at the foot of steeply sloping green hills is one of the busiest and most beautiful in the world. Its industries are in a period of explosive growth, and its population has soared from about 600,000 at the end of World War II to nearly 4,000,000 today.

The Land.   The Crown Colony of Hong Kong is made up of the island of Hong Kong, Kowloon peninsula on the mainland, and the New Territories, which extend north from Kowloon peninsula to the China border. The New Territories also include the island of Lan Tao and many smaller islands. Most of the colony's area of 398 square miles (1,031 square kilometers) is in the New Territories. The island of Hong Kong has an area of just 29 square miles (75 sq. km.).

The busiest and most crowded parts of the colony are Victoria, the capital, on the island of Hong Kong, and Kowloon, which faces Victoria across Victoria harbor. In the harbor, large ships lie at their moorings, while thousands of small craft—junks, sampans, water taxis, and ferryboats —scurry around them. Miles of docks and wharves line the harbor. Hong Kong has its modern sections, with wide streets, hotels, apartment and office buildings, theaters, and shops; and nearby, its old sections, with narrow streets and poor, overcrowded buildings. People jam the sidewalks, spill over into the streets—even live on rooftops. By contrast, in the New Territories there are still hundreds of quiet farm villages with few inhabitants. About 100,000 people spend their whole lives afloat, in villages of junks and sampans moored in several fishing harbors.

Hong Kong has cool, dry winters and hot, humid summers with occasional typhoons. The average annual rainfall is 85 inches (216 centimeters). Land, especially flat land, is so scarce it has to be reclaimed from the harbor. Hills are leveled and the soil, rock, and gravel used to fill in a bay. The runway of Kai Tak Airport is built on reclaimed land.

The People.   In 1941, during World War II, when the Japanese occupied Hong Kong, the population was estimated at 1,600,000. Large numbers of Hong Kong's population went to the mainland of China during the occupation. Only about 600,000 people were left when the occupation ended in 1945. After the establishment of the People's Republic of China in 1949, thousands of people crossed the border into Hong Kong, and the population soared. (Immigration has since been limited.)

Caring for a vastly increased population has raised serious problems. One of the most serious is housing. The government has built housing units for 1,000,000 people, but there are still 400,000 squatters living in flimsy shacks on the hillsides. A second serious problem is the shortage of water. A number of reservoirs have been built to help solve the problem of water storage, and Hong Kong buys part of its water from China. A third problem is educating the huge school-age population. There is no system of public education, but the government hopes to provide by 1971 places in government or government-aided primary schools for all who are likely to seek these places. Many schools are already on a schedule of three shifts a day, and new schools are opening fast. The University of Hong Kong and the Chinese University provide higher education. Most of the people in Hong Kong are Chinese. Instruction in schools is in Chinese, and English is taught as a second language.

As they go about their work, both men and women wear trousers and a loose pajamalike jacket. It is common to see a chubby baby in a cotton sling tied to his mother's or older sister's back. The *cheongsam*, a fitted dress with a high collar and slit skirt, is fashionable, and many men and women wear Western dress.

Almost every religion is represented in Hong Kong, and the year is brightened with traditional Chinese festivals. Chinese New Year, which falls in January or February, is a 3- to 5-day celebration with firecrackers and feasting. People exchange presents and children receive "lucky money" in red envelopes. Going to the movies is a favorite form of entertainment all year long. Many of the films shown are produced in Hong Kong, where film making is an important industry.

Economy. Hong Kong's economy grew up around its harbor, which is a free port. (Duties are charged on only a few items.) Goods from all over the world are sent to Hong Kong, unloaded, and then shipped on to other countries. The port also has excellent modern warehouses for storing goods, and dockyards for repairing and building ships. Since the end

Victoria, capital of Hong Kong, has both modern buildings and refugee shacks.

Part of the harbor of Hong Kong, crowded with sampans.

of World War II a number of industries have grown to tremendous importance in the economy: cotton and synthetic textiles; the making of clothing; plastic products (flowers, dolls, toys); and electrical and electronic products (radios, TV sets). The supply of skilled labor, low wages, and low taxes helps explain the growth of industry. Each year more and more tourists are attracted by Hong Kong's colorful sights, its shops, luxurious hotels, and lovely beaches, such as Repulse Bay, Shek O, and Clear Water Bay.

Farmers, who are just a small percentage of the population, raise mainly vegetables, pigs, and poultry, but the farmers cannot raise enough for the people's needs. Hong Kong buys much of its food from China. A fishing fleet of over 6,800 vessels supplies most of the colony's fish.

History. The island of Hong Kong was ceded to Great Britain by China in 1842, after the Opium War. The tip of Kowloon peninsula and Stonecutters Island were ceded in 1860, and in 1898 the New Territories were leased to Great Britain by China for a period of 99 years. As a crown colony, Hong Kong, a member of the Commonwealth of Nations, is ruled by Queen Elizabeth II, who is represented by a governor.

At the moment, Hong Kong is experiencing a boom. But the picture could change quickly. Hong Kong's future is linked to the attitude of the People's Republic of China. Since 1949 China has tolerated the British colony on its coast, but no one can predict how long that toleration will last.

Reviewed by ROBERT SUN, Hong Kong Trade Development Council

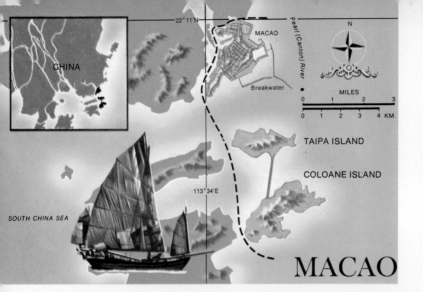

MACAO

MACAO

There is a certain sense of excitement about Macao, for this Portuguese territory in Southeast China has a history of adventure and intrigue. Its location on the South China Sea made it vital to the fortunes of traders, who amassed great wealth smuggling opium and tea. Now the illegal traffic has been greatly reduced, and Macao's economy today is based mainly on shipping, fishing, and tourism.

The Land and the People.   Macao (Macau in Portuguese) is located on the Pearl (Canton) River and comprises the city of Macao and the islands of Taipa and Coloâne—a total of 6 square miles (16 square kilometers). Originally, most of the inhabitants were Portuguese traders and missionaries. Today, Chinese make up the great majority of the population, which is estimated at 260,000. Due to the influx of people from neighboring regions, the ranks of the Chinese community are continually swelling. This has created a serious economic problem, and Macao is forced to import most of its food from China. In addition to the permanent population, some 400,000 Chinese live on sampans on the Pearl River. The mixture of Portuguese and Chinese cultures is striking. Roman Catholic churches stand side by side with Chinese temples. Streets with Portuguese names are lined with the shops of Chinese merchants. One of the most impressive sights in Macao is the Praia Grande, a beautiful promenade along the sea.

History.   In the 16th century Macao was an important trade center between China and Europe. The Chinese Emperor permitted Portugal to establish a settlement there, and for centuries the Portuguese paid a yearly rental to China. In a treaty signed between the two countries in 1887, the Chinese agreed to recognize Portuguese sovereignty over the area. After Portugal's revolution in 1974, the Lisbon government granted greater internal authority to Macao and changed its status from overseas province to territory.

Government.   The territory of Macao is administered by a governor appointed by the Portuguese Government, a cabinet of five secretaries, and a legislative assembly. The cabinet secretaries are selected by the governor. Of the 17 members of the legislative assembly, five are appointed by the governor, six are elected by direct vote, and the remaining six are elected by professional and workers' groups. Macao sends one elected representative to the Portuguese parliament.

Reviewed by LEÃO LOURENÇO, Permanent Mission of Portugal to the United Nations

TAIWAN

In the late 1500's a group of Portuguese seafarers and traders became the first Europeans to visit the island of Taiwan. The Portuguese were so impressed by the rugged beauty of the island that they named it Ilha Formosa—"beautiful island." Following its discovery by the Portuguese, Taiwan (also known as Formosa) was ruled successively by the Dutch, a Chinese pirate-general, the Manchu emperors of China, and the Japanese. In 1949 Taiwan became the seat of the Chinese Nationalist Government (the Republic of China), which had been driven from the mainland by the Communists. Today, Taiwan is a thriving country of about 16,000,000 people, who have turned this mountainous island into both a prosperous farming land and a growing industrial nation.

THE LAND

Taiwan is located about 100 miles (160 kilometers) off the coast of mainland China, across the storm-swept seas of the Taiwan (Formosa) Strait. It is the largest of the 14 islands of the Taiwan group, which along with the Pescadores, Matsu, and the Quemoy group make up the territory of the Republic of China. All of the islands together comprise

Terraced fields cover the slopes of Taiwan's rolling hill country.

an area of nearly 14,000 square miles (36,000 square kilometers). The island of Taiwan alone is about 13,800 square miles (35,740 sq. km.).

Much of Taiwan is mountainous, with high-rising peaks and tree-lined ridges. Mountain ranges run the entire length of the island, covering the eastern two-thirds. The remaining third in the west is a level plain, which contains most of the island's farmland and most of its population. An east–west highway, completed in 1960, connects the highlands of the east with the fertile western lowlands.

Climate. Taiwan has a subtropical climate, with extremely hot and humid summers and mild winters. Monsoon winds bring heavy rains to the south during the summer and to the north during the winter months. Violent storms called typhoons lash the island during the late summer and early autumn, often resulting in serious damage and flooding.

Economy. Only about one fourth of the land on Taiwan is suitable for farming, but agriculture is still the main activity, involving about 45 percent of the population. As a result of a land reform program begun in the 1950's, roughly 70 percent of the farmers own the land they till. By means of modern techniques and the use of chemical fertilizer and insecticides, crops have been increased. Each year the farmers grow and harvest two crops of rice. In between rice crops they use the fields to grow soybeans, melons, and sweet potatoes, along with some wheat, jute, and tobacco. On drier land or in less fertile fields, pineapples, tea, sugar, and bananas are planted. Because they are no longer dependent on rice alone, the farmers now feel more economically secure.

According to many experts, Taiwan has nearly reached the limit of its agricultural productivity. Therefore, the country is moving to develop further its small but growing industries. During the past decade, Taiwan's industrial production has risen almost 14 percent each year. Among Taiwanese products are electrical appliances, textiles, processed foods, plastics, motor vehicles, radios, and ceramic dinnerware. Lumbering and fishing are important industries, and the country also manufactures iron and steel, aluminum, fertilizers, and machinery.

These are aircraft maintenance workers for CAT, a Taiwanese airline.

FACTS AND FIGURES

REPUBLIC OF CHINA is the official name of the Nationalist Chinese Government on Taiwan.
CAPITAL: Taipei.
LOCATION: Island about 100 miles off the coast of mainland China. **Latitude**—21° 54′ N to 25° 18′ N. **Longitude**—120° 03′ E to 122° E.
AREA: 13,885 sq. mi. (35,961 sq. km.), not including islands of Quemoy and Matsu.
PHYSICAL FEATURES: Highest point—Sinkao Shan (Mount Morrison) (13,113 ft.; 3,997 m.). **Lowest point**—sea level. **Chief rivers**—Tanshui, Choshui.
POPULATION: 16,000,000 (estimate).
LANGUAGE: Mandarin Chinese (official), plus many regional dialects.
RELIGION: Buddhist, Taoist, Confucian, Christian, Muslim.
GOVERNMENT: Republic. **Head of government**—president. **Legislature**—national assembly. **International co-operation**—Sino-U.S. Mutual Defense Treaty, 1954.
CHIEF CITIES: Taipei, Kaohiung, Taichung, Keelung, Tainan.
ECONOMY: Chief minerals—coal, limestone, salt, petroleum, natural gas, silver, gold, pyrites. **Chief agricultural products**—rice, sugarcane, sweet potatoes, bananas, pineapples, citrus fruits, soya beans, groundnuts. **Industries and products**—textiles, cement, glass, electrical products, refined oil, paper, fertilizers. **Chief exports**—textiles, metals, machinery, wood and wood products, bananas, chemicals, canned foods, sugar. **Chief imports**—machinery and tools, ores and metals, vehicles and parts, electrical equipment, raw cotton, crude oils, wheat.
MONETARY UNIT: New Taiwan dollar (yuan).
NATIONAL HOLIDAY: October 10, National Day.
NATIONAL ANTHEM: *San Min Chu I* ("Three Principles of the People").

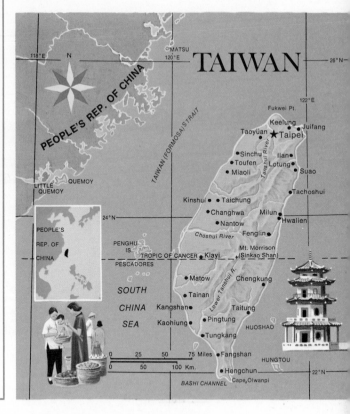

THE PEOPLE

The people of Taiwan are mainly Chinese in origin, though there are some 200,000 aborigines belonging to Malayan and Polynesian ethnic groups. Over 80 percent of the Taiwanese are descended from Chinese people who came to Taiwan during the 17th, 18th, and 19th centuries. An additional 2,000,000 Chinese fled the mainland after the Communists took over in 1949.

Mandarin Chinese is the official language of the country, but many other Chinese dialects are spoken by the people. The dominant religion is a mixture of Buddhism and Taoism, which also includes ancestor and spirit worship. There are about 400,000 Christians and a scattering of Muslims on the island.

Way of Life. In the larger cities of Taiwan, people live, dress, and work in much the same way as the people in big cities everywhere. Their way of life is determined mainly by the occupations they follow. Family structure is based on ancient Chinese tradition, and it is common for households to contain members of three generations—grandparents, parents, and children. Generally speaking, the old customs are not as well kept by city people as they are by the rural Taiwanese.

In the farm villages and smaller towns, people live much closer to the traditional Chinese way of life. When doing business in village markets, farmers follow the ancient practice of dealing first with their closest relatives. The circle then extends to include relatives by marriage, villagers bearing the same name, fellow villagers, schoolmates, and foster brothers, in that order of preference.

Among rural people rice farmers are the richest and the most

Outdoor markets such as this one are a common sight in Taiwan's cities.

socially prominent group. Usually they live in comfortable red-brick houses with tile roofs. These homes, which are located in the center of the farmers' rice paddies, are spacious by Taiwanese standards. Most have electricity, and it is now common to find modern electrical appliances in many farmhouses. The family's water buffalo—still a useful domestic work animal—usually occupies one room in the house, even though this means some crowding for the rest of the family.

The farmers sell their rice to the farmers' association, an organization that serves as a credit union and a co-operative. If a farmer wishes, he can deposit his savings with the association or borrow money to buy seed and tools when he needs them. The farmer can also buy chemical fertilizer from the association's general store and attend illustrated lectures on the latest farming methods.

The most important holiday celebration on Taiwan is the Lunar New Year. The occasion is marked by parades, fireworks, and family gatherings. During this period, the Taiwanese put on red-colored clothing, give red-wrapped packages of gift money to children, and post good luck mottoes on red paper on the doors of their homes. The color red signifies the hope for a happy and prosperous year to come. New Year festivities end with the coming of the first full moon of the lunar year. The final event is a great parade featuring children carrying handmade lanterns shaped in the form of animals.

On significant religious festival days, the farmers offer animal sacrifices to their favorite Buddhist gods. The most revered of the idols is the sea goddess Ma-tsu, the patron saint of the fisherman. Each year the more prosperous farmers hold an extravagant festival called the Ghost Festival, which is centered around a village feast (*pai-pai*). Next to the Lunar New Year the Ghost Festival is probably the biggest social event of the countryside.

Education. Free education is provided by the national government for all children between the ages of 6 and 15, and over 80 percent of the people on Taiwan can read and write. Higher education is provided

Bikes, scooters, and cars crowd the avenues of Taipei, capital of Taiwan.

by nine universities and about 70 colleges and technical institutes. The largest of the universities is the state-run National Taiwan University, which is especially noted for its medical and agricultural schools. The two best-known private universities are Tunghai and Soochow.

Cities. Taipei, the island's largest city, is located in the northern part of Taiwan, on the banks of the Tanshui River. Besides being the capital of the Nationalist Republic of China, the city is also a center of commerce, industry, and cultural life. Taipei has a population of about 1,200,000. The most important southern city is Kaohiung, one of the country's major ports (population about 660,000). Other urban centers include Taichung, Keelung, and Tainan.

HISTORY

Chinese settlers first arrived on Taiwan in the early 7th century A.D., but large-scale migration did not take place until the 17th century. By that time Europeans had also discovered the island, and the Spanish and the Dutch had set up trading posts. The Dutch drove out the Spaniards in 1642 and controlled Taiwan until 1661, when a Chinese army under a former pirate named Cheng Ch'eng-kung (Koxinga) invaded the island. Cheng and his followers defeated the Dutch and established a Chinese government opposed to the Manchus on the mainland, who had recently toppled the Ming dynasty (1368–1644).

In 1683 the Manchus took over Taiwan. They ruled the island for 2 centuries, until they were defeated in the Sino-Japanese War of 1894–95. The Treaty of Shimonoseki, which ended the war, provided that Taiwan, along with other Chinese islands, was to be ceded to Japan.

After 50 years of Japanese rule, Taiwan was returned to China at the end of World War II (1945). In 1949 the Nationalist Government of the Republic of China was overthrown by the Communists, and the Nationalists were forced to flee to Taiwan. Since then, the Nationalists have maintained their government on Taiwan, which they consider a province of the Republic of China.

A Taiwanese craftsman puts the finishing touches on a porcelain vase.

GOVERNMENT

Taiwan has two governments—the national government of the Republic of China and the provincial government of Taiwan. The national government is based on a constitution adopted in 1946. Under this constitution, the government is divided into five Yuans (branches), according to the teachings of Dr. Sun Yat-sen, the founder of the Chinese republic. These include the executive, legislative, judicial, examination, and control Yuans. The president of the republic is the head of state. He is elected by the National Assembly, the legislature. The office of President of the Republic of China was held by Chiang Kai-shek from 1948 until his death at the age of 87 in 1975.

Operating under the president are the five Yuans. Besides the courts and the National Assembly, there is an executive Yuan consisting of a premier, appointed by the president, and a cabinet of ministers. There are also an examination Yuan to administer the civil service and a control Yuan, which is responsible for maintaining standards of honesty and efficiency among public officials.

Local affairs on the island are administered by the Taiwan Provincial Government. It is headed by the provincial governor, who is appointed for an indefinite term by the national government. A provincial assembly, elected by popular vote, is the legislative body.

Over the years Taiwan has been able to achieve political and economic stability. Economic progress was helped considerably from 1950 to 1965 by United States economic and technical aid. As a result, the people of Taiwan enjoy one of the highest standards of living in Asia.

OTIS H. SHAO, Dean of the Graduate School, University of the Pacific

KOREA

Korea has long been known as the "land of morning calm," a poetic translation of the Korean word *choson,* an early name of the country. But the history of Korea—marked by wars, invasions, and foreign domination—has not been very calm. Both China to the north and Japan to the east have fought in the past to control the Korean peninsula. The veil of morning mist that rises each day over the rugged peaks and broad ridges of Korea's numerous mountains covers a land still struggling for survival, as it has for centuries. Today, as before, this struggle is economic and political. Korea has relatively few natural resources, and only about 20 percent of the land is suitable for farming. In 1945, after World War II, Korea was divided into two states, the Communist Democratic People's Republic of Korea (North Korea) and the Republic of Korea (South Korea). Tension between the two states has remained high since the Korean War (1950–53), when North and South Korea fought each other in a long and bitter conflict.

Modern Korea is a mixture of the old and the new. Farmers work their small plots of land dressed in the simple white cotton clothing that has been traditional in Korea for hundreds of years. They till the soil as

A woman prays atop Wang Mountain, overlooking Seoul.

A farmer with his plow and bullock near modern buildings outside of Seoul.

their fathers and grandfathers did, using wooden plows drawn by bullocks, and they still use the old A-frame (a wooden pack device at least 1,000 years old) to carry heavy loads. But while the farmer continues to follow ancient ways, a new urban Korea is emerging. In the cities, commuters in Western-style clothes climb aboard buses that take them to modern office buildings and factories. The white-clad farmer, the businessman in his Western-style suit, the housewife in her *chima* (a long robe), and the young girl in a miniskirt illustrate the mingling of the ancient and the modern in today's Korea.

THE LAND

Korea is a peninsula surrounded by three seas—the Yellow Sea to the west, the Sea of Japan to the east, and the East China Sea to the south. To the north lies Manchuria and a small strip of Soviet Siberia. Much of the land is covered by hills and mountains, through which flow numerous rivers. Korea has a total area of about 85,000 square miles (220,000 square kilometers), which includes more than 3,000 offshore islands. Under the present division of the country, North Korea occupies more than half of the peninsula, although it has less than one third of the total population (about 13,300,000 out of a total of some 44,000,000 people).

North Korea is more mountainous than the south, and its central section is an almost continuous mass of rolling hills and rugged peaks. The Taebaek Range, Korea's largest chain of mountains, runs along the east coast below the city of Wonsan, and includes the Diamond

Dry rice fields are in the foreground in this view of the Korean countryside.

Mountains, which are noted for their forest-covered ridges, waterfalls, and splendid views. Along the west and southeast coasts, the mountains taper off into narrow coastal plains, which contain most of Korea's fertile land. Most of the good farmland lies in the southern part of the peninsula, while the north is more favored with mineral resources. Two of North Korea's major rivers, the Yalu and the Tumen, mark the northern border with Manchuria. Both rivers begin on the steep slopes of Paektu-san, Korea's highest mountain. These rivers are an important source of hydroelectricity for North Korea's industries. South Korea's major rivers include the Kum and the Han, which provide water for crops in the western lowlands, and the Naktong, which irrigates the southeastern coastal plains. Korea's climate is generally continental, but winters are much colder and harsher in the north, and summers are warmer and more humid in the south. Rainfall is much heavier in the south than in the north. About three fifths of the total annual rainfall comes during the summer months.

Economy.   South Korea has been primarily an agricultural country, while North Korea has been more industrialized. Rice, barley, corn, wheat, fruit, potatoes, and tobacco, however, are grown in both the north and the south. South Korean farmers own and work small plots of land, usually a few acres. In the north, farmers are organized into co-operatives where they live together, share farming equipment, and work under government supervision. North Korea has substantial deposits of coal, iron ore, copper, lead, and gold. Iron and steel production is a major industry in the north, as is food processing and the manufacture of chemicals and textiles. Assisted by other Communist nations, North Korea has increased its production of heavy machinery, tractors, electrical equipment, and other items. North Korea's economic growth, which was rapid until the 1960's, has now slowed.

There has been a remarkable growth of industry and other economic activities in South Korea during the past decade. Fishing and the manufacture of textiles remain the leading industries, but many new plants have been constructed. They include oil refineries, synthetic fabric plants, and an automobile factory, which produces the first Korean-assembled cars. South Korea's economic expansion, however, has been helped by aid from the United States, the United Nations, Japan, and Western European countries. While the south has tungsten and small deposits of coal, iron, and gold, its most important natural resource is its people.

THE PEOPLE

The Koreans are descended from various Tungusic peoples (related to the Mongolian people), who migrated into the Korean peninsula from Central Asia several thousand years before the birth of Christ. According to Korean folklore, the Tungus were joined (around 1100 B.C.) by a group of Chinese led by a legendary figure named Kija. They are credited with introducing Chinese culture to the early Koreans.

Language. Although divided politically and geographically, North and South Koreans are united by their language, Korean. Until the 15th century A.D., Koreans used the Chinese alphabet, since they had no written language of their own. But under the guidance of King Sejong, Koreans invented a new system of letters called *hangul,* a very advanced and scientific alphabet.

Hangul was officially adopted in 1446, but until the 20th century Chinese was more widely used by literary people and the nobility.

Korean women in traditional dress with their children.

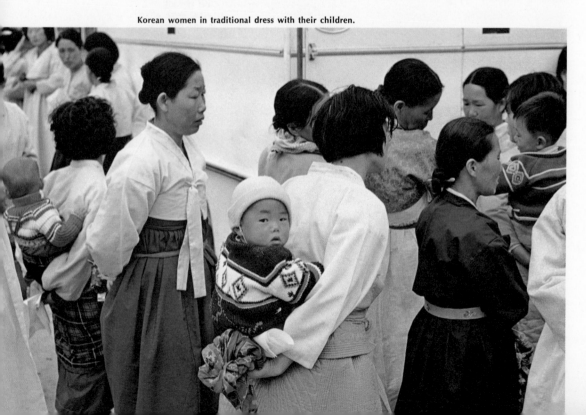

FACTS AND FIGURES

DEMOCRATIC PEOPLE'S REPUBLIC OF KOREA—Choson Minjujuui Inmin Konghwa-guk—is the official name of North Korea.

CAPITAL: Pyongyang.

LOCATION: Northeastern Asia. **Latitude**—38° N to 43° 01′ N. **Longitude**—124° 18′ E to 130° 39′ E.

AREA: 46,540 sq. mi. (120,538 sq. km.).

PHYSICAL FEATURES: Highest point—Paektu-san (9,003 ft.; 2,744 m.). **Lowest point**—sea level. **Chief rivers**—Yalu, Tumen, Imjin, Taedong.

POPULATION: 13,300,000 (estimate).

LANGUAGE: Korean.

RELIGION: Buddhist, Confucianist, Christian, Chondokyo, animistic beliefs.

GOVERNMENT: Communist republic. **Head of government**—president. **Legislature**—Supreme People's Assembly.

CHIEF CITIES: Pyongyang, Wonsan, Chongjin, Hamhung, Najin, Kaesong.

ECONOMY: Chief minerals—coal, iron ore, copper, lead, zinc, tungsten, gold, graphites, phosphates, silver. **Chief agricultural products**—rice and other grains, meat, milk, vegetables, hemp, tobacco, flax, fruit. **Industries and products**—iron and steel, food processing, textiles, fishing, electrical machinery, tractors and other farm equipment, cement, mining and construction machinery, glass. **Chief exports**—iron and steel, agricultural products, minerals, chemicals. **Chief imports**—machinery, fuel, textiles, rubber.

MONETARY UNIT: Won.

NATIONAL HOLIDAY: September 9, Founding of the Democratic People's Republic.

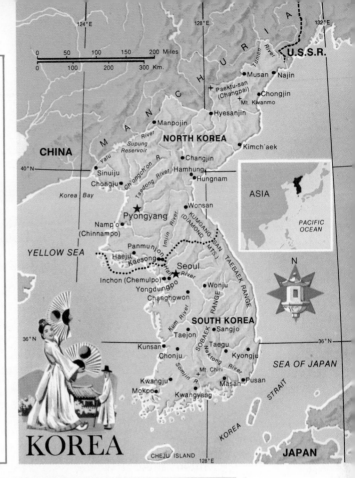

KOREA

FACTS AND FIGURES

REPUBLIC OF KOREA—Taehan Min-guk—is the official name of South Korea.

CAPITAL: Seoul.

LOCATION: Northeast Asia. **Latitude**—34° 17′ N to 38° N. **Longitude**—124° 58′ E to 129° 35′ E.

AREA: 38,022 sq. mi. (98,477 sq. km.).

PHYSICAL FEATURES: Highest point—Mount Chiri (6,283 ft.; 1,915 m.). **Lowest point**—sea level. **Chief rivers**—Naktong, Han, Kum, Somjin.

POPULATION: 31,000,000 (estimate).

LANGUAGE: Korean.

RELIGION: Buddhist, Confucianist, Christian, Chondokyo, animistic beliefs.

GOVERNMENT: Republic. **Head of government**—president. **Legislature**—national assembly.

CHIEF CITIES: Seoul, Pusan, Inchon, Taegu, Kwangju.

ECONOMY: Chief minerals—coal, tungsten, iron ore, gold, copper, graphite, silver. **Chief agricultural products**—rice, wheat, barley, soybeans, potatoes, fruit, cotton, tobacco, silk. **Industries and products**—fishing, textiles, chemicals, cement, electrical products, rubber tires, shoes, diesel engines, food processing, paper and paper products, glass, plywood. **Chief exports**—clothing, plywood, electrical and electronic goods, textiles. **Chief imports**—machinery and transport equipment, chemicals, mineral fuels, livestock, foodstuffs.

MONETARY UNIT: Won.

NATIONAL HOLIDAY: August 15, Liberation Day.

NATIONAL ANTHEM: *Aegug-ga* ("The National Anthem").

Recently, however, both North and South Korea have taken steps to make *hangul* the only written language used.

Religion. Buddhism and Confucianism, the country's major religions, came to Korea from China. In the earlier periods of Korean history, Buddhism was the main religion and a source of inspiration for much of Korean art. Confucianism was introduced after the 7th century A.D. and was made the official state religion in the 14th century. South Korea also has a large Christian population, about 7 percent Protestant and 3 percent Catholic. A local religion called Chondokyo ("religion of the heavenly way") is practiced by many people.

Way of Life. The majority of Korea's people live in rural areas, since farming is still the major occupation. Most of South Korea's farming population lives in the country's "rice bowl"—the fertile lowlands along the coast. Because of the heavy concentration of people in this area, South Korea has one of the heaviest densities of population per square mile of farmland in the world (more than 800 people per square mile). North Korea is much less densely populated, and more people are employed in industry than in the south.

Whether in cities or in rural areas, most Koreans lead a simple, often hard life. The average farmer's cottage has walls made of earth or baked clay and a thatch roof, while the homes of wealthier families are constructed of stone or brick and have tile roofs. Glass is used in the windows of modern houses, but more often in traditional homes the windows are made of an oily, translucent paper. Lacquered paper is also used to cover the floors (which are heated by ducts carrying hot air), and people must therefore remove their shoes before going inside. Outside the house, in the courtyard, the family's basic food supply is stored in earthenware jugs standing on a raised platform called the *changdokte*. Many Koreans still maintain the old custom of separate tables for men and women at mealtime. Rice is the basic food and is served with every dish. Beef, pork, poultry, and dried fish are also part of the diet, along with such Korean specialties as *kimchi*, a hot pickle, *kuksu*, a spaghettilike dish, and a variety of soups. Chopsticks are most commonly used for eating, except in the case of soups, when long-handled brass or silver spoons are used.

A custom that has survived through the centuries, particularly in the small farming villages, is that of arranged marriages. Usually the parents of the boy contact a matchmaker and have him find a suitable girl for their son. Then the two families agree to the marriage by signing a contract and announcing the couple's engagement. However, the boy and girl do not actually meet until the wedding day. In former times it was quite common for marriages to take place when the bride and groom were only 13 or 14 years old, but such early marriages are now very rare.

Education. Both North and South Korea have made great strides in education since Japanese rule ended. In both sections most people can read and write, and the rate of illiteracy is under 10 percent. Korean children are required to attend primary school for 6 years, and in the north 3 years of middle school (seventh to ninth grade) are also compulsory. Because of a shortage of funds, middle schools and high schools (10th to 12th grade) in the south are overcrowded. North Korea's schools also are believed to suffer from this problem. Korean culture stresses the value of education and most families want their children to go to a university. But university facilities and enrollment are limited by the government, and only a relatively small number of Koreans are university educated. In recent years technical and scientific training have been stressed in both northern and southern universities so that graduates will be able to find jobs in the growing industrial field.

Arts and Sciences. Although greatly influenced by Chinese culture, Korea developed an artistic and scientific tradition of its own. Examples of traditional Korean art and architecture can be seen today. The palaces of Seoul, modeled after those in the Chinese city of Peking, are a reminder of the old splendor of Korea's royal court. Ancient Buddhist temples, which dot the mountainsides or are hidden in remote valleys, continue to provide monks with a quiet atmosphere for meditation. Within these temples are beautifully sculptured figures of Buddha, made from stone, iron, and gold. One of the outstanding achievements of Korean art is the magnificent pottery created during the great dynasties of past centuries. In particular, the gray-green bowls and other ceramic products of the Koryo period (A.D. 935–1392) are highly regarded throughout the world. Even today pottery craftsmen in many other countries use Korean ceramics as a model for their own work.

Traditional Korean literature was written in Chinese and was often based on stories and legends from Chinese folklore. However, Koreans also developed a literature of their own. Much of Korean poetry deals with the natural beauty of the country and with life in the royal court, where many of the early poets were employed. Some Korean poetry deals with universal emotions, as in the case of the love poems of Hwang Chini (1506–44), Korea's most famous woman poet. Korea has a rich musical tradition, which includes both royal court music and ballet and folk songs and dances. The folk dances are very spirited and are often accompanied by heavy beating of drums. Delicate movement of the arms, shoulders, and hands are features of Korean folk dancing.

Koreans have also made important technical and scientific advances. Movable printing type was first used in Korea (the date is variously given as 1234 or 1403), before the German printer-inventor Johann Gutenberg developed it in Europe. One of the world's earliest astronomical observatories was built near the ancient city of Kyongju in the 7th century. The Korean admiral Yi Sun-sin used history's first iron-clad warships (the "turtle boats") to defeat a Japanese invasion fleet in 1592.

Cities

Most people in the cities live in fairly cramped quarters in crowded residential districts. Unemployment is a serious problem in the south. A considerable number of people in the work force are without jobs, while many others are employed only part of the time. Many farmers hoping to earn extra money come to the cities looking for work—especially in winter, after the farming season—and this adds to the unemployment problems.

Seoul. Korea's largest city is Seoul, the capital of South Korea, which has over 4,000,000 people. Seoul is an example of the changing face of Korea. Gleaming new office buildings and hotels rise up next to centuries-old palaces and Buddhist temples. Seoul is located on the banks of the Han River, in a narrow valley framed by low mountains. Founded in 1392, the city has many historic shrines, including the majestic Kyongbok Palace.

The country's second largest city and major port is Pusan (population about 1,400,000), which is located on the southeastern tip of the peninsula. Two other important ports on the west coast are Inchon (about 500,000), which serves Seoul, and Kunsan. Other large urban centers include Taegu (about 850,000) and Kwangju (about 400,000).

Pyongyang. North Korea's principal city is Pyongyang, its capital, which has an estimated population of 653,100. Pyongyang, like Seoul, was almost completely destroyed during the Korean War and has been largely rebuilt. Situated on the Taedong River, it is an important industrial and commercial center. Other leading northern cities include the ports of Wonsan, Hamhung, and Chongjin. (Population figures for most of North Korea's cities are not available.)

HISTORY

Korea has one of the longest continuous histories of any nation (within the same boundaries) in the world, going back some 5,000 years. The early history of Korea is shrouded in legend and revolves around the ancient kingdom of Choson, which came into being about 2300 B.C. The Chinese established the colony of Lolang in Korea in the 2nd century B.C., but the Koreans drove them out 5 centuries later. By then three small kingdoms had emerged in Korea. The kingdom of Silla eventually defeated its two rivals and united the peninsula under its own rule. The united Silla dynasty that followed (A.D. 668–935) brought in an era of great cultural and scientific achievement, as well as relative peace and prosperity. Internal troubles led to the decline of the Silla, and in the 10th century A.D. the Koryo dynasty arose in its place. During this period (935–1392) Korea was plagued by invasions. The Mongols led by Genghis Khan swept across the peninsula, and Korea became a part of the Mongol Empire.

After the Mongol Empire collapsed in the late 14th century, various aristocrats and military men attempted to seize power in Korea. Finally a Korean general named Yi Sung-gy drove out corrupt officials and established the Yi dynasty (1392–1910). Confucianism was introduced as the official religion, social and political reforms were begun, and the nation's capital was moved from Kaesong to Seoul. But Korea still found itself threatened by China and Japan. Both countries wanted to control Korea in order to expand and to protect their own borders from attack. Following the unsuccessful Japanese invasion of 1592–98, Korea came under partial control of the Manchu from the north. For the next few centuries Korea shut itself off from the rest of the world, becoming a hermit nation. In the 1800's Russia, Japan, and China competed for control of Korea. After the Russo-Japanese War (1904–5), the victorious Japanese moved into the peninsula and annexed Korea in 1910. The Japanese modernized Korea, developed industries, and built new railroads. But the improvements were made to benefit the Japanese conquerors, who held all important government and industrial posts, and Koreans resented them. The Koreans staged peaceful demonstrations for independence in 1919, but Japanese police broke up the protest and killed many Koreans.

In 1945, at the end of World War II, Russian troops occupied the northern half of the peninsula and American troops the southern half, and by agreement between the United States and the Soviet Union the country was divided along the 38th parallel. The division, which was intended to be temporary, remains in effect. An American military government controlled the south until August, 1948, when the Republic of Korea was established on the basis of a United Nations-supervised election. Dr. Syngman Rhee, a leader of the resistance movement against Japanese rule, was elected South Korea's first president and continued in that office until 1960. United Nations observers were not permitted in North Korea, where the Soviet Union set up the Communist Democratic People's Republic of Korea in September, 1948.

On June 25, 1950, North Korean troops invaded South Korea in a forcible attempt to unify the peninsula under Communist rule. The poorly armed South Korean Army was overwhelmed by the Soviet-equipped North Korean troops, and only the arrival of military forces from the United States and other nations—brought together under the leadership of the United Nations—saved South Korea from defeat. The United Nations forces counterattacked, and drove the North Koreans almost to the Manchurian border. But then the Communist Chinese entered the war and the United Nations Army had to retreat south. Two years of bitter fighting followed, until an armistice was worked out in July, 1953. Technically, however, there is still a state of war between North and South Korea.

GOVERNMENT

South Korea has a presidential form of government. Changes to the constitution made in 1972 gave the president much stronger powers, including the power to appoint one-third of the legislature, the National Assembly. The balance is elected by the people. The president is assisted by a newly-formed body called the National Conference for Unification.

Under the new constitution of 1972, North Korea's Communist government is headed by a president and a cabinet. According to law, the Supreme People's Assembly, the legislature, is North Korea's most powerful political body. But in practice policy decisions are made by the president and cabinet and automatically approved by the Assembly.

DAVID I. STEINBERG, Representative, the Asia Foundation
Author, *Korea: Nexus of East Asia*

Military Armistice Commission at Panmunjom. North Korean delegates are at right.

JAPAN: AN INTRODUCTION
by H.I.H. Prince Takahito MIKASA

Mount Fuji is a symbol of Japan. In the days before air transportation, the first impression of Japan a foreign visitor received as his ocean liner approached the port of Yokohama was Fuji's magnificent peak towering over the land. But even today, if the weather is clear, a traveler arriving in Japan by plane can see Fuji's peak in the distant west as his plane descends toward Tokyo's international airport. The mountain looms like a huge inverted fan, its slopes extending far on either side.

Tens of thousands of years ago Japan's Mount Fuji and the area around it was a volcanic zone. This volcanic zone was erupting, pouring out volcanic ash. The ash accumulated to form a stratum, or layer, of loam. Then, gradually, another stratum of earth covered the layer of loam to become the land on which the Japanese people live today.

The loam stratum was laid down during the latter part of Japan's Diluvian Age—about 50,000 to 10,000 years ago. In this period Japan was connected by a land bridge to the Asian mainland. But for many years no one could prove that man had used the land bridge to cross into Japan. For years scientists thought that while Old Stone Age (Paleolithic) man developed in Europe, Africa, and inland Asia, Japan was uninhabited. Yet within the past 25 years, scientists have pushed back by hundreds of thousands of years the dates of Japan's known prehistory.

For a long time the earliest known traces of man in Japan belonged to the Jōmon culture. The people of this culture seemed to have appeared in Japan after it was cut off from the mainland by a rise in the sea level. Why man appeared in Japan only after it was separated from the rest of Asia was a mystery. The answer to the mystery began to unfold in 1949. One day a young peddler of sweets with a deep interest in archeology discovered in the loam stratum formed from the ash of Mount Fuji stone implements that seemed to have been made by man. Archeologists found them to be relics of the Paleolithic Age. Soon other relics were found, proving that man had lived in Japan in Paleolithic times.

Meanwhile anthropologists were also at work. In 1957 pieces of human bone were unearthed in a limestone quarry. These bones were thought to belong to a man who had lived about 100,000 years ago. The man was named Ushikawa man, after the place where he was found.

A few years later another discovery of human bones was made, these belonging to what came to be known as Mikkabi man. Since fragments of the skull were found, anthropologists could tell that he belonged to the same species we do—*Homo sapiens*. Mikkabi man was thought to have lived in Japan more than 20,000 years ago. Scientists now concluded that both Ushikawa man and Mikkabi man had entered Japan when it was part of the mainland. The Jōmon-culture people, whose origins had been unclear, were now thought to be the descendants of Mikkabi man.

Much of Japan's history after the Jōmon period has long been known. But scientists are refining our knowledge. Several thousand years after the height of the Jōmon culture, a great development, which we can now trace in detail, changed Japanese life. During the 4th and 3rd centuries B.C., the Japanese learned how to grow food—mainly rice. Rice cultivation was first conveyed from the Asian continent to the southwest of Japan. It slowly spread northward, and by A.D. 300 it had reached the northern tip of the main island of Honshu. Copper, bronze, and iron implements were imported too, and soon they were being made by Japanese hands. This culture is called the Yayoi culture. According to Chinese official history, by the time the Yayoi culture had spread throughout Honshu, the many small Japanese states were about to become one country.

During the 4th and 5th centuries, a unified Japan and a new civilization were developing. Japan was ruled by a line of emperors from which the present ruling house descends. The early emperors built their palaces where the city of Osaka stands today. The size and solidity of the mausoleums of two 5th-century emperors, Ōjin and Nintoku, show that even in that far-off time the prestige of the emperor was very great.

Japanese civilization was influenced by neighboring China and Korea. The importation of kanji (Chinese word characters) and of Buddhism was particularly important. The introduction of writing in the 5th and 6th centuries was a great advance. In the same period many priests, scholars, and artisans came to Japan from China and Korea. The influence of these adopted Japanese was an additional element in the development of Japan's ancient civilization.

Japan's relations with its neighbors grew closer. But diplomatically speaking this took the form of Japan's paying tribute to China. When Prince Shōtoku became Prince Regent of Japan, at the end of the 6th century, he altered Japan's relations with China. The two nations thereafter dealt with each other on an equal footing. In 604 the Prince set forth Japan's first written laws, based on the ethics of Buddhism. The 8th century was the high point of Japan's ancient period. A great capital was built at Nara and the earliest history of Japan was compiled. The famous statue of Buddha at Nara was completed. And the culture of West Asia (Sassanian Persia) was introduced into Japan by way of China.

Although Japan's ancient civilization was built in part upon cultures introduced from the Asian mainland, during the last 1,000 years the Japanese have absorbed these elements and have re-created them as their own. Japan's climate and landscape have also played a great part in forming its unique culture. The mountains covered with green trees, the lakes and rivers filled with clear water, the gently changing seasons, the plains resplendent with flowers—all these have affected the arts and every aspect of daily life. Flower arranging, the tea ceremony, tanka and haiku poetry, Japanese painting, the kimono—these have been developed in close harmony with the changes of the seasons.

But now that Japan has transformed itself into an industrial society, it is in danger of losing the lovely natural surroundings it has been so proud of. Still, it is inconceivable that Japan's natural environment will be destroyed. The Japanese and their scientists are already seeking ways to preserve the beauty of the country they love so well.

As the reader turns the following pages, it is hoped that he will come to understand the Japanese a little better. Such an understanding should contribute to goodwill among nations and to world peace.

Mount Fuji, Japan's most famous landmark.

JAPAN

The ancient Egyptians told a legend about a huge and very beautiful bird called the phoenix. It lived for hundreds of years and then burnt itself up; from its ashes, however, another phoenix was born, just as big and just as beautiful. Japan has been likened to the phoenix. At the end of World War II much of the country lay in ruins, its economy almost completely destroyed. But in the relatively short span of time since then, the Japanese have rebuilt their land so that economically it now ranks third among the nations of the world, after the United States and the Soviet Union, and its beauty once again astonishes its foreign visitors.

The comparison to the phoenix breaks down at one point, however. Japan is not a very big land compared with the Soviet Union, China, Canada, or the United States. It is, in fact, only about $\frac{1}{20}$ the size of the United States. Furthermore, less than a fifth of its land is suitable for farming. Yet Japan supports the basic food needs of a population of over 100,000,000—half as large as that of the United States.

One of the most striking aspects of Japan is the contrast between ancient and modern ways, between Oriental and Western styles. The Japanese are among the most progressive people in the world, yet they want to keep some of their traditional customs. Thus the old and the new mingle in Japan. The fastest long-distance train in the world, traveling at speeds of 130 miles (210 kilometers) per hour and controlled by computers, races through countryside where farmers in paddy fields cultivate rice much as their ancestors did. In the midst of one of the most glittering entertainment districts in Tokyo stands an ancient

People from many lands visited Japan's world's fair, EXPO '70, outside Osaka.

Buddhist temple, Senso-ji (usually called Asakusa Kannon), said to have been founded in the 7th century. In the cities young Japanese wear the latest Western fashions, though elderly Japanese women may be seen clad in the traditional kimono. Many young women, after a day's work in one of the large electronics plants, practice ceremonial tea making. Many men and most girls practice the graceful, centuries-old tea ceremony from time to time. Japan, the greatest shipbuilding nation in the world, still turns out almost to a man to admire the cherry blossoms in the spring.

Probably the contrasts will become less and less noticeable in time. Industrialism seems to erase the differences among nations. So far, however, modernization has not greatly changed certain old manners and ways of thinking in Japan. The old Japan may still be seen in the quiet and courteous manners of the people, in their respect for old people, in their habit of hard work, in their love of learning and sensitivity to beauty, and in their great pride in their country.

One way of thinking, however, has definitely changed. The Japanese experience in World War II has persuaded them that they never again want to see such terrible suffering and destruction. It is vital in understanding Japanese beliefs today to realize that they are the only people ever bombed by nuclear weapons. The Japanese Constitution outlaws war. Article 9 of the Constitution states that "the Japanese people forever renounce war as a sovereign right of the nation, and the threat or use of force as a means of settling international disputes. . . . land, sea, and air forces, as well as other war potential, will never be maintained. The right of belligerency of the state will not be recognized."

In spite of this famous article, the Japanese Government did in fact begin to re-arm during the Korean War (1950–53). But what Japan spends on weapons is still small compared with what other powers spend. Nevertheless, many Japanese are critical even of this expense, and any suggestion that Article 9 be changed meets angry outcries from many of the Japanese people.

Japan's years of peace since the end of the war in 1945 and its comparatively low military expenses help account for the amazing growth of the Japanese economy. To comprehend that growth more fully, we must first look at the Japanese land. Surprisingly, it offers few of the natural resources that have helped other nations become prosperous.

THE LAND

The islands of Japan form a long, relatively narrow chain, lying about 100 to 500 miles (160–800 km.) off the east coast of Asia. Four of the islands—Kyushu, Shikoku, Honshu, and Hokkaido—are quite large, but there are over 3,000 smaller islands as well, some of them unpopulated. The distance from the southern to the northern tip of the four main islands is approximately the same as that from southern Spain to northern Germany.

The landscape varies widely, but most of Japan is mountainous. Mountains run like a backbone down all the islands. Some are worn and rounded old mountains, but more are jagged and young. The most famous mountain peak in Japan is the cone-shaped Mount Fuji, which the Japanese call Fuji-san. Many volcanoes also mark Japan's landscape. Over 50 of them have been active in historical times, and nine have erupted since 1958. There are also numerous hot springs, which are favorite holiday spots for the Japanese.

While the mountains and hot springs are fine for sportsmen and vacationers, the coastline and the small plains are more important to the livelihood of the Japanese people. Japan has a long coastline indented by many bays and inlets. (If a string following the coastline were stretched out to its fullest, it would reach almost three fourths of the way around the world.) In Japan one is never very far from the sea. Thus many Japanese are seafaring people—fishermen, sailors, shipbuilders, and merchants in international trade. Japanese shipbuilders meet the needs not only of their own country, but of many others as well. The fisherman is almost as important to the Japanese dinner table as the rice farmer. The waters around Japan, where cool currents from the north meet warm currents from the south, abound in tuna, pike, bonito, and other fish. Japanese fishermen and whalers also sail to many other parts of the world for their catches.

Most of Japan's farmlands are in the various lowland plains scattered across the island. The largest of these lowland areas is the Kanto plain, surrounding Tokyo, on the island of Honshu. Honshu is the largest of the Japanese islands and the most heavily populated; over three fourths of the people live there. Since land is so precious, it is intensively cultivated. Even mountainsides are farmed, some of the gentler slopes being terraced into plots for orchards and vegetable crops. The northern island of Hokkaido is next in size after Honshu, but it is much more thinly populated. In contrast to the predominantly small farms of the south, those in Hokkaido are relatively large. Farms in Kyushu and Shikoku benefit from a warm climate and long growing season. Japan's rivers are generally short and swift. They are usually too swift to be used for transportation but some have been harnessed to produce hydroelectricity.

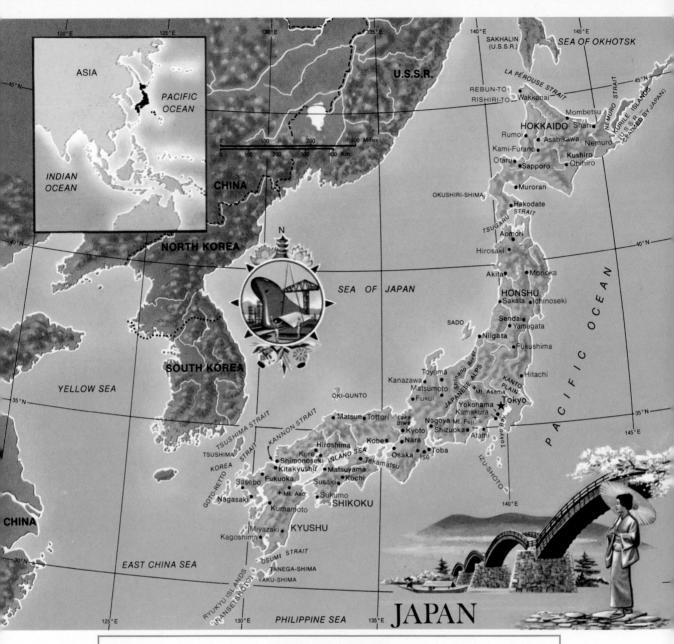

FACTS AND FIGURES

JAPAN—Nippon or Nihon—is the official name of the country.

CAPITAL: Tokyo.

LOCATION: East Asia. **Latitude**—20° 59′ N to 45° 33′ N. **Longitude**—122° 56′ E to 153° 58′ E.

AREA: 145,737 sq. mi. (377,459 sq. km.). Includes Okinawa and southern Kuriles.

PHYSICAL FEATURES: Highest point—Mount Fuji (12,389 ft.; 3,776 m.). **Lowest point**—Tokyo (about 11 ft.; 3.4 m.). **Chief rivers**—Shinano, Ishikari, Tone. **Major lake**—Biwa.

POPULATION: 104,650,000. Includes Okinawa.

LANGUAGE: Japanese.

RELIGION: Shintoist, Buddhist, Christian.

GOVERNMENT: Constitutional monarchy. **Head of state**—emperor. **Head of government**—prime minister. **Legislature**—national Diet. **International co-operation**—United Nations, Colombo Plan.

CHIEF CITIES: Tokyo, Osaka, Nagoya, Yokohama, Kyoto, Kobe, Kitakyushu, Sapporo, Kawasaki.

ECONOMY: Chief minerals—coal, iron, manganese, zinc, copper, lead, gold. **Chief agricultural products**—rice, wheat, barley, tea, sweet potatoes, soybeans, tobacco, oranges, apples. **Industries and products**—heavy machinery, textiles, steel, optical and electronic products, fishing, shipbuilding. **Chief exports**—iron and steel, ships, textiles, fish, heavy machinery, cameras, optical instruments, radios, TV sets, cultured pearls. **Chief imports**—petroleum, iron ore, coal, wheat, lumber, soybeans, sugar, ginned cotton.

MONETARY UNIT: Yen.

NATIONAL HOLIDAYS: New Year's Day; May 3, Constitution Day; April 29, the Emperor's Birthday; plus nine other national holidays.

NATIONAL ANTHEM: *Kimigayo* ("The Reign of Our Emperor").

Climate. Weather as well as land determines where the people live. The Japanese climate ranges from cold to warm. The climate of the areas along the Sea of Japan is influenced by the fiercely cold air from Siberia, but the Pacific coast is warmed by the Japan Current, a body of warm water that flows up the southwest. Winters in the region that are influenced by the Japan Current are not as severe as in the other parts of Asia that are located equally far north. Luckily, Japan is blessed with abundant rain, for rice, its most important crop, needs a great deal of water. One of the first things one notices is that Japan is a very green land. The areas that receive the heaviest rainfall, which comes in summer, are on the Pacific coast. Winds from the Asian mainland bring snow in winter to the Japan Sea coast and to the mountains. The Takada region in Honshu receives the heaviest snowfalls in Japan.

Nature usually is fairly generous and very regular in its ways in Japan. Although the climate varies from place to place, in any given region the four seasons are clearly marked, and a Japanese can predict pretty accurately how the weather will be in any month. Much Japanese art and literature is based on this regular rhythm of the seasons, and many popular festivals and customs mark the different times of the year. But nature is not always kind. Typhoons sometimes strike causing great damage and loss of life. An enormous number of earthquakes are registered every year in Japan. Very few cause damage, but some do, particularly if they are followed by tidal waves and fires. Until recently, most Japanese buildings were made of wood, so the Japanese are very careful about fire. Their history has forced them to be so. In 1923 over 100,000 people lost their lives in fires following an earthquake in the Tokyo-Yokohama area.

The backbreaking work of rice farming. Rice is Japan's staple food.

Natural Resources. Japan has not been well endowed with natural resources. Like a salesman's sample case, Japan has a small amount of almost every kind of mineral, but not enough of each to be worth much. There is some coal, but coal is becoming less and less important as a source of power in Japanese industry, which finds it cheaper to import petroleum. As a result of diminishing demand, production has declined and some mines have closed. Some zinc, copper, and lead are also mined. A more important resource is timber. Almost 70 percent of Japan is covered with forests, and Japan is among the world's leading producers of such products as wood pulp and paper. But Japan's economic successes, both before and after World War II, can hardly be explained by its natural resources. Japan's greatest resource is its people, and they are the way they are very largely because of their history.

THE JAPANESE PEOPLE AND THEIR HISTORY

Nobody knows where the ancestors of the present-day Japanese people came from. But Korea, some 100 miles (161 km.) from the southwest coast of Japan, is the part of Asia closest to Japan and the most likely route that migrating people moving east might have taken to reach Japan. Others may have traveled north from the island of Okinawa, or south from Sakhalin island. Historians believe that in prehistoric times people from Siberia, China, Korea, and Southeast Asia used these "bridges" to Japan. Already living in Japan at this time were the Ainu, a people with some physical characteristics more like those of Europeans than Asians. (For example, the Ainu has more hair on his face and body than the typical Asian.) The new Japanese, wherever they came from, pushed the Ainu farther and farther north until they became a minority, small in number and weak. Today, only a few thousand Ainu remain. They live on the island of Hokkaido.

These people from the mainland of Asia gradually evolved into a distinct people, although clearly related to other Asians. Since then the Japanese have remained physically the same. Probably no other important nation in the world has a population that has mixed so little with other peoples.

Japan's position off the east coast of Asia is somewhat similar to that of Great Britain off the west coast of Europe. Japan is close enough to the Asian mainland to have exchanged ideas with its peoples, as Britain has with Europe. But the 100-mile straits that separate the mainland of Asia and Japan are much wider and rougher than the English Channel, so that the Japanese have been able to keep invaders from the mainland away from their shores. There is no record of a successful invasion of Japan until World War II. Japan's history is unique in this way. The Japanese have been able to exchange ideas with foreigners when they wanted to and to keep them away when they wanted to.

By the 1st century A.D. the Japanese had begun to create a form of political order. Between the 1st and 4th centuries this order was maintained by councils of families and clans. One of them, the Yamato Court, became powerful enough during the 4th century to control much of Japan, so it can be said that the country has had a more or less unified existence for over 1,500 years. During the reign of the Yamatos, formal contacts with the Asian mainland began, and thus started the introduction of new ideas and techniques into Japan—Buddhism from

China, India, and elsewhere; Chinese art, learning, and political organization; the Chinese form of writing. Each of these, however, was made into something different and distinctly Japanese.

We need not follow closely the complex history of Japan, from the time the emperor began to lose his power in the 9th century (eventually he became a figurehead) to 1868, when the imperial system, greatly changed, was restored. At times a feudal ruler (called a shogun) would acquire enough power to unify Japan. (The Edo Period, from 1603 to 1868, was generally a time of peace and political stability.) At other times, particularly during the 15th and 16th centuries, Japan was torn by civil wars as rival clans competed for power. Japanese history during these times of strife is in some ways similar to the struggle of Italy to regain the unity it once had under the Roman Empire.

Cultural Achievements. Many of the courts of European nobles during the early Middle Ages were very rude places, and art and learning

Kabuki performers. The Kabuki theater dates from 17th-century Japan.

stayed alive only in the monasteries. But many of the Japanese noblemen during the same period maintained centers of culture at Nara, Kyoto, and Kamakura. The Fujiwara Court at Kyoto, for example, from the 9th century to the 12th, was the center for learned Buddhist monks and literary ladies. "Chinese learning, Japanese spirit," was the slogan of one of the leading statesmen of the time. The monks changed Chinese Buddhism into forms better suited to Japan, and in the 11th century one of the ladies of the court, Lady Murasaki, wrote the long *Tale of Genji,* sometimes said to be the first novel. Another period, the Muromachi Period (1338–1573), was marked by great political disorder, but it was also the time when many purely Japanese art forms were invented. A form of drama called No, the tea ceremony, a new kind of architecture, and a style of landscape painting marked by simplicity and serenity, in the mood of Zen Buddhism, were all developed during this warlike period. Again a comparison to the West is interesting. About this same time Italy was torn by wars within and among its city-states, but the same period produced the literature of Dante, Petrarch, and Boccaccio, the paintings of Leonardo da Vinci and Michelangelo, and the great cathedrals, such as St. Peter's.

The Japanese were very proud that no invader succeeded in conquering Japan. The emperor gave military leaders the title Seii-Tai-shogun—"barbarian-quelling generalissimo"—a title that remained in use until the middle of the 19th century. The powerful Kublai Khan, the Mongol emperor of China, twice tried, unsuccessfully, to conquer Japan —in 1274 and 1281. Each time the Khan's ships were destroyed by a typhoon that suddenly blew up, which the Japanese called *kamikaze*— "the divine wind."

Japan Isolates Itself.   For over 2 centuries, between 1639 and 1853, during the Tokugawa Period, Japanese rulers excluded from Japan all foreign missionaries and all foreign traders except a few Dutch and Chinese, who were permitted only at the port of Nagasaki. The Japanese themselves were forbidden to travel abroad. It was a period of peace, rich culture, and the accumulation of great wealth among some families. However, these years of isolation meant that Japan remained a feudal society longer than many European nations and did not share in the rapid advances in science and technology made during that period in the West. It also meant that some Japanese, like some of the ancient Greeks, found it hard to break the habit of thinking of all foreigners as barbarians, that is, people without culture. Living alone for such a long time, the Japanese worked out a complex and subtle code of manners, a way of smoothing out relationships among people crowded together. Some of this code is still maintained. It makes daily life easier and pleasanter for the Japanese, although it sometimes makes them seem mysterious to foreigners.

Another important change took place during these years of isolation. Art and learning, which earlier had been a pleasure known only to monks and people at the warlords' courts, became available to the newly wealthy merchants in the cities. These townspeople developed their own art forms, livelier than those of the Muromachi Period. The gaudy and melodramatic Kabuki theater grew up alongside the quiet No. Architecture took on more color and ornamentation. At the same time, a very concise form of poetry, called haiku, which was to become popular in the West in later years, developed among these newly rich townspeople.

Beginnings of Modern Japan. The Japanese could not keep to themselves forever, for the Western nations, competing for more and more foreign trade, kept pressing Japan to open its doors. Finally, in 1853, an American naval officer, Commodore Matthew Perry, sailed into Tokyo Bay and overawed the Japanese with his fleet of warships. The following year a peace treaty was signed with the United States, and in 1858 a trade treaty was signed between the two countries. The narrow wedge that Perry opened in Japan widened, bringing great changes to the lives of the Japanese. The emperors of Japan had long been figureheads, without real power, but in 1868 the 15-year-old Emperor Meiji, supported by new political groups, was made the symbol of a new Japan unified under the emperor. These men soon made Japan the most modern nation in the Orient.

The most able young men in Japan were sent to the United States and Europe to study. Europeans and Americans were invited to Japan as teachers, and foreign trade was increased. The government was re-made, with a constitution, a two-house legislature (the Diet), and a modern legal system. The feudal system was abolished. The Army was re-organized to fit European models. Most important, perhaps, was the great change in the educational system. A ministry of education was established to organize a school system that would be the same for all Japanese children throughout the country, and this new system made use of many ideas of Western educators.

This period of change, called the period of the Meiji Restoration, also opened Japan to ideas of democracy and industrialization. Then the Japanese copied something else that seemed to be the mark of a successful modern nation—military power. Japan was successful in wars against China in 1894–95, against Russia in 1904–5, and against German colonies in Asia and the Pacific during World War I. Military leaders came to be widely honored and influential. Joining with other men who were suspicious of the new democratic ideas in Japan, they made plans that eventually led to the invasion of Manchuria and northern China in the 1930's and of Southeast Asia in 1940, and then, finally and fatally, on December 7, 1941, to the bombing of Pearl Harbor in Hawaii. After 4 years of terrible war, Japan was ruined.

Young people in Japan are often heard to say, these days, that they are glad that their country lost the war. At any rate, Japanese now have a dread of war, and their Constitution, written during the time when the Allied forces occupied Japan after the war, provided many political and economic reforms that most Japanese like.

JAPAN'S CITIES

Three fourths of the Japanese live in cities. Three of the four most densely populated areas of Japan lie along the southeast coast of Honshu. They include the Tokyo-Yokohama metropolitan area, Nagoya, and the cluster of cities that includes Osaka, Kyoto, and Kobe. The fourth population center includes the cities of Kitakyushu and Fukuoka on the northern tip of Kyushu, and Shimonoseki City on the westernmost part of Honshu. The cities of Kitakyushu and Shimonoseki are connected by an undersea tunnel.

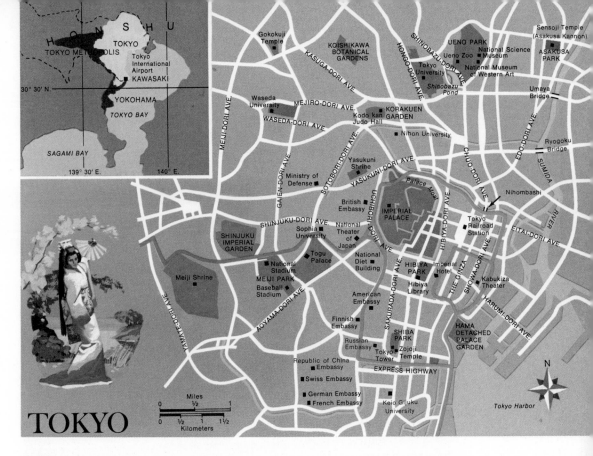

Tokyo

A city as huge as Tokyo has many faces. It is the capital, a manufacturing center, a center of international trade, and the seat of the main offices of most of Japan's big businesses. It is the center of Japan's culture, with many colleges and universities, publishing houses, museums, and art galleries. Most of all, though, Tokyo is many, many people. Over 11,000,000 people live in Tokyo and its near suburbs. One fifth of all the Japanese people—20,000,000—live within a circle about 30 miles (50 km.) from its center.

New York is Tokyo's nearest rival in size. Most major cities of the world are old, but these two cities were founded comparatively recently, at about the same time, early in the 17th century. By the beginning of the 19th century, New York was still a city of fewer than 100,000 people, while Tokyo had a population of over 1,000,000—more than London or Paris. The reasons for this very rapid growth are unusual. In 1590 Ieyasu Tokugawa, who became the most powerful ruler of Japan (he became shogun in 1603), decided to set up his headquarters in a little fishing village named Edo (it was renamed Tokyo, meaning "the eastern capital," following the Meiji Restoration in 1868). Tokugawa wanted to get away from the plots and political struggles of the old government at Kyoto, and he knew that the wide plains around Edo would grow rice enough to feed more soldiers than the smaller plains around Kyoto. Then, in order to maintain control over the strong men who followed him, at a time when a ruler needed all the help he could get, Tokugawa thought of a clever scheme. The first year he made half the warlords of Japan come to his new capital and work in his government; the next year he made the other half come, but then he insisted that the wives, children, and servants of the first year's warlords move to Edo, as a sort of insurance that the husbands would not plot against him while they were away. These warlords and their families built grand palaces, which required many servants, builders, and other workmen. Merchants arrived to meet the needs of the new capital. They were followed by other people seeking their fortunes, and Edo grew rapidly. By the middle of the 18th century it was the largest city in the world. But these were hardly sound economic grounds upon which to build a city, and eventually Tokyo was outstripped by New York. Only after World War II did Tokyo regain its leadership as the world's largest city, this time for sound economic reasons.

One need not read statistics to appreciate Tokyo's economic vigor. Signs of its dynamism—its intense activity—are everywhere. Old buildings are constantly being torn down and new ones being built to take their place. During the rush hours on commuter trains, "pushers" are employed to shove all the passengers into the cars, so that no space will be wasted. (Students are hired for this job.) People from other parts of Japan stream into Tokyo, for the average income there is higher than in other parts of the country—almost three times as much as in the poorer farming districts—and it is going up every year. At graduation time, tens of thousands of young people from senior high schools and colleges come to Tokyo to get acquainted with the city and to find jobs of the sort they are trained for.

A Changing Tokyo.   A person gone from Tokyo for a year is always surprised when he returns, for the city is always changing. Many of Tokyo's buildings were put up hastily after World War II, when almost two thirds of the city was in ruins, and now these old buildings are being replaced. Replacement is not enough, however, with the tremendous growth in population, and additional buildings are being constructed. But the Japanese want these new buildings to be beautiful too. To fill

Morning rush-hour crowds at Shinjuku station in Tokyo.

The Ginza, Tokyo's commercial, business, and entertainment center.

A section of the port of Tokyo, where trade fairs are held.

A side street in Tokyo. The gateway is called a torii.

these needs, some of the best architects in the world have developed in Japan in recent years. These men have not only designed original and beautiful buildings, but have changed methods of building as well. Formerly, because of the old fear of earthquakes, the Japanese built only low buildings, and wood was the only material most of them could afford. New wealth has made more materials available, and engineers have discovered ways of building safe skyscrapers. Today, Tokyo is a mixture of the old, narrow, winding streets crowded with small shops and homes, and of wide avenues with handsome new steel-and-concrete office buildings and stores.

The Ginza, the famous shopping, business, and entertainment district in the center of Tokyo, is always crowded with people. First there are the people going to work, then the shoppers and tourists, and finally, under the neon lights, the pleasure-seekers at night. A very modern and efficient subway system, connected with private railway systems, takes these people to and from their homes. More and more, these homes are big concrete apartment buildings in the suburbs. The price of land is so high in and near Tokyo that a single-family home with a little plot of grass around it costs too much for most people. Tokyo, like most large cities, has plenty of poor housing as well as new, modern apartment buildings; sometimes old and new exist side by side.

The City's Problems.   Tokyo's problems are much the same as those

Hibiya Park, one of Tokyo's many parks.

of other big modern cities. The struggle to keep the air and water clean is a constant one. Following a severe water shortage in 1964, Tokyo began to reach out farther for its water supply; it is now many miles away, but even this source will probably not be sufficient in a few years. New incinerators have been built to get rid of garbage and other waste, but the amount of waste, roughly 10,000 tons a day, increases faster than the incinerators can handle it. Crowding on the subways and suburban trains also seems to increase faster than the new subways can take care of it, and with the number of new automobiles skyrocketing, the streets are jammed too, in spite of new highways. These are problems all big cities face, but they are more serious in Tokyo than elsewhere because of the high price of land.

The cost of land in a Tokyo suburb is many times that of New York or London suburbs. So a man in Tokyo must put aside a higher part of his wages—sometimes as much as half his earnings—in order to buy land for a house for his family. And land prices rise so fast that the amount of land he can afford gets smaller and smaller. When he can finally build his house, it is likely to be small and of poor quality. Renting a place to live is also very expensive. (Because of the high cost of land in the suburbs, there is a growing tendency toward renting apartments in the new, tall apartment buildings in central Tokyo.) Since land prices rise so rapidly, people who already own land are reluctant to sell it.

This makes it difficult for the city government to plan any changes in the zoning of buildings in Tokyo, and it also means that it is difficult to get land for parks, new schools, hospitals, and other public places necessary for the growth of a modern city.

None of this seems to discourage people from swarming to Tokyo. It is more than better jobs that attracts so many people. Perhaps another reason is that life there is so varied. A Tokyoite can enjoy the quiet old gardens and temples, the gay night life, and the glittering stores; he can visit the huge fish market, with its daily catch that feeds millions; he has a choice of plays, movies, concerts, art exhibits, and sporting events at the big arenas; he can savor the fine restaurants with dishes from all nations; and, not least, he has the excitement of being where the important decisions are made that affect the lives of the whole nation and much of the rest of the world.

Other Cities

A ride on the new Tokaido bullet train, southwest from Tokyo, takes a traveler to the Kansai region where the cities of Osaka, Kyoto, Kobe, and Nara are located. **Osaka**, the second largest city in Japan, with about 3,000,000 people, is an industrial center where textiles, appliances, and other goods are manufactured. Almost 13,000,000 people live in the cities and towns within a circle of about 30 miles (50 km.) of Osaka. The people take great pride in the fact that until the Meiji Restoration

Osaka, Japan's second largest city.

The cenotaph in Hiroshima's Peace Memorial
Park, dedicated to Japan's war dead and to peace.

The resort city of Atami.

in 1868 their region was the commercial and cultural heart of Japan. While Osaka was the commercial center, **Kyoto** was the cultural center and also served as the capital until 1868. The small nearby city of **Nara** was the first permanent capital of the country. It is famous for its beautiful ancient temples dating from the 8th century. **Kobe** is a port city through which much of Osaka's trade moves. It has an international flavor something like that of **Yokohama**, the port that handles much of Tokyo's trade. The economy of the whole Kansai district has suffered in comparison with that of the Tokyo-Yokohama area because Kansai formerly depended heavily on trade with China and the rest of the Asian mainland, and that trade is now largely cut off.

Nagoya, another manufacturing city, with almost 2,000,000 people, lies about midway between Tokyo and Osaka. The people of Nagoya call their city Chukyo, or "central capital," and they pride themselves on its open, well-planned downtown area and its progressive spirit. There are also the cities that are regional centers, such as **Sapporo**, on Hokkaido, which has a population of over 1,000,000; **Sendai**, the chief city on northern Honshu; **Fukuoka** in northwest Kyushu; and **Hiroshima** in southwest Honshu, once again a thriving city of 500,000 after being devastated by an atomic bomb in World War II. All are proud of their local history and their position as cultural centers of their regions. And there are little resort cities, such as **Atami** and **Beppu**, whose size depends on the tourist season. Both are famous for their hot springs.

ECONOMY

The Japanese economy, almost totally destroyed by 1945, has returned to health with amazing speed. Japan is far ahead of any other nation in shipbuilding. It is the world's third largest producer of raw

A tanker being built in Nagasaki. Japan is the leading shipbuilding nation.

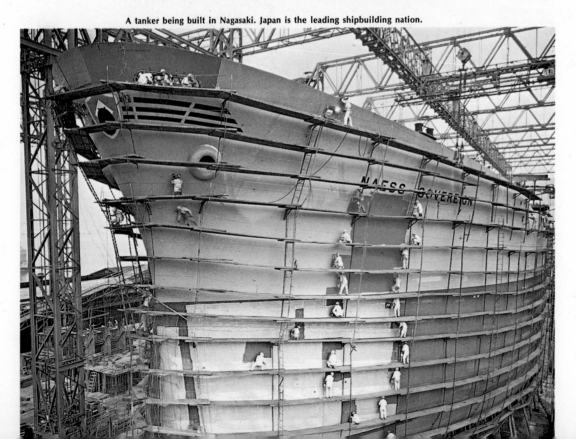

steel. It is the second largest producer of motor vehicles and of synthetic cloth. The rate at which the total annual income of Japan has been growing has for some time been the highest in the world, almost matched by West Germany, but far ahead of the other nations of the world.

The Japanese economy had been strong before World War II, but much of the wealth then went to maintaining a strong army and navy. Furthermore, a very large part of the country's wealth was controlled by five or six families called the *zaibatsu*. The power of those families was broken up during the Allied occupation.

Before the war much of what Japan produced was exported. Today many more of the goods produced in Japanese factories are meant for domestic use. Trade with other nations is still necessary, however; the economy would fall apart without it. Since Japan has so few natural resources of its own, much of the country's economy depends on its buying raw materials from other countries, transforming these materials in its factories into goods people want to buy, and then selling the goods at home and abroad. This means that the health of the Japanese economy depends largely on the skill of its people. The most important reason for Japan's amazing economic growth since 1945 is its large supply of well-educated and hardworking people and their high level of scientific and technical knowledge. Reforms in agriculture and the replacement of plants and equipment destroyed in the war also helped Japan toward its economic "miracle."

Other reasons for Japan's success have to do with its political policies. The amount of money Japan spends on its army and navy, which are hardly more than a police force, is much less than that of other major nations. At the same time, Japanese manufacturers and merchants have made money, directly or indirectly, from other people's wars, in Korea, the Middle East, and Vietnam. Also, Japan's labor costs were comparatively low, especially in the years after the war. Finally, a comparatively small amount of Japan's tax money is spent for governmental programs such as public housing and public payments to the aged and jobless. The amount spent for education and health, however, compares well with other countries. One criticism many Japanese make of their country's economic policies is that some businessmen and political leaders think too much about the growth of the economy for growth's own sake, emphasizing more and more production while putting too little emphasis upon improving the standard of living by raising the workers' pay or by improving hospitals, parks, and other public works.

A Split-Level Economy

The Japanese economy is sometimes called a split-level economy, because some of the production is organized along the most modern lines, while more of it is carried on in a rather old-fashioned way. This means that working conditions are very different at the two levels.

Work, for most Japanese people, is organized differently than for most Americans and Western Europeans. Perhaps the fact that Japan's feudal period lasted longer than Western Europe's affects the relationship between a worker and his manager. There is still a strong feeling in Japanese offices and factories that the worker owes his manager obedience, loyalty, and a hard day's work, and that the manager owes the man who has worked for him long and loyally certain things besides

424

The Toyota plant. Japan ranks second in production of motor vehicles.

his pay. In other words, the relationship between worker and manager is still somewhat more personal, and less a matter of pure bookkeeping, than in the West. Most workers stay with one company throughout their lives.

When a Japanese finishes the amount of schooling he plans to get, at the end of junior high school, senior high school, or college, he may take an examination given by one or more of the big businesses. The company that hires him, if it is a big one, will keep him on for the rest of his working life, as long as he is loyal and obedient. His wages will be small, but he can expect many fringe benefits that make his income seem larger. One of the most important fringe benefits is free housing (or housing at greatly reduced rents) offered by most of the large companies and the government agencies. He might be given lunch at work and perhaps free or cheap transportation to and from work. He might be sold clothing at reduced rates, or given the product of the factory at cost. The company probably has some athletic equipment so that he can play volleyball or some other sport during lunch hour.

As his seniority grows, he can expect promotions and more fringe benefits. If he is ready to get married (a very expensive undertaking in Japan), he may be able to use the company's wedding chapel, and perhaps the company will give his bride a gift. Every year, the company may organize a trip to a resort for him and his fellow workers. It may provide him with some sort of insurance policy and an annual bonus or two, often equal to several months' pay, at New Year's and perhaps in the summer.

If he is lucky enough to rise to a managerial job, the benefits may become quite extensive; as well as free or very inexpensive housing, they

A steel plant. Japan is one of the leading iron- and steel-producing countries.

may include a car with a chauffeur. He might be given an expense account, so that he can go to costly nightclubs, and membership in some fashionable golf club. And all along the line he can be sure, if it is a large company and if he remains loyal to it, that he will not be fired. He usually retires before the age of 60 with a single retirement payment equal to several years' salary.

If, on the other hand, the young man out of school cannot get a job with a big firm and takes a job with one of the little companies that still produce more than half of Japan's goods, he can be sure of nothing. Many big companies arrange with smaller firms to make certain parts they need in their product. These smaller companies then arrange with even smaller companies to make parts of that part. If a depression comes, the big companies will probably survive, but the smaller ones will be badly hit. Typically, the very small companies hire fewer than 10 workers. They pay lower wages and cannot offer either the fringe benefits or the security that the big companies can. Thus, in a split-level economy the difference between good jobs and poor jobs is a wide one.

Japanese Traditions and Work Habits

The arrangement of work, even in the big companies, may strike many Westerners as inefficient. Some Japanese experts on business feel that faster production should be more rewarded than loyalty and obedience. Japanese tradition, however, honors hard work, and men and women work hard of their own accord because they feel they ought to. So the system undoubtedly works very well.

About 35 percent of all Japanese workers belong to unions, a higher percentage than in the United States, lower than the United Kingdom,

Stringing necklaces of cultured pearls, an important export for Japan.

Women divers. Once they dived for pearls. Today they harvest shellfish.

and about the same as West Germany and France. Japanese industrial productivity has not been seriously impeded by strikes and labor strife. The Japanese tradition of loyalty to one's superior and of hard work may help to explain this. However, labor shortages in recent years have increased the bargaining power of the labor unions, and wages have been increasing at a faster rate in Japan than in the countries of North America and Europe.

The labor shortage is having an effect on women workers too. Traditionally, the women's place has been in the home. Formerly, women worked in offices and factories after they left school, but it was always expected that they would marry someday, so they were thought of as temporary workers, given less pay and fewer benefits. Now, however, more of them are allowed to work after they are married, and their jobs are slowly becoming better. A few women, in fact, have begun to move into important jobs in management.

All workers, men and women, must be very thrifty, for the government's social security system is much less generous than that of the United States, the United Kingdom, or Sweden, and the company retirement pensions are never sufficient. Again, old Japanese traditions help. Thrift, like hard work, has been as powerful a rule of behavior in Japan as it was in Puritan England and New England. Buying things on credit, which became fashionable in the United States in the 1920's, is only now becoming popular in Japan, and people generally still save for what they buy. Nature was always hard on the lazy and spendthrift farmer in old Japan, and fairly kind to the thrifty and hardworking one, and the tradition of thrift is still strong in modern Japan.

Another old tradition, that of the family, serves a useful purpose in Japan's economy. The Japanese family has always been very close. Families followed strictly a very complicated system of duties and responsibilities toward one another, young to old, women to men. Among other duties, the family took care of its members when they were too sick or old to work—even distant cousins if no one else would care for them. People are less exact about following this tradition today, especially in the cities, but the system still works well enough to relieve many people of the worry of being unable to work.

In another way, too, Japanese workers are more secure than workers in other industrial nations. Because the economy is growing rapidly and because workers are badly needed, only a little over 1 percent of Japanese workers are without jobs—less than in most Western industrialized countries. Some Japanese, however, are underemployed; that is, they have jobs that do not allow them to make full use of their abilities or do not pay them enough to live decently.

The Work of the Farmers

Life is often harder, though, in the small towns and on the farms. In a small town a man might have three jobs. He might farm a bit when the crops need his attention, fish a bit or cut some timber, and work in a factory occasionally. Many companies in these small towns schedule their work so that they can get farmers and fishermen when the men are not busy. Another man might farm during certain parts of the year, then go to the city to work at a construction job. The Japanese wife has always helped in the fields during the busy seasons, but nowadays she may also go to a resort at vacation time to work as a maid or waitress. Almost one fifth of the Japanese men, however, continue to spend most of their time working the farms and the forests.

The Rice Farmer.   If the Japanese farmer happens to be a rice farmer, his job is vital, for rice is even more important to the Japanese than bread and potatoes are to the Westerner. The government recognizes this and buys almost the whole rice crop of Japan every year, at quite a high price. After paying the farmer, the government sells the rice to housewives at a somewhat lower rate. The original purpose of this agricultural aid policy was to maintain Japan's ability to feed her population, but it also served to keep the farmers satisfied with the government. Recently, however, it has resulted in rice surpluses and the government has been forced to curtail subsidies at the cost of losing farm votes.

But even government aid did not keep the rice farmer's life from being a hard one. Growing rice has always been backbreaking work, and modern farming methods have not made it much easier. A small tractor can help work the ground and then help fertilize it, an important step that the Japanese make much of. But the most important steps are planting and transplanting, and this work must be done by hand, after the fields have been flooded. The second step, transplanting, must be done delicately. It is so difficult that some factories give their workers time off to return to their farms at replanting time to help with the work. Rice farmers must work bent over for hours, day after day. One can see old rice farmers whose backs are so bent that they cannot straighten up. Nor does all this work and government aid make them rich. Japanese farms are tiny by American or even French standards. If a family has 2 or 3 acres (1–2 hectares) of land, it is considered well-to-do, and the father might be a community leader.

The social order of a rice-growing community has always been very close by necessity. Water must be on hand at certain times or the crop will be ruined. Long ago, the farmers gathered together to make irrigation systems, which are still, with some changes, in use. Leaders in each of the neighborhoods make certain all co-operate. Everybody knows everybody in these communities, for many of the families have had their own small farms for centuries. They take great pride in them, and it is a sad day for a family, especially since there is a labor shortage on the farms, when for economic or other reasons one of the children leaves the farm to go to the city.

Other kinds of farming are done with almost as much care. It is said that a silk farmer gets little sleep, for he must be up, off and on, all night long, feeding the silkworms their mulberry leaves. (Silk has, of course, greatly declined in importance with the invention of synthetic cloth, but those who can afford it still prefer it to other kinds of cloth.) In the apple or pear orchards, the fruit is wrapped in paper before it is harvested. A kind of beef cattle raised near Kobe is fed a special diet, which includes beer, and is then massaged by hand to make the meat more delicious. It is this great care taken by Japanese farmers that accounts for the fact that a country with so small an amount of fertile land can feed so many people. Only in Hokkaido, where there is more land and where farming methods have been influenced by American experts brought in during the late 19th and early 20th centuries, is farming more free and easy, less intensive.

In all the work of Japan, from the long automated assembly lines to the farms, we can see the importance of skilled and intelligent workers, managers, and engineers. The Japanese have for centuries given education a high place in their value system. The importance of this to their economy cannot be overstated.

EDUCATION

In the 6th century A.D. the Japanese had enough respect for learning to begin to import ideas and arts from the Chinese, who then had the most highly developed culture in the world. For centuries after that, the Japanese remained in awe of Chinese scholars and borrowed much from them. By 701, Japan had created a governmental office of education, the beginning of its national educational system. In Europe during the Middle Ages, learning was mainly the province of monks; in Japan at this time education was the concern not only of priests, but also of the ladies and gentlemen of the imperial court; and by the 17th century the samurai, the old warrior class that by then had little military function, began to take up scholarship. By the 18th and early 19th centuries, learning became popular also among the merchant class. They sent their children to schools called *terakoya*, run either by Buddhist priests or samurai, to learn reading, writing, and arithmetic. Feudal lords set up their own schools for their children.

After the Meiji Restoration, the government organized compulsory schools. The old schools were brought into the new system, which was controlled from Tokyo. By the early 20th century all children received at least 6 years of free schooling, and plans were made for a system of higher schools, trade schools, professional schools, and universities. The plans were very ambitious and worked out well, but they led to a debate about the purpose of education that has not yet completely ended. Some Japanese believed that the purpose of an education was to help each student develop his talents to the fullest; others felt that the purpose was to make students into obedient soldiers and workers. Some teachers accepted the first idea, but more accepted the second. At the end of World War II, however, the Japanese school system was remodeled after the American system, and the idea that schools were supposed to teach superpatriotism was thrown out. The central national office of education, Mombusho, was left in control of the system, but teachers were given much more freedom in the classroom than they had had before the war. Young people were now required to go to school for at least 9 years. They would then either go to work, at the age of 15, or take examinations to try to qualify for senior high school. About three fourths of the students get into senior high school, but after they graduate they must take further examinations, which are even more difficult, to qualify for a college or university. About one fourth of the high school graduates enter college.

Since World War II, the number of junior colleges, colleges, and universities has greatly increased, but they cannot begin to handle all the young people who wish to attend. Many older private colleges have also grown, but not fast enough. In sheer numbers of institutions of higher learning, Japan is second only to the United States, but more schools are needed.

This all-out effort to try to give every student as much education as

A young student prepares her lessons. Education is highly prized in Japan.

he wants has had both good and bad results. Japan has achieved almost 100 percent literacy, rare in the world. The award of the Nobel prize to two Japanese physicists, Hideki Yukawa (in 1949) and Shinichiro Tomonaga (in 1965), and to Yasunari Kawabata for literature (in 1968) may be some indication of the success of Japanese education at its highest levels. Perhaps the most important advantage of the Japanese educational scheme is that all students, rich or poor, who can pass the series of difficult examinations that mark every important step in their advancement through school can have an equally good education. The finest universities in the country are free or almost free to the students who do well on these examinations.

However, some Japanese believe that this system is too much like a hurdle race and that it puts excessive pressure on the students—the examinations often determine how successful their lives will be—while they are too young to cope with the strain.

The examination of 12-year-olds to get into junior high school are not so important, since all children must finish junior high school. But the pressure on 15-year-olds to get into senior high school is severe. Not only do some fail to get into any high school at this time, but the students who do pass try to get into the best high schools. For the next stage of their education is vital: since these high schools train students for the competition for entrance into a college or university, a graduate of one of the better senior high schools is more likely to be admitted to one of the better colleges or universities.

The competition is keenest for admission to college. Some colleges and universities have more prestige than others, and businesses and government offices tend to give job examinations only to graduates of these universities. If a person is admitted to Tokyo University, for example, he is said to have a "diploma for life." His chances of graduating are almost 100 percent—few students once admitted to a college are failed. The chances are good that the student then will get a job with a large company or a government office. Some businesses and government agencies prefer students from particular universities, and students from other schools have little chance. One result of this is that some students try to gain admittance only to certain of the famous universities; if they fail their examinations one year, they may try them again, and then again and again. "Cram schools," private unofficial schools that do nothing but prepare such students for these examinations, are a big business.

THE JAPANESE LANGUAGE

Spoken Japanese, which dates from prehistoric times, is not particularly difficult, but written Japanese is extremely difficult to learn. Written Japanese began about 1,300 years ago, when the Japanese began to borrow Chinese writing. At first this was a picture language, like Egyptian hieroglyphs. The Japanese, however, used the Chinese characters, called kanji, to represent images and sounds in their own spoken language. These kanji are the basis of the Japanese written language. Scholars know thousands of kanji, and 1,850 of them are used in everyday reading of newspapers and magazines. A student must learn nearly 900 kanji by the end of sixth grade.

In addition to kanji, there is a group of 48 characters called hiragana, which are used mainly to represent the sounds of prepositions, participles, and other less important parts of speech, and also 48 katakana, used mainly to represent the sounds of foreign words that have come into the Japanese language. Japanese also learn one or another system of Romaji, which are Japanese words represented in the Roman alphabet. (Usually in the seventh grade, students also begin to learn English.)

A few people have suggested that this difficult language be simplified, perhaps reduced to an alphabet such as Western nations use. But most Japanese would rather not change, for the whole rich literature of the centuries would then be largely lost. A number of foreign words, particularly scientific terms, have entered the Japanese language. But Japanese is rich in the ways in which it can express fine shadings of personal feeling and delicate differences in the sights, sounds, and smells of things.

HOME LIFE

Patterns of life in the Japanese home began changing in the 1920's during the Taisho Period, but the old ways still show clearly through the new ones. The center of the old way of life was the close family group with the father in control. In very poor families, however, this might be hard to manage. If the father could not earn much money the mother might have quite a bit of authority. Also, four or five poor families might share a small house. But ordinarily the male gave orders to the female, and the older member of the family to the younger. If the husband's mother was a member of the household, the wife had to respect her, and an older sister had to defer to a younger brother. Some work was thought to be proper for men, and some for women; thus a wife never bothered her husband in his life outside the home, and he never bothered her about her household duties (nor did he help wash the dishes). The family waited until the father returned from work in the evening and had his bath before they began supper. After the father finished his bath, the boys had theirs, then the girls, and finally the wife.

The design of the house emphasized cleanliness, tranquillity, and simplicity. One entered through a little alcove, took off one's shoes, and stepped onto the soft and fine-smelling straw matting (tatami) that covered the floor. The walls inside the house were paper-covered screens, which could be arranged to make small or large rooms. Ideally, as few objects as possible cluttered the room. For example, though a family might have any number of paintings on scrolls to hang in the little nook (tokonoma) meant for them, they never used more than one at a time, one that was appropriate to the season. Nor was the room crowded with chairs, beds, sofas, and other furniture.

Meals were simple. Breakfast might consist of rice and some soup and tea; lunch, of rice or noodles, vegetables, and tea; and the evening meal of fish, rice again, and several vegetables, some of them pickled. There was more variety than this menu suggests, however, because of the wide variety of fish, vegetables, and fruits that were available. Bathing was eagerly looked forward to. The big wooden tub, fragrant when wet, was big enough so that when one sat down in it the water came up to one's neck. It was filled once, with very hot water, for the whole family. One used the tub for soaking and relaxing, however, not for washing; that was done outside the tub, before one got in. In winter, warmed by the bath, one might hurry to the hibachi, a pot filled with a little charcoal fire. Here sometimes the meals were kept warm and the kettle heated. Then the family might sit, with their legs under quilts, by the *kotatsu,* a small heating unit in the center of the livingroom floor, until bedtime. Bedding was stored in wall cupboards and rolled out on the tatami at night. During the summer, the house could be opened up to let breezes blow through, and perhaps play the chimes hanging in the doorway, giving a sense of coolness. Most houses had a garden, if only a few square feet in size, and possibly containing only small pebbles and a few artistically arranged rocks. Life for most people in Japan has always been meager, but they have been clever at making it as pleasant as possible.

Changes in Japanese Life

For some of the Japanese people this traditional picture of home life has changed very little. But for most Japanese it has changed sharply. Many forces have been at work to cause these changes. Most important, perhaps, has been the great appeal of Western ideas and manners since the Meiji Restoration. Ideas about the equality of man and woman and of the aristocrat and the poor man, and ideas about individual freedom, all began to cut into older beliefs. Industrialism made its own demands upon old ways of life. Since jobs are better in the cities, people began to leave the old family farms, starting the breakup of the tight family order. People also began to want new things, not necessarily because they were better, but because they were new. The design of Japanese houses, for example, has taken on more and more new, Western lines. These changes in housing reflect changes in the whole Japanese style of life.

Very few of the new houses and apartment buildings put up in recent years omit Western-style kitchens, including the modern refrigerators that by themselves have helped change Japanese life styles. The use of tatami is decreasing, and Western beds, chairs, and other furniture are becoming more popular with families who can afford them and who wish to seem modern (the Japanese have borrowed this word—*modan*). Many Japanese are even giving up their traditional baths for Western-style showers. Practically every Japanese home has a television set and a radio, and now the homes are beginning to lose their spare, uncluttered look.

At the same time, changes have occurred within the family. The father of the house began to lose some prestige and authority following World War II. More and more wives began to work, not necessarily to earn enough money so that the family could afford to live, but sometimes to buy the new Western-style luxuries that advertisements and commercials were telling them about. Successful businessmen often work very late, and many workingmen often take second jobs at night in order to buy more of the new products. Since no family could wait until these late-workers returned home to eat their evening meals, the old tradition of waiting for father is beginning to die out. If the mother is at home

Shoppers have a wide variety of stores to choose from in the city of Sasebo.

A marketplace in a small village.

during the day and watches some of the excellent educational television shows that are available, she may become more knowledgeable than her busy husband. Thus she may feel more and more impatient about keeping quiet when the men talk about important economic and social problems, although the old tradition holds that she should be silent on such matters. Diet, too, has begun to change. To the old menus are now added more meat, fish, and bread, vegetables and fruits not formerly grown in Japan, peanut butter, cheese, breakfast cereals modeled on the American brands, and Italian-style spaghetti.

Formerly, the Japanese housewife had to spend much of her day shopping at various small meat, rice, fish, and vegetable shops in the neighborhood. Daily shopping meant that everything was fresh—particularly important with fish. Now supermarkets are almost everywhere in the cities, stocked with food from all over the world, and the steady increase in refrigerators means that fewer shopping trips are required.

But acquired tastes for new foods and other goods means that more money is needed in the home. Because of this, two changes are taking place in the cities. One is the growth of credit buying. The other is that more wives are taking jobs, and they are now protected by law from being fired simply because they are married. This, in turn, means that more and more Japanese children are becoming "latchkey" children, who have their own apartment keys because neither parent is home when the children return from school.

Perhaps because of the wide gap between what people want and what they can afford, Japanese do not marry until comparatively late in life. The average age at marriage is 27 for men (compared with 23 in the United States) and 24 for women (compared with 20½ in the United States). College graduates may be even older when they marry. Families are small, usually with two or three children.

The marriage itself may be different from that of the couple's parents. What are known as love marriages are more and more replacing the traditional marriages, which were arranged by the two families through a go-between. Today in the cities, the courtship may take place without any discussion with the parents at all. The wedding ceremony itself may be different, too. It may be traditionally Japanese or thoroughly Western. Many are a mixture of the two: the bride and groom may wear kimonos and be married by a Shinto priest according to the old ceremony, then change into Western dress for a reception and a wedding trip. This mixture of styles probably indicates the mixture of Oriental and Western patterns in the lives the couple will live.

Love marriages are only one cause of differences among parents and

A newly married couple. The bride wears traditional garb, the groom Western clothes.

their children in Japan. Since ways of thinking about most important matters have been changing so rapidly in Japan, especially since 1945, young and old often find it very difficult to get along. Thus many young people, like their fathers, now spend a good part of their spare time away from home.

At the same time, however, certain old attitudes seem to have changed little. The same modern young couple that might have disobeyed their parents' wishes about their marriage most likely still see it as their duty to care for the old people in their family. Old rules of hospitality, too, seem to last. The young couple might live from week to week on a monotonous and thrifty diet, but when they have guests, they will entertain them grandly. And for New Year's and in June, the traditional gift-giving times, their gifts might be very expensive ones.

Holidays

This sharp difference between day-to-day thrift and openhanded generosity on special occasions makes the rhythm of Japanese life different from that in most industrial nations. Japanese expect most days to be drab and difficult, but they look forward to the few times a year when it is exciting.

The Japanese worker has about 20 days of vacation a year, besides 12 national holidays. About half the workers must work until noon on Saturdays, and about half have all of Saturday off. The workers and their families spend the weekend going to the movies, working around the house, watching television. But a worker can bunch his vacation days around clusters of national holidays and thus have two or three good-sized vacations a year. One of the most popular vacation times is around the New Year's holidays, when, traditionally, little work is done by anybody. The Japanese now celebrate Christmas too (although only about 1 percent of them are Christians), since it comes at about the same time as their own gift-giving time. The other favorite time for vacations is the first week in May, called Golden Week, which is preceded by the Emperor's Birthday (April 29) and includes Constitution Day (May 3) and Children's Day (May 5). During these long vacations, Japanese like to get into the countryside, perhaps on an excursion, perhaps to visit their old family homes. The big companies often sponsor organized workers' parties and trips on chartered buses.

LEISURE ACTIVITIES AND THE ARTS

As the workweek for many Japanese becomes shorter, and as the wealth of Japan slowly spreads among more people, the Japanese are thinking more and more about how to spend their leisure time. In fact, they have even adopted the English name for it—pronounced "reja." Some writers have said that too much leisure has put a strain on many Westerners. But the Japanese feel no such strain; instead, they find much of the meaning of their life in hobbies and sports. In the West, when a man is asked "What do you do?" he answers by naming his job. In Japan he tells you what his hobby is.

A man's hobby may be traditionally Japanese or it may be modern; it may be a mass activity or a private one. Playing and watching baseball is a passion for many. Going to the movies is an active occupation for some Japanese; films are seen as art rather than mere entertainment,

Cheerleaders at an inter-city non-professional baseball championship game.

A Tokyoite ponders a choice of movie theaters.

A very popular spectator sport in Japan is sumo wrestling.

A favorite spot for sightseers: the Itsukushima Shinto Shrine.

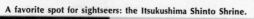

and they are scrutinized closely. This is not to say that the Japanese do not watch many films that have no great artistic merit—American or Italian Westerns, Japanese films about feudal warriors that are not very different from Westerns, and Japanese and American crime movies. But the Japanese also make many fine films, which have won international prizes, and they import the best films from other countries. Japanese commercial television shows material of spotty quality, including much that is imported from the West. However, a government-sponsored radio and television station, NHK, produces shows of a very high quality, without commercials. Programs include interviews with Japanese and foreign scholars, concerts of Western and Japanese music, and folk songs and dances from the various regions of Japan. Practically every home in Japan has a television set. Radios, too, are everywhere, including the streets, where people listening to their transistors can always be seen.

Other leisure-time activities have also been borrowed from the West. Golf is a current fad. Besides the fine courses throughout the country, many office buildings have practice tees on their roofs, where business-men may practice during their lunch hours. It is an expensive sport in Japan, played mostly by men on expense accounts. Bowling, another imported sport, is much less expensive and has become increasingly popular. The Japanese version of the Western pinball machine is the pachinko. There are a million such machines in Japan, and pachinko parlors employ about 100,000 people. From early morning until late at night, thousands of people can be seen gambling at pachinko. Coffee-houses are numerous in the cities. They are meant mainly for young people and often bear French or Italian names. Each specializes in a certain sort of music—classical, folk, or jazz—and for the price of a cup of coffee one can listen to music for hours. Skiing and mountain climbing are very popular sports. During all seasons, the railroad stations are crowded with people carrying climbing gear, heading for the mountains; and in winter the skiers are even more numerous. Swimming, tennis, and track are all popular, and so is volleyball, which workers often play during their lunchtime.

It would probably be hard to find a Japanese who did not count traveling and sightseeing among his favorite pastimes. The natural beauty spots, the ancient Buddhist temples and Shinto shrines, the cities, the hot-springs resorts, the beaches, the famous gardens, all have a constant flow of Japanese tourists. Many schools make arrangements to send their students on tours in chartered buses, and some business concerns reward their employees with sightseeing tours. (Hawaii is considered the grandest prize, but Japanese will gladly settle for visits to Atami or Beppu or Tokyo.) Women's associations in rural villages save their money until they have enough to hire a bus for a tour. Although their main purpose may be to visit a sacred shrine to pay their respects, they do not let this pious purpose spoil their fun. The chartered buses all have charming young women who serve as hostesses, describing the regions through which the bus is traveling and perhaps singing some of the local folk songs.

Traditional Japanese pastimes are also plentiful. Some men spend hours playing go, a game said to be even more complex than chess, or mah-jongg, another complicated game of Chinese origin. The championship sumo wrestling matches are attended by thousands of people and watched by millions more on television. Sumo is formal in the ceremonies that surround it and very fast and subtle during the actual matches. The action might be over in a few seconds, but those familiar with the sport can detect very delicate differences in kinds of attack. The wrestlers are very large men, fed on a special diet. Other sports of a rather violent nature that are becoming familiar to Westerners are judo, karate, and a kind of fencing called kendo in which wooden sticks are used instead of swords.

Traditional Japanese Arts

More gentle are certain arts that many Japanese, young and old, still enjoy. Flower arranging (*ikebana*) is practiced by both men and women, and many department stores have exhibitions of both traditional and new forms of this art. Many Japanese have a strong feeling for the koto, a long, 13-string harp played on the floor, and the samisen, a three-string, mandolinlike instrument.

Many Japanese attend No, Kabuki, and Bunraku drama. No is an ancient form of drama in which the players, like those in the ancient Greek plays, wear masks and move in a careful, solemn dance; the story is chanted with great formality. At first glance, No appears slow, but then one sees that great power and passion lie just beneath the surface of the careful movements and speeches. Kabuki is a more recent form, which flourished among the merchant classes during the time of peace and prosperity of the late 17th century. It is more popular, very colorful, and full of movement and sentimental plots designed for the newly

A demonstration of "ikebana," or flower arranging, a traditional Japanese art.

wealthy merchants of that period. Bunraku, which dates from the same period and is still centered in Osaka, uses puppets. The puppets are almost life-size and act out their roles with such dignity and strong emotion that the viewer often forgets that he is watching puppets. People have been known to bring box lunches and stay all day watching good performances of Kabuki or Bunraku.

The beauty of the tea ceremony (chanoyu) is a bit difficult for some Westerners to feel. Basically, it is a way of serving tea to guests and a way the guests have of showing their appreciation for the hospitality. But it has become very formal, with great emphasis on the beauty of the implements used, the grace of the host or hostess, the delicacy of the flavor of the tea itself, and the courtesy of the people in the ceremony. Like so many of the Japanese arts, the tea ceremony celebrates serenity, grace, and discipline of movement, a way of achieving peace in a troubled world. Traditionally, during the 16th century, warriors resting from battle would practice this ceremony. It is good, the Japanese say, to find a few minutes of peace and serenity even in the midst of battle.

Very many Japanese write poetry, or even invent it on the spot to fit the occasion. Mostly, they prefer the old forms like the haiku, which attempts to capture in 17 syllables a whole vision of the world. To do this in so few words takes the kind of control and sharp insight that the Japanese admire in all their arts. Calligraphy, a way of combining art and penmanship, is a required course in schools, and Japanese students study it in the 3rd to 9th grades.

The tea ceremony, or chanoyu, one of the most graceful Japanese arts.

The garden of a Japanese inn.

Gardening is a form of constantly changing sculpture. The object here, too, is to suggest a vision of the world, usually with plants, but also with stones and small pebbles. Besides tiny home gardens, there are large and famous public ones. Good restaurants often boast fine gardens, and *ryokan*, the old-style Japanese inns, have gardens that can be seen from every room. It is characteristic of the Japanese that in their gardens and parks they manage to combine several art forms in one object: a sculptured stone might bear a short poem carved in handsome calligraphy.

Art in Everyday Life

The fact that many Japanese follow these arts suggests a difference between their culture and some Western cultures. Art in Japan is not thought of as the special province of a few people who are specially educated for it. This more widespread interest in art reflects not only tradition and a good educational system, but also the fact that in Japan art is close to nature, religion, and ordinary life. Gardening and the tea ceremony, for example, are only ways of formalizing ordinary matters of farming and hospitality. A formal meal is laid out in many little dishes, carefully arranged by shape and color, so that the diner sits down to a sort of fine-smelling, delicious painting. The spirit of the greatest Japanese painting and literature is closely related to the spirit of Buddhism, which emphasizes harmony, serenity, and restraint. This Buddhist spirit is also found in the way Japanese form their manners, arrange their dress, and decorate their homes. The language itself contains many words with which to describe feelings of beauty.

This popular interest in the arts extends to all kinds of art, old and new, Western or Oriental. All of the large department stores in the big cities reserve their top floors for art galleries. They may show, from time to time, painting and sculpture from all periods and all places. Among the people who may be seen attending these art exhibitions are not only fashionably dressed men and women and students in their school uniforms, but also workers on their lunch hours, with the characteristic sweatbands (*hachimaki*) wrapped around their heads, and wearing aprons showing the emblems of their trade. When, a few years ago, the famous statue of Venus de Milo was borrowed from Paris and put on display in Ueno Park in Tokyo, lines of people a mile long waited for as long as 4 hours to see her, day after day for almost a month.

A similar popular interest in the arts may be seen in the production of newspapers, books, and magazines. A fantastic number of words come off the printing presses in Japan each day. Roughly one newspaper for every two people in the country is circulated every day. The number of books published in Japan is also increasing. Many novels by Japanese writers are published in serial form in newspapers and magazines, and the proportion of serious books of all kinds to the total number published is very high.

Western Influences

Japanese have seen and heard Western arts in great quantity for a century now, and many Japanese artists have mastered the Western forms of art. For example, the well-known orchestra conductor Seiji Ozawa is Japanese. More interesting than the Japanese skill in Western arts, how-

Gymnasium in Tokyo, designed by Kenzo Tange for the 1964 Olympic Games.

ever, is the way in which many of the best artists are joining Japanese and Western styles to make a new, modern art that is still clearly Japanese. Kimio Eto, the greatest koto player in Japan, has played with some of the best Western symphony orchestras and is constantly working on koto music in which Eastern and Western styles meet. The famous author Junichiro Tanizaki (who died in 1965) wrote, as many modern authors do, on psychological subjects, but his manner is typically Japanese. Rather than depending on the sharp climaxes that often mark Western fiction, he slowly examines his characters' most subtle feelings and the way in which they change. The busy architect Kenzo Tange, who designed, among many other buildings, the National Gymnasium for the 1964 Olympic Games, which were held in Tokyo, has been trained in all the methods of the so-called international school of Western architecture. While studying, however, he was reminded of principles of ancient Japanese architecture, which were very similar. As a result, his buildings are modern, in the international style, but they are also distinctly Japanese, some of them resembling ancient Shinto shrines. Rudyard Kipling, the English poet, once wrote "Oh, East is East, and West is West, and never the twain shall meet." Many 20th-century Japanese artists are proving him wrong.

This meeting of East and West is taking place in areas other than the arts, too. Western interest in the religions of Asia is centuries old, but recently this interest has become livelier than ever. Zen Buddhism, for example, one of several Buddhist sects in Japan, has acquired particular interest among Westerners.

RELIGIONS OF JAPAN

The Japanese have usually taken a view of their religions different from that of most Christians, Jews, and Muslims. The latter have thought their god to be the only god, perfect, all-powerful, and sometimes angry when his commandments are disobeyed. Christians, Jews, and Muslims have all fought long and bloody wars in the name of their own gods. For the ancient Japanese, there were innumerable gods, most of them having human characteristics, like the gods of the ancient Greeks. These ancient Japanese gods inhabited all the forces of nature—the rivers, the winds, fire, mountains, and especially the sun, from which all life flows. Belief in these gods came to be known as **Shinto**, which means "the way of the gods." Shinto, as it developed, came to teach respect not only for the forces of nature, but for one's ancestors as well. Furthermore, since Shinto taught that all men are basically good, it taught men to believe in the impulses that come from their hearts. Much later, Shinto combined worship for the sun goddess with the tradition that the Japanese emperors were descended from the sun and declared that the emperors were divine. Emperor Hirohito, in January, 1946, renounced this idea of imperial divinity.

Buddhism was introduced to Japan from the Asian mainland about the 6th century. The interest that Japanese scholars found in the many forms of Buddhism gradually spread among all the Japanese people. Shintoism became less powerful, but it did not disappear as Buddhism rose in influence. Nothing in either religion required a person to give up the other, so both religions existed side by side, and even today it is usual for a Japanese to be married in a Shinto ceremony and buried

A Shinto priest at prayer.

The Great Buddha at Kamakura. The bronze statue is over 700 years old.

The great Todaiji temple at Nara houses a giant 8th-century statue of the Buddha.

with a Buddhist ceremony. The point is that religion in Japan has strongly influenced morals and manners and has always had important ceremonial purposes, but it has not ordinarily been thought of as the complete and final answer to all questions. Generally, Japanese religions have made slighter demands on the lives of the people than Western and Middle Eastern religions have.

In times of great stress, however, when the land was troubled by strife and disorder, religion has been stricter. During the Mongol invasion of Japan in the 13th century and the civil wars of the 15th and 16th centuries, Buddhism took on deep spiritual meaning for more people, and strict piety was thought to be of great importance.

Christianity was brought to Japan by Catholic missionaries in the 16th century. It made a deep impression, particularly in the western part of the country. The shoguns during the Tokugawa Period feared the power of this foreign religion and, hoping to strengthen their own position, massacred the Japanese Catholics in 1637 and then closed the country to all foreigners, except for the Dutch and Chinese, who were restricted to the port of Nagasaki.

During the Meiji period the influence of Shinto increased. Following World War II, governmental support of any religious institution became illegal. Today, with some exceptions, organized religion is still comparatively unimportant to most people. The exceptions are a few very pious people of all sects and the **Soka Gakkai**. The Soka Gakkai is a well-disciplined lay organization of one of the sects of Buddhism. The millions of Japanese who have become interested in this movement are mainly people in small businesses, new to the cities and therefore lonely, with no other powerful voice to speak for them. They have also gained some political strength.

Christianity survived the old laws forbidding it, and has had some importance in Japan even though few people have joined its churches. During the Meiji Restoration, Christian ideas, particularly Protestant ideas, were popular among certain scholars and writers. Today, the fact that there are a number of Christian colleges and universities in Japan indicates that the influence of Christianity continues. The strength of this influence is hard to measure, because of the old habit of the Japanese of borrowing from foreigners what he wants and throwing out what he does not want, and also because of the easygoing way in which most Japanese have usually taken religions.

The Japanese have inherited many customs, festivals, and holidays from the various religions. They like to celebrate such occasions, and they enjoy visiting historic shrines and temples. A man might feel like a Christian on Christmas Day, a Shintoist at a wedding, and a Buddhist at a funeral. The rules that a Japanese is most used to following, however, come not from the priests or monks, but from custom, from the family, and from government.

GOVERNMENT

The lives of most Japanese are a blend of old Japanese ways and newer Western ways, and government is no exception. The Constitution of 1946 kept the emperor as symbolic head of state, but put real power into the hands of the Diet, the Japanese legislature or parliament. As in the British system, a prime minister is the head of government. He is responsible to the Diet, and he appoints the Cabinet from among members of the Diet. The system of law courts, established during the Meiji period, was fashioned after French, German, and British models. Japan, however, has far fewer judges and lawyers in proportion to its population than most Western countries. It was an old custom in Japan to ask some older man in the neighborhood to settle minor disputes before they came to trial. But this tradition is dying out, except perhaps in some rural areas, and Japan badly needs more judges and lawyers. The Japanese Supreme Court, like the American, is the final court of review in all important questions of constitutional law.

Lawmaking has been, since 1945, almost entirely the work of the Liberal-Democratic Party (the Liberal and Democratic parties joined in 1955), which has usually won a good majority in national elections. This party gains its strength largely from businessmen and from farmers who are interested in keeping the price of rice high. The Liberal-Democrats maintain friendly relations with the United States and have not pushed hard for a reduction in the number of American military bases in Japan or for increased trade with China. A number of political parties that wish faster change in Japan gather strength from teachers, some laborers, and students. These parties would like to make Japan more independent of the United States and to work out ways for spreading the wealth more equally among all Japanese. Recently, the political party representing Soka Gakkai has made some gains in the number of representatives elected to the Diet.

JAPAN'S FUTURE

Throughout their history the Japanese have shown themselves capable of making the changes in their way of life necessary to meet new conditions. Many Japanese these days are studying life in Japan and elsewhere and asking themselves what adjustments they must now make. Some of the most important questions they are asking are these:

What can be done to close the growing gap in the split-level economy? Big business is getting bigger all the time, as companies merge to make profits larger and more certain. But the small businesses are insecure, easy victims of an economy that, because it is booming, is also liable to rapid changes.

The rapid growth in Japan's economy creates other problems as well. How can the farmers and fishermen be assured a fair share in the rising standards of living brought by industrialization? How can more adequate housing, transportation, and sanitation be provided for the increasingly urbanized population? Pollution of air, water, and land is already a serious problem. Can the gross national product again be doubled without creating intolerable living conditions? Growth is impossible without foreign trade. Can Japan be confident that her access to both raw materials and markets will not be restricted?

For 25 years Japan has been relatively free to concentrate on urgent economic needs because her security was assured by the United States. With the return of Okinawa in 1972 most of the political problems carried over from World War II will have been settled. But the question of the Kurile Islands, occupied by the Soviet Union, remains unresolved, and treaties of peace have not yet been concluded with either the Soviet Union or the People's Republic of China. The United States has announced that it is reducing its role in East Asia. Japan, with the world's third strongest economy, seems destined to play a more active role in international affairs. Is it possible to do so most effectively by continuing its reliance on American military strength, by breaking those ties, or by increasing its own expenditures on defense?

This question of military security in a troubled world leads to another question. Most people in Japan still favor the anti-war clause in the Constitution. If the alliance with the United States remains close, pressure to change that clause remains slight, even from people who believe that strong military power is necessary to ensure safety. Others, however, feel that the alliance itself makes the danger to Japan greater, from enemies of the United States; and in any case, many of these same people have little faith that a strong military organization ensures the country's security. At any rate, one tradition of the Japanese is gone forever—isolationism. People wish that the gap between the rich and the poor, whether on the farm or in the city, would also become a dead tradition. Finally, can the traditions that the Japanese *want* to keep survive in a world of mass production and mass society? Can they be maintained strongly enough so that Japan can fulfill the promise to make East meet West?

THOMAS SHIBATA, Tokyo Metropolitan University
CHARLES G. CLEAVER, Grinnell College

AUSTRALIA: AN INTRODUCTION

by Sir Robert Gordon MENZIES
Former Prime Minister of Australia

Australia, where I was born and bred, and of which I was prime minister for a total of over 18 years, is a paradoxical country. Geologically it is an old continent, in which obvious traces may be found of the shellfish of old seas, the Ice Age, and the Volcanic Age. Ethnically its aboriginal inhabitants are the not very numerous survivors of a race of people whose recorded history is scanty. They are not, it is generally thought, comparable with the Maori of New Zealand, though they have produced a number of people of talent and achievement, and will produce more in the future.

Australia's history as a nation in the modern world began with the first colony established by England nearly 2 centuries ago. Before that there had been far-ranging navigators, Dutch and French in particular, who explored and charted its shores without discovering its inland possibilities. For a long time it was a *terra incognita*—"unknown land"—in the South Seas. The early Dutch navigators recorded it as New Holland.

And now, after only about 180 years of colonization, it is a predominantly European nation of over 12,000,000 people, active in world affairs, creative in talent, significant to the Old World powers of the Western World and to the New World power, the United States of America.

You will learn much of a factual and objective nature about Australia in the following article. But I would wish to make a few more general observations.

What sort of people are we? Well, we are overwhelmingly European, chiefly of British descent and with an inherited and deep-seated loyalty to the British Crown. We have inherited and practice parliamentary democracy on the British pattern. We have inherited and practice, with suitable modifications, the English common law. We adhere to the rule of law and the impartial administration of justice. We speak "the tongue that Shakespeare spoke," although our prevailing accent might have fallen strangely upon his ears; and we are the active and inventive heirs of English, Scottish, and Irish literature. In recent years we have received hundreds of thousands of migrants from Europe, who brought with them backgrounds and elements that are enriching and diversifying our own culture. So far you might conclude that we are no more than a physically transplanted European people living in a once primitive land. But we are much more than that. Our geographical situation, in the Southwest Pacific close by Southeast Asia, detaches us, to an extent, from Europe, and involves us more and more in the politics and economic arrangements of the emergent Asian nations, as well as Japan and the United States.

This does not mean that our old loyalties have died, that we no longer have a special feeling for what is for most of us still our mother country. A royal visit to Australia still evokes an enthusiasm that surprises and disappoints the small but no doubt growing band of republicans. Their great theme, as it was in India and Pakistan and the new African nations, is that complete independence must be signified and proved by cutting off the allegiance to the British Crown. To my mind this is sorry and unperceptive nonsense.

Australia is as independent politically, socially, and economically as any nation in the world. We receive and observe no instructions from Great Britain or from any other power. But we are not so immature as to think that for all purposes we are independent of all others. Our common allegiance to the Queen, shared with Canada, New Zealand, the West Indian nations, and a few others, not only appeals to our emotions but makes real our sense of community and makes the Commonwealth of Nations something significant, particularly in periods of world crisis.

We have learned, as most nations have in the 20th century, that just as no civilized man can live by himself, so can no nation live by itself. Even the great United States is no exception to this rule, for it has the great responsibilities of great power.

Australia has great and expanding resources, a vigorous and growing population, a high standard of living, and a highly developed educational system. It has great scientific and technological achievements to its credit. The Australian is a person of instinctive though occasionally obscured independence of mind and action.

Yet when we turn to the problem of national security, we are willing to admit that we could not successfully resist armed attack by hostile hordes from a vast Asian country unless we had powerful allies willing to come to our aid. The fact of interdependence prompted us to join the mutual defense organization known as SEATO (Southeast Asia Treaty Organization), along with the United States, New Zealand, Britain, France, Thailand, and the Philippines. Although SEATO has been phased out of existence, Australia is still a member of the ANZUS Pact, a mutual defense organization whose members are Australia, New Zealand, and the United States.

But although there has been, in view of the policy of and military developments in Communist China, a preoccupation with the problems of defense, there is a wide realization of our responsibility for encouraging the economic growth of the Southeast Asian countries. Australia expects to have growing contacts and co-operation with Malaysia, Singapore, the Philippines, and Indonesia. And it will continue to aid and assist Papua New Guinea, an independent nation that was formerly an Australian-administered territory.

In short, given peace, the nations of this section of the world should see an exciting period of development, rising standards of living, and the creation of free institutions of self-government that will change the rest of the world's old conceptions. These nations were for many decades thought of as relatively unknown but romantic countries to be visited by novelists and travelers, part of their charm being that they were seen by the Western world as primitive and picturesque, with such profitable activities as tin-mining, rubber-planting, and oil-drilling primarily controlled by outsiders. But in the near future they will emerge as a series of modern nations, with great population and resources, of notable international significance. Their future, provided that the great industrial powers play their part in the building up of trade and of the new technical skills, will be one of the exciting things in the world that my grandchildren, and millions of others of today's grandchildren, will inherit.

The long waves challenge water sports enthusiasts off the Australian coast.

AUSTRALIA

"Down under" in the Southern Hemisphere is a unique island, country, and continent—Australia. Australia is the earth's smallest continent and the earth's largest island. It is the only continent occupied by a single nation. Australia, which comprises the continent of Australia and the island state of Tasmania, is an English-speaking nation, part of the Commonwealth of Nations; yet physically it is closest to Asia. Australia lies southeast of Asia between the Indian Ocean on the west and the Coral and Tasman seas of the South Pacific Ocean on the east.

Formed in an early period of the earth's development, Australia is one of the oldest landmasses on earth. Yet it was not discovered until 1606, more than a century after the discovery of the Americas, and not settled by Europeans until 1788, almost a century after the settlement of North America.

On January 1, 1901, 113 years after the settlement of Australia, the six self-governing colonies—New South Wales, Victoria, South Australia, Queensland, Tasmania, and Western Australia—joined to become the Commonwealth of Australia. Australia also administers the Northern Territory, a huge, sparsely settled mainland area; the Australian Capital Territory, which includes the capital city, Canberra; and a number of island territories.

AUSTRALIA

G EQUATOR
H
F
G
E
D
C
B
A

EQUATOR

0° 170° 160° 150° 140° 130° 120° 110° 100°

10° 20° 30° 40° 50° 160° 170° 180°

1 **2** **3** **4** **5**

TROPIC OF CAPRICORN

PACIFIC OCEAN

INDIAN OCEAN

CORAL SEA

TASMAN SEA

ARAFURA SEA

BANDA SEA

TIMOR SEA

JAVA SEA

NEW HEBRIDES

NEW CALEDONIA
Nouméa

SOLOMON ISLANDS
BOUGAINVILLE IS.

PAPUA NEW GUINEA
Port Moresby
Lae
Rabaul
NEW IRELAND
NEW BRITAIN
IRIAN JAYA
Torres Strait
NEW GUINEA

INDONESIA
REPUBLIC OF
Djakarta
Bandung
Surabaja
Surakarta
Jogjakarta
Palembang
Banjermasin
Balikpapan
Macassar
Dili
BORNEO
CELEBES (SULAWESI)
MOLUCCAS
AMBOINA
SUMATRA
JAVA
BALI
LOMBOK
SUMBAWA
FLORES
SUMBA
Macassar Strait
Sunda Strait
Karimata Strait
Lombok Strait

NEW ZEALAND
Auckland
Hamilton
NORTH ISLAND
Wanganui
Palmerston North
Wellington
Cook Strait
SOUTH ISLAND
Christchurch
Mt. Cook
Dunedin
Invercargill

AUSTRALIA

WESTERN AUSTRALIA
Perth
Fremantle
Broome
Kalgoorlie
Cape Leeuwin
North West Cape
Steep Point
King Sound
HAMERSLEY RANGE
GREAT SANDY DESERT
GIBSON DESERT
GREAT VICTORIA DESERT
GREAT WESTERN PLATEAU
KIMBERLEY PLATEAU

NORTHERN TERRITORY
Darwin
Alice Springs
ARNHEM LAND
Cape Arnhem
Ord River
MACDONNELL RANGE
MUSGRAVE RANGE
Mt. Woodroffe
Ayers Rock
CENTRAL RANGE
GULF OF CARPENTARIA

QUEENSLAND
Brisbane
Ipswich
Lismore
Cape Byron
Bundaberg
Rockhampton
Townsville
Cairns
Toowoomba
MORETON BAY
CAPE YORK PENINSULA
Cape York
GREAT BARRIER REEF
GREAT DIVIDING RANGE
THE GREAT ARTESIAN BASIN
Flinders River
Diamantina R.
CENTRAL LOWLANDS

SOUTH AUSTRALIA
Adelaide
Port Pirie
Port Augusta
KANGAROO ISLAND
Yorke Pen.
SPENCER GULF
GULF ST. VINCENT
MT. LOFTY RANGE
Lake Eyre
Palmer River
NULLARBOR PLAIN
GREAT AUSTRALIAN BIGHT

NEW SOUTH WALES
Sydney
Newcastle
Wollongong
Lithgow
Canberra
Broken Hill
Wagga Wagga
BOTANY BAY
Mt. Kosciusko
Darling River
Lachlan River
Murrumbidgee River
Murray River
GREAT DIVIDING RANGE
AUSTRALIAN ALPS

VICTORIA
Melbourne
Geelong
Bendigo
Ballarat
PORT PHILLIP BAY
Bass Strait

TASMANIA
Hobart
Launceston
Mt. Ossa
Southeast Cape

TROPIC OF CAPRICORN

NATIONAL CAPITAL
STATE CAPITAL
DECIDUOUS FOREST
MEDITERRANEAN SCRUB FOREST
TROPICAL FOREST
STEPPE (SHORT GRASS)
DESERT

©1956, JEPPESEN & CO. DENVER, COLO., U.S.A.
ALL RIGHTS RESERVED
REVISED 1977
PUB BY THE H.M GOUSHA CO.

Miles
0 200 400 600 800 1000
0 200 400 600 800 1000 1200 1400 1600
Kilometers

INDEX TO AUSTRALIA MAP

FACTS AND FIGURES

COMMONWEALTH OF AUSTRALIA is the official name of the country.

CAPITAL: Canberra.

LOCATION: Southern Hemisphere between Pacific and Indian oceans. **Latitude**—10° 41' S to 43° 39' S. (including Tasmania). **Longitude**—113° 09' E to 153° 39' E.

AREA: 2,967,910 sq. mi. (7,686,900 sq. km.).

PHYSICAL FEATURES: Highest point—Mount Kosciusko (7,305 ft.; 2,227 m.). **Lowest point**—Lake Eyre (39 ft.; 11 m. below sea level). **Chief rivers**—Murray, Darling, Murrumbidgee.

POPULATION: 12,550,000 (estimate).

LANGUAGE: English.

RELIGION: Anglican (Church of England), Roman Catholic, various Protestant denominations.

GOVERNMENT: Independent, self-governing federation within the Commonwealth of Nations. **Titular head of state**—governor-general. **Head of government**—prime minister. **Legislature**—Parliament. **International co-operation**—United Nations, Anzus Pact, Colombo Plan.

CHIEF CITIES: Sydney, Melbourne, Adelaide, Brisbane, Perth, Newcastle, Wollongong, Hobart.

ECONOMY: Chief minerals—coal, lead, iron, copper, zinc, gold, rutile, tungsten. **Chief agricultural products**—wheat, meat, oats, butter, cheese, wool, livestock, fruits. **Industries and products**—meat, wool, and dairy products; mining; food processing; manufacture of iron and steel; motor vehicles; heavy equipment; electrical appliances. **Chief exports**—wool, cereals, chemicals, hides and skins, meat, ores and metals, fruits and vegetables, sugar and sugar products. **Chief imports**—transportation equipment, machinery, textiles, petroleum products.

MONETARY UNIT: Australian dollar.

NATIONAL HOLIDAY: January 26, Australia Day.

NATIONAL ANTHEM: "God Save the Queen," "Waltzing Matilda," "Advance Australia Fair," "Song of Australia."

AUSTRALIA'S STATES AND TERRITORIES

STATE	AREA (Approx.) (Sq. Mi.)	(Sq. Km.)	POPULATION	CAPITAL CITY
New South Wales	309,400	801,300	4,566,900	Sydney
Victoria	87,900	227,700	3,443,900	Melbourne
Queensland	667,000	1,727,500	1,799,300	Brisbane
South Australia	380,100	984,500	1,164,700	Adelaide
Western Australia	975,900	2,527,600	979,700	Perth
Tasmania	26,200	67,900	392,500	Hobart
Northern Territory	520,300	1,347,600	71,300	Darwin
Australian Capital Territory	940	2,430	133,100	Canberra

Today Australia is emerging as a world power, but it is still a country in the process of growing. Its relatively small population, challenging geography, vast mineral wealth, and stable government all combine to make Australia attractive to people from Europe and America who are looking for new lands to settle and new frontiers for investment.

Because Australia was separated from other landmasses and was relatively untouched by human influence, animals, birds, trees, plants, and flowers have evolved there that are seen nowhere else on earth. The fascination of the country is greatly increased by the remarkable variety

AUSTRALIA

★ State capital

Perspective map by J. Donovan

of its scenes. Modern, busy, populous cities thrive on the coast. Inland, seemingly endless, often arid, desertlike plains stretch for hundreds of miles. There are pastoral areas that resemble the English countryside; enormous stock stations (ranches) as in Texas in the United States; vast wheatfields as in Canada's Saskatchewan; tropical rain forests as in Brazil; snowy uplands as in Norway; and still other unexpected sights.

THE LAND

The Australians are the only people who have a continent to themselves. It is a large, extraordinary land. From Cape York Peninsula in Queensland on the north to Southeast Cape in the island of Tasmania on the south is approximately 2,000 miles (3,220 kilometers), and from Steep Point in Western Australia on the west to Cape Byron in New South Wales on the east is 2,400 miles (3,860 km.).

The more than 12,000 miles (19,310 km.) of coastline is remarkably even, because it has been smoothed out by water washing against it for millions of years. There are two large indentations, the Gulf of Carpentaria on the northern coast and the Great Australian Bight on the south. Along the northeast coast is one of the most spectacular natural wonders of the world—an enormous underwater garden. The Great Barrier Reef, the longest coral formation in the world, runs for 1,250 miles (2,010 km.). Separated from the mainland by from 10 to 100 miles (16–160 km.) of shallow water are some 600 little islands that lie in this gigantic marine playland.

Millions of years of wear and tear by wind and rain have worn away the ancient mountains of Australia, making this the flattest of all the continents. Almost half of the land is a huge plateau that does not rise very high above sea level. Despite this, the face of Australia does present dramatic contrasts. There are three great natural regions: the Great Dividing Range, or Eastern Highlands, the Central Lowlands, and the Great Western Plateau.

The Great Dividing Range is a system of plateaus, hills, and low mountains that parallel the east and southeast coasts. The Great Dividing Range includes the Blue Mountains and the Australian Alps.

Along the coast, on the eastern and southern sides of the Dividing Range, are the great cities of Australia—Sydney, Melbourne, and Brisbane —as well as many other smaller towns and cities. Along the western slope is the "fertile crescent," a well-watered area where many large farms and sheep stations are located.

Water is a constant concern in Australia, for it is the driest of all the continents. The Great Dividing Range is Australia's main watershed. There are several short, swift rivers that flow eastward or southward to the coast. The long inland rivers that supply water for the fertile lands flow away from the coast and, because of the flatness of the terrain, are leisurely and slow-moving.

The 1,600-mile-long (2,570 km.) Murray, the greatest of Australia's rivers, forms the major part of the boundary between the states of New South Wales and Victoria before it enters the sea in South Australia. The Murray, with its tributaries, the Darling, Murrumbidgee, and Lachlan (flowing into the Murrumbidgee), forms the main watering system of Australia. Floodwaters in the upper reaches of these long, lazy-moving rivers may take weeks to reach distant parts of the country.

Mount Olga rises majestically in Australia's vast Central Lowland region.

The Central Lowlands contain some of Australia's finest pastureland. An unusual feature of this area is the Great Artesian Basin, 670,000 square miles (1,735,300 square kilometers), where water is found deep underground. It has been estimated that more than 2,500 bores have been sunk to make this artesian water available for irrigating grazing land and pastureland. The water contains too much salt and other minerals to be used for watering croplands.

Lake Eyre, in the state of South Australia in the southern part of the Central Lowlands, is generally a vast, dry depression. Most of the time, since droughts often last for years, the lake is a giant basin of dried, salty mud. Farther south, between the Murray and Lachlan rivers, is some of Australia's richest farmland and best grazing land.

The Great Western Plateau covers nearly one half of the entire continent. Archeologists think that the southwestern part of the plateau may be 3,000,000,000 to 6,000,000,000 (billion) years old. Much of the interior of the plateau is desert—the Great Sandy Desert, the Gibson Desert, the Great Victoria Desert—and the so-called "gibber" plains, vast stretches covered with pebbles or consisting of barren land with some grass and spiky bushes. There are dramatic outcrops such as the majestic Ayers Rock, jokingly called the largest pebble in the world.

Mineral wealth is found in all parts of Australia, but this dry western area is literally a treasure house of minerals. In the Hamersley Range there are important iron ore deposits. Valuable minerals such as lead, zinc, silver, copper, uranium, nickel, gold, and others are found in many parts of the Great Western Plateau.

Some areas on the edge of the Western Plateau do receive regular rainfall. They are being developed, with the aid of scientific irrigation projects, into croplands and sheep and cattle stations.

Climate and Rainfall

Since all of Australia lies in the Southern Hemisphere, the seasons may seem to be reversed to people who live in the Northern Hemisphere. Australians, who are enjoying summer in December, very often celebrate Christmas by having beach picnics.

Almost everywhere in Australia one may enjoy sports such as tennis and surfing nearly all year. And the cattle that roam Australia's ranches do not need to be sheltered even during the winter. Australia enjoys a more temperate climate over most of the continent than do other countries in the same latitude. The reasons for this are found in Australia's global position, the pattern of the prevailing winds, and the fact that the continent is an island with no great land barriers.

More than one third of Australia, in the north, does suffer the extreme temperatures of the tropics, since it lies within the tropical zone, but the rest of the continent is within the temperate zone.

Despite the inviting climate, Australia suffers from insufficient rainfall. About one sixth of the continent—a belt along the eastern seaboard, the northern coast, and parts of the southern coast—receives more than 40 inches (102 centimeters) of rain annually. Enormous interior areas of Australia have an average annual rainfall of less than 10 inches (25 cm.). These arid regions are of little use to man, and predictably the major cities of Australia have grown up in the areas where rainfall is comparatively plentiful.

Unique Flowers and Trees

An amazingly varied and abundant plant life flourishes on this dry continent. It is believed that many of the 13,000 native plants are unique to Australia. Native wildflowers grow in many places, but Western Australia often is called the wildflower state. There on the heathlands in the southwest and in the western coastal regions are more than 2,000 rare flowers, which after a rainfall give the land the appearance of a brilliantly painted canvas. Among the unusual native wildflowers are kangaroo paws, predominantly red and green plants that seem to resemble the paws of the kangaroo.

Along the east coast one finds the waratah, a plant that grows taller than a man and has clusters of vivid red blooms; the honeyflower; and the lovely native fuchsia. The christmas bush, christmas bells, and coachwood burst into bloom at Christmastime.

Boronia, a beautiful flowering shrub, grows in more than 80 varieties in Australia. Its leaves supply aromatic oils for perfume. Orchids grow wild, but many of the more than 600 species are also cultivated for export.

Although vegetation varies with climate, national and state parks with extraordinary flowers and trees are within an hour's drive of every capital city in the country. And everywhere in Australia one finds the eucalyptus, or gum tree. There are more than 500 varieties of this native tree, and they grow in all kinds of climate and soil. They vary from the small, stunted eucalyptus of the dry areas to the tallest hardwood trees in the world.

The coat of arms of Australia depicts sprays of the sweet-smelling acacia, commonly known as the wattle tree. This name is derived from the fact that early settlers found acacia saplings just right for interlacing, or wattling, the framework for mud-plaster walls and roofs. The acacias range from dwarfs that are almost hidden by natural grasses to trees rising more than 80 feet (24 meters).

Strange Animals and Birds

The isolation of Australia from the rest of the world since prehistoric times permitted animals that disappeared elsewhere to survive and develop on the island continent. Australia is the home of the marsupials, animals that produce small, immature offspring that are carried in the mother's pouch until the babies are fully developed. Nearly half of the Australian native mammals are marsupials, and the kangaroo is the most famous of them. Among the more than 40 types of kangaroo are the red kangaroo, which may grow taller than a man; the wallaby, often the size of a large dog; and the rat kangaroo, less than a foot (.3 m.) long.

The koala, which looks like a child's teddy bear, is the best-loved marsupial. In its natural habitat, the koala spends its day sound asleep curled up in a eucalyptus tree. It drinks no water and lives entirely on the leaves of these trees. The platypus, Australia's strangest mammal, has four webbed feet and a long snout shaped like a duck's bill. The platypus represents a stage in evolution between reptile and mammal. It is a mammal that lays eggs.

Australia's dingo, a strong, handsome, usually light-brown animal with a bushy tail, is now believed to have descended from the Asian wolf. Other exotic animals that roam the sparsely settled areas of the continent are descended from buffalo, horses, donkeys, camels, deer, pigs, and dogs that were brought to Australia, escaped, ran wild, and multiplied.

Birds as strange and unique as the native animals abound in Australia. The emu, Australia's largest bird, like the smaller cassowary, cannot fly but is a fast runner. Equally distinctive is the black swan, its black

The best-known of Australia's marsupials (pouched mammals) is the kangaroo.

The koala bear's favorite habitat is a fragrant eucalyptus tree.

plumage highlighted by a red beak. The lyrebird, perhaps the world's greatest mimic, has been known to sing more than 40 different calls in one concert.

THE PEOPLE

In the latter part of the 18th century, England was faced with the problem of what to do with its large convict population. For some years considerable numbers had been transported to America, especially to the colonies of Virginia and Maryland, but the War of Independence quickly stopped this outlet. Soon the English jails were overflowing, and the government looked around the globe for other places for penal settlement.

It was Sir Joseph Banks, the famous botanist who accompanied Captain Cook on his voyage of discovery in 1770, who suggested Botany Bay as a place for settlement. On May 13, 1787, Captain Arthur Phillip set sail from Portsmouth to be governor of New South Wales, a new English colony. He commanded 11 vessels, including six convict transports, and 1,487 people, of whom 759 were convicts. Among the first settlers were officials of the government, Royal Navy marines and seamen, merchant seamen, children, and men and women prisoners—some had been held as political prisoners, and many others had been convicted of very minor offenses. Eight months later, the First Fleet, as it is known, anchored in Botany Bay after completing the arduous 15,000-mile (24,135 km.) voyage.

Captain Phillip, disappointed in Botany Bay, found in Port Jackson, a few miles north, "one of the finest harbours in the world, in which a thousand sail of the line might ride in perfect security." On the 26th of January, 1788, permanent settlement began on one of the bays of Port Jackson, which Captain Phillip named Sydney Cove. This day is now commemorated throughout Australia as Australia Day. A government was established and formal possession was taken on behalf of the British Crown of the whole of the eastern half of the Australian continent and the island of Tasmania. The aborigines, natives of the country, simply observed but did not protest the activities of the white men.

The Aborigines

It is believed that between 20,000 and 30,000 years ago, aborigines, the first inhabitants of the continent, gradually migrated from southeast Asia. When the white man first began colonizing the continent, the Australian aborigines are believed to have numbered about 300,000. They were nomads who hunted, fished, and lived simply, although they had developed a highly complex social organization. They had no knowledge of agriculture. The dingo, a dog, was their only domestic animal. They had almost no contact with the outside world. In the dry areas of Australia, they managed to exist where Europeans would have died from hunger, thirst, or exhaustion.

The settlers looked with contempt upon these people, with their dark skin, dark hair, and deep-set eyes, who wore no clothing and spoke strange words. The newcomers made no attempt to understand the customs, tribal organization, and religious beliefs of these primitive peoples. Many aborigines died from diseases brought by the settlers; some were killed in clashes with them. Other aborigines gradually lost the will to live as they lost their tribal territories and saw the breakdown of their customs. Today about 116,000 persons of more than 50 percent aboriginal blood survive.

Aborigine boys proudly display their day's catch of crabs.

The aborigines gave to Australia many place-names, some words, vivid cave paintings and rock engravings, and the boomerang. The boomerang was used by the Australian aborigine in hunting and fighting. They threw the heavy weapon so that it was able to hover briefly motionless in the air. The "return" boomerang, which is more curved than the hunting and fighting stick, is really only a sort of toy.

Today the Australian government is attempting to improve the status and prospects of the aborigines, and their number is actually increasing. Some still live the life of nomads; some work as miners and as stockmen on stock stations; some live in supervised reservations; and some have moved to the cities and lead the everyday life of the average Australian in the poorer suburbs.

Pioneer Settlers

Australians sometimes have been unhappy about the origins of their country and the predominance of convicts among the first immigrants. But recently the Fellowship of First Fleeters was formed, with membership open to descendants of anyone who arrived with Captain Phillip, regardless of their original status.

The early settlers, convicts or freemen, came from the British Isles. They found their new country vast, harsh, and awe-inspiring. Despite their inexperience in pioneering, many newcomers displayed great enterprise. Some drove sheep or cattle across the mountains to the plains beyond. Others made their way through the forests along the coast, cutting timber, which was in great demand. And still other groups settled down to farm on the land that had been cleared.

In the early days the governors of the colonies had the right to make free grants of land to anyone who would employ convicts and be responsible for their food and clothing. In addition, as settlements spread to the western plains beyond the mountains, ownership of land was established by living, or "squatting," on the land.

Sheep raising has always been an important part of the Australian economy.

The settlers needed additional land for the sheep industry. The first sheep had been brought to Australia from the Cape of Good Hope in 1797. Captain John Macarthur, who is credited with sparking the development of this great Australian industry, obtained some of these sheep. Before the beginning of the 19th century, he brought in some fine merino sheep from Spain. By skillful breeding he and others developed a hardy type of sheep that produced an exceptionally good quality of wool. The first wool was exported in 1807, when 245 pounds were sent to London. This industry was perfectly suited to the dry climate of Australia's vast interior.

The widespread sending of convicts to most of Australia ended because of public opposition in 1850. In Western Australia, however, because of extreme shortages of labor, quotas of convicts were accepted until 1868.

The great rush to Australia by immigrants did not come until the discovery of gold by Edward Hargraves in the Bathurst area on the central plains of New South Wales in 1851. Immigrants came from many parts of the Western world but mostly from Great Britain. This early gold discovery was followed by finds at Ballarat and Bendigo in Victoria, and later by even richer deposits at Kalgoorlie, in Western Australia.

The continent was opened up by the arrival of vast numbers of prospectors who went deep inland. They were followed by adventurers and tradesmen. In 10 years the population, though still sparse, had almost tripled. Many new settlements became flourishing cities as settlers turned from prospecting for gold to agriculture and other ways of making a living. As always, roads, railroads, commerce, and educational institutions developed to fill the needs of the new settlers.

New Australians

The original British settlers were joined during the gold rush by Americans and many Chinese. The Americans were active in the movement to create an Australian republic and in the fight against unwise official regulations in the goldfields. There was considerable conflict between the Chinese and the Australians on the goldfields, and this was reflected in the adoption by the colonies of restrictions on Chinese immigration. These restrictions later developed into what was in practice a prohibition on most Asian and other non-white migrants.

Since World War II, in one of the most exciting chapters in Australia's history, 2,000,000 new settlers have come to the country. More than one out of six persons now living in Australia is a "new Australian." In 1945 the Commonwealth Government decided to launch a planned program of population expansion by assisting settlers from Great Britain to make the expensive trip to Australia. It was soon extended to include would-be settlers from other Western European countries. Two thirds of the new settlers have come from Great Britain, but large numbers have come from Italy, the United States, Greece, Germany, the Netherlands, Yugoslavia, and other countries.

Australia's immigration policy has often been attacked because of its discrimination against people of non-European stock. The policy is directed toward maintaining a predominately homogeneous population. Provisions exist for the entry of some non-Europeans, but they must have special skills or professional qualifications.

Players leap for the ball in an Australian Rules football match.

Athletic Heroes

The typical Australian seems to be the product of a hard, healthful, out-of-door existence. People think of Australians as lean, sun-tanned, independent in outlook, laconic in speech, friendly especially to their own kind. The increasing move from country to city life has changed many Australians from outdoor workers to office and factory workers. Still there is a noticeable lack of formality, an admiration for toughness and loyalty, and a great interest in the outdoors. Most people live well in comfortable homes they own, and there are no apparent extremes of wealth and poverty.

Australian heroes tend to be athletes. The climate permits year-round outdoor sports in almost all areas. The youth and good health of the population and the fact that sports are easily available and cheap make Australia seem to be a country of sports enthusiasts. They are not just spectators, but participants. Many Australians seem to have been born in the water, they take so naturally to swimming, surfing, skin diving, sailing, and yachting.

Tennis, lawn bowling, track, golf, and cricket are all very popular. There are four varieties of football, Australian National Rules, Soccer, Rugby Union, and Rugby League. Australian Rules, which is especially popular in the southern states and Western Australia, draws the biggest crowds, and on occasions more than 100,000 people have attended the grand finals of this code. Interest in horses is great and horse races draw large crowds, as do car-racing events. For many years Australian players have dominated the world tennis events. Many city-dwellers play tennis at night on brightly lit courts in nearby suburban areas.

EDUCATION

The Australian colonies early introduced free, compulsory education. Today children must attend school between the ages of 6 and 15 or 16, depending on the particular state in which they live. The state government operates schools, but in addition to free schools, Australia has many independent schools, mainly operated by religious denominations. Some of these are modeled after the British boarding schools, and, as in England, are known as public schools. The curriculum and graduating examinations are set by the state and are the same in all schools.

The Australians have had to improvise to meet the problems of educating children in the remote areas. On the fringes of the metropolitan centers buses bring children to nearby schools. In some cases, however, the children travel almost 50 miles (80 km.) each way. Farther out, schools are established wherever there are 8 or 9 children, but in the outback children and teacher speak to one another on radio.

The Australian Government plays a major role in making university education available to students. Thousands of scholarships are awarded on the basis of merit, determined, at least in part, by the results of the qualifying examination. About half of the students receive some form of public assistance.

There are 15 universities in Australia. Each state has at least one university. The Australian National University at Canberra is a graduate and postgraduate institution with faculties in arts, economics, law, medicine, oriental studies, and science.

Language

When John Masefield, the great English poet, saw the Australian and New Zealand soldiers in World War I, he was moved to write, "They are the finest body of young men ever brought together in modern times. For physical beauty and nobility of bearing they surpassed any men I have ever seen. . . ." But it is doubtful that John Masefield could have understood them had he heard them as they marched during World War II singing about a "jolly swagman" who boils his "billy" at a "billabong" where a "jumbuck" comes to drink. The soldiers were singing the great marching song, "Waltzing Matilda," about the tramp (swagman) carrying his bundle (swag), who boils water in his pan (billy) at a waterhole (billabong) when a sheep (jumbuck) comes to drink.

Many English-speaking people traveling to Australia are confused by the local vocabulary spoken rapidly in the typical accent, which is flat and nasal. Soon they learn that to take a sheila to a shivoo is to take a girl to a party, where they may be served plonk, or lunatic soup (wine). If an Australian warns you of a willy-willy, he means a windstorm is threatening. Or he may tell you not to be a drongo (fool) or you will end up in a picnic (real mess) or up a gum tree (in trouble).

Australian speech is also spiced with aboriginal words. Aborigines have contributed names of animals (kangaroo, koala, dingo), trees (mulga), and birds (kookaburra). Idiomatic words also came out of early experiences in Australia. The Aussies, as Australians are commonly called, refer to the English as Pommies. This possibly comes from P.O.H.M., which was on the convicts' uniforms when they arrived from England, for they were "Prisoners of His Majesty."

The early days in Australia produced words such as "bush," which now means the country, in contrast to the city, and "bushed," which means "lost." A "digger," originally a miner, now refers to a soldier. The language of the sheep and cattle stations includes "jackeroo," meaning a man who works as a stockman but hopes to manage his own property eventually; "ringer," the leading shearer in the shearing shed; "poddy-dodger," a cattle thief; and many others. And the vast inland of Australia produced descriptive phrases such as "outback," "way out," and "west of the sunset," among others.

THE ARTS

Literature. The first Australian writers wrote in a wholly British style. True idiomatic Australian writers began to emerge in the late 19th century with "bush balladists" such as A. B. Paterson, who wrote "Waltzing Matilda," and Henry Lawson, who wrote short stories as well as verse.

They were followed by writers who told, in fiction, of the life of the convicts and life in the bush. In more recent times Australia has produced important prose writers such as Henry Handel Richardson, Alan Moorehead, Patrick White, Jon Cleary, Ruth Park, Morris West, Shirley Hazzard, and many others.

Australian publishing has expanded greatly since World War II, although many of the books read are still produced overseas. The Australians are among the biggest buyers of books in relation to their population in the English-speaking world.

Performing Arts. In the earlier days in Australia, promising writers, artists, and performers often went overseas to escape cultural isolation and to have greater opportunities for financial reward. Since World War II, with the increasing prosperity, the arrival of many new Australians with wide cultural backgrounds, and the support of the Australian Government, the picture has changed.

The Australian Broadcasting Commission, a government agency, has a generous program to encourage Australian composers, performers, dramatists, and writers. Facilities for the arts have been greatly improved with the building of concert halls in the leading cities. Ballet companies, theater, and opera all receive government support.

Many Australian singers have achieved international fame. Dame Nellie Melba (1861–1931), the great soprano, born Helen Mitchell, took her stage name from her native city, Melbourne. Joan Sutherland is one of the most celebrated sopranos in the world today. Among the Australian composers who are best-known outside their own country are Percy Grainger (1882–1961) and Arthur Benjamin (1893–1960).

Many Australians are known for their stage and screen performances. Judith Anderson, Cyril Ritchard, Diane Cilento, Peter Finch, Zoe Caldwell, and others have traveled to London, New York, and Hollywood to act on stage and in films.

Painting. Australia has produced many well-known artists. The earlier painters, such as Arthur Streeton, Hans Heysen, Tom Roberts, and Charles Conder, painted vivid scenes of the Australian landscape, of sheep stations, the lonely outback, and eucalyptus trees. In recent years, Sidney Nolan has painted expressionist interpretations derived from colonial legends and Australian history; while William Dobell has done fine portraits.

A whole group of contemporary realistic as well as abstractionist painters are at work in Australia. There has also been great interest in sculpture and mural art since many artists have been commissioned to decorate the spendid new buildings.

Aboriginal artists have attracted much attention in recent years, and the national museums have become interested in the aboriginal art. Albert Namatjira, an aboriginal artist who died in 1959, gained fame for his work in watercolors, a form of painting that has attracted many aboriginal artists in recent years. Public art galleries are now found in all the state capitals as well as in some smaller cities.

NATURAL RESOURCES

Australia is still sparsely populated and is seeking additional human resources to develop her great natural resources. Australia's most valuable agricultural resource is the vast grassland in the eastern and central areas where the sheep graze. Sheep are so important to the economy that it has often been said that Australia travels on the sheep's back. Even today the export of wool continues to be a primary source of foreign currency.

Every state in Australia has important mineral resources. The minerals are often found in the exposed rocks that occur over much of the continent. In Western Australia, South Australia, the Northern Territory, and parts of Queensland and western New South Wales there is iron ore, lead, zinc, silver, copper, uranium, nickel, and gold. In a belt along the eastern portion of the continent from western Victoria to northern Queensland are mineralized rocks that once contained major deposits of gold. The gold is now largely worked out, but some large copper, lead, zinc, and silver deposits remain. In both areas there are smaller amounts of tin, tungsten, mica, manganese, cobalt, and other metals. Bauxite for aluminum is also an important Australian resource.

An offshore oil rig, 10 nautical miles off the Australian coast.

Very large coal reserves are found in Australia. They vary from poor grade brown coal in Victoria to the highest quality coals in New South Wales and Queensland. Opals, lovely iridescent gems, are found in the sediment of the Great Artesian Basin in Queensland, in New South Wales, and in South Australia.

In the 1960's important commercial oil fields were discovered in Queensland, west of Brisbane. Offshore fields in the area of Victoria, and on Barrow Island, Western Australia, have proved of great value. Natural gas was also found there.

Australia's eucalyptus trees provide a fine supply of hardwoods and also help to fill the country's newsprint and paper needs. The supply of softwoods is being replenished by extensive planting.

Australia does lack a most important natural resource—sufficient water. Much is being done to use the available water in the most scientific ways to provide irrigation. The greatest single development is the Snowy Mountains Hydro-Electric Scheme—a vast complex of dams, tunnels, power stations, and transmission lines, which trap, store, control, and distribute the snow-fed waters of the Snowy River and its tributaries. Formerly these waters flowed down the southern slopes of the Great Dividing Range and poured largely unused into the Tasman Sea. Today the waters are diverted through the mountains down the western slopes, spinning mighty turbines and finally pouring the life-giving waters into the Murray and Murrumbidgee Rivers. The Snowy Mountains Scheme has been called one of the seven engineering wonders of the world.

An opencut iron ore mine at Iron Knob near Whyalla on the southern coast.

INDUSTRIAL EXPANSION

Australia's industrial expansion during the past two decades has been spectacular. Manufacturing now accounts for a fourth of the gross national product and employs a quarter of the workers. Since World War II expansion has been in the highly skilled and technically advanced industries. Even before the war, Australia was known as the producer of textiles, metal and electrical goods, and processed foods.

With its enormous wealth of coal and iron ore, it is not surprising that Australia has a tremendously successful and greatly expanding iron and steel industry. Automobiles are produced by subsidiaries of the great companies of the United States, Germany, and Great Britain. Australia also produces all sorts of highly sophisticated machinery such as diesel locomotives, power turbines, earth-moving equipment, television sets, and computer equipment.

There has been a huge expansion of the petroleum and petrochemical industry since the large oil and natural gas deposits were discovered. Petroleum products and related chemicals, explosives, plastics, fertilizers, drugs, and cosmetics are important new industries.

The pulp, paper, and paperboard industry is also being enlarged. Shipbuilding is a significant industry. Commercial fishing is important to the economy, and canneries, particularly of salmon and tuna, have been developed.

Perhaps the most exciting story in Australia's industrial expansion has been the growth of Australia's mineral industry. Coal has always been important, but Australia is now one of the leading producers of lead and one of the largest producers of zinc in the world. Copper and gold are also produced in large quantities. In recent years the discovery of large deposits of bauxite has made possible an aluminum industry. Uranium and a number of other minerals, including the lovely Australian jewel, the opal, are exported.

The expanding Australian economy and markets, its mineral treasures, and its stable government, which welcomes foreign capital, have all combined to attract foreign investors. Australia was one of the first countries to establish independent industrial tribunals for compulsory conciliation and arbitration of industrial disputes.

CITY LIFE

Most Australians, more than 80 percent, live in urban centers. The chief cities are the capitals of Australia's states, all of which are ports, and Canberra, the federal capital. One third of all Australians live in two of these capitals, Sydney, the capital of New South Wales, and Melbourne, the capital of Victoria. They and the other state capital cities —Brisbane, Adelaide, Perth, Hobart, and Darwin—have grown rapidly since World War II, keeping pace with the expansion of manufacturing and trade. Canberra is a comparatively new city that is also growing by leaps and bounds.

Australian cities tend to grow outward rather than upward since more than 70 percent of Australians own their own homes. One does find apartment buildings, of course, but most people have their own house and garden. Since the work week is from 35 to 40 hours, the average worker has some time left every day for sports and gardening when he returns from his job.

Australian homes and apartments are generally well equipped with modern labor-saving devices. Most houses have one story of four or six rooms and are set in gardens planted with flowers and vegetables. Nearby are easily available and inexpensive facilities for tennis, cricket, golf, and lawn bowling. Residents of most coastal cities have magnificent, free beaches within easy reach.

Many Australian cities, with tree-lined streets, lovely parks, and traffic driving on the left side of the road, are often compared to such cities as Edinburgh, but, despite many British customs, life is quite different. The temperate climate in the areas where most Australians live is congenial. The pace of life is slower than in Great Britain or the United States. Although manufacturing is growing rapidly in Australia and many of her people are city-dwellers, its pastoral and pioneering background gives a different flavor to life.

Because of Australia's healthy economy, the people live well. Their wages are generally lower than in the United States, their taxes higher, and the prices they must pay for many necessities are higher, but they do not have extremes of poverty and wealth. Health services are good and available to all under a voluntary contribution insurance plan sponsored by the federal government. All school systems have health services.

COUNTRY LIFE

Most immigrants tend to settle in cities, but some seek their fortune in the country. Despite Australia's rapid industrialization, wool is still its single most important export item. Sheep are run in all the states, but the larger sheep stations are found in Queensland, New South Wales, and Western Australia.

Cook Station on the Nullarbor Plain is one of the stops on the Trans-Australian Railway near the border of Western and South Australia. The railway provides a vital link for the immense and varied parts of the nation.

Sheep farmers, called graziers, take care of thousands of sheep, and the shearer may shear 10 pounds or more from 150 to 200 sheep a day at the annual shearing. Some sheep farmers lead fairly isolated lives. Others are part of holdings so large that they form little communities of their own.

The sparsely settled interior of Australia is known as the outback, and that is the area of many of the great cattle ranches or stations. They have been so successful that Australia now ranks second only to Argentina in the export of beef and veal.

The outback is a vast, largely flat, dry, lightly populated land. There are cattle stations which vary from a few thousand acres to one that is about 6,300 square miles (16,300 sq. km.) in area. The Australian stockman, or cowboy, often uses trucks, jeeps, and planes these days rather than depend on horses.

Aircraft and radio do much to relieve the isolation of the outback. Children learn their lessons via a radio educational system called School of the Air. They are taught by two-way radio communication and mail in their tests and homework. The radio gives neighbors a chance to talk together, although they may be many miles apart. It brings news and advice to the homestead. It also summons medical aid.

The doctor and dentist arrive by plane under a program supported by the government and called the Royal Flying Doctor Service. Flying doctors operate a chain of bases, each serving a part of the vast Australian continent. Altogether they provide medical assistance for some two thirds of the entire country.

Many Australians live on large wheat farms. The country is also a

A typically large, mechanized wheat farm in the Wimmera district.

major supplier of other cereals, dairy products, sugar, and fruit. Maize (corn) has been cultivated since the first year of settlement; and wheat, barley, and oats are grown in all the states. Sugarcane, tobacco, and cotton are also important Australian crops, grown in the warmer areas.

Smaller farms, found in areas near cities, provide fruits and vegetables as well as dairy products. Many of the market gardens, fruit farms, and vineyards are run by Australians newly arrived from similar farms in Europe.

AUSTRALIAN STATES AND TERRITORIES

New South Wales. The mother colony is the fourth in size among the Australian states, but it has the largest population. It leads the country in industry, volume of shipping, and in agriculture; and it has more factories than any other state. They range mostly along the coastal plain and make everything from textiles to motor vehicles and plastic toys to agricultural machinery.

New South Wales has great iron and steel centers and a considerable mining industry. It also produces sugar and timber in the coastal areas; while inland, sheep range and wheat is grown. Fruit and grapes for wines are grown inland, often in irrigated areas.

In New South Wales is the Kosciusko National Park, one of the largest in Australia, which is an important winter and summer playground, and the Snowy Mountains Scheme, the ambitious engineering undertaking.

Most visitors who come to Australia enter the country through **Sydney**, the capital of New South Wales. Arriving by ship, one sails into Port Jackson, the harbor of Sydney, one of the largest and most beautiful in the world. If you come by air, you see below you a city that spreads out from the blue Pacific to the Blue Mountains beyond.

Sydney, with a population of about 2,800,000, is not only Australia's oldest and largest city, but also its chief manufacturing center and busiest port and the largest center for selling wool in the world. It is also a city with 30 golden beaches nearby. In Sydney if you say you are going skiing, it often means water-skiing. If you are going to Mount Kosciusko, about 270 miles (434 km.) away, you are probably going snow skiing.

The harbor of Sydney is spanned by the famous arch bridge and commanded by the dramatic new Sydney Opera House. Described by many architects as one of the great buildings of the 20th century, the Opera House has great white curving roofs built to resemble wind-filled sails.

There are old, classic buildings in Sydney as well, such as St. James Church, Hyde Park Barracks, the Old Mint, and Parliament House. Many of these were designed by the convict architect Francis Greenway. There are also modern, towering skyscrapers.

It is estimated that by the end of this century there will be 5,000,000 people living in a continuous metropolitan complex along the 200 miles (322 km.) of coastline north and south of Sydney. Along this coastal strip are industrial centers such as Newcastle, Port Kembla, and Wollongong.

Victoria. In southern Australia is the smallest mainland state, Victoria, which has an importance in the country's economy far greater than its size might indicate. Almost one fourth of the people of Australia live there.

The Harbour Bridge frames the skyline of Sydney, with its modern Opera House.

Victoria has snowcapped mountains in the Australian Alps, lovely fertile valleys, and volcanic plains. Cattle graze in the eastern highlands of the state, while sheep and wheat are important in the west. On the lands along the Murray River, fruits are grown and trucked to the markets in the cities.

The gold rushes of the 1850's transformed Victoria from a quiet, pastoral colony into a thriving, vital place. Substantial towns were established, many of which continued to grow after the diggers no longer found mining lucrative. Ballarat and Bendigo, both born as goldfields, have become agricultural and manufacturing centers.

Coal is mined in Victoria. Brown coal from Latrobe Valley, the world's largest known continuous brown coal deposit, is used to make cheap electricity for the cities. Oil and natural gas found off the coast of Victoria may eventually overtake coal as the great source of Victoria's mineral wealth.

Melbourne, Australia's second largest city, was established in 1835, 16 years before the separation of the colony of Victoria from New South Wales. John Batman, after purchasing 600,000 acres (242,915 hectares) of land from a wandering band of aborigines for some blankets, tomahawks, flour, and other supplies, decided that the lower reach of the Yarra River was "the place for a village." From 1901 until 1927 Melbourne was the seat of the federal government of Australia.

Melbourne, on Port Phillip Bay, was the natural port for all the men and goods that came to and from the goldfields. It flourished accordingly, and it continues to be the natural outlet for the products from southeastern Australia.

The most English of Australian cities, Melbourne has attracted many European settlers. Stately stone buildings rise along wide, tree-lined streets. There are lovely parks, with statues of people notable in Australian history, and fine shops in lanes between main street arcades. Around the city is a green parkland belt, which includes the Royal Botanic Gardens.

Melbourne is regarded as the cultural and sports center of Australia. The Victoria National Gallery in Melbourne has the finest collection in Australia and is known throughout the world. Melbourne is the scene of Australia's richest horse race, of numerous golf matches, of cricket matches played against teams from England, and of many exciting Davis Cup (tennis) finals.

Queensland. The tropical northeastern corner of Australia is occupied by the second largest and youngest state. Queensland is about two and a half times the size of Texas and more than three times the size of France. Yet less than 2,000,000 people live in this vast area. Most of them are clustered along the narrow coastal plain in the east.

It is off this coast that the Great Barrier Reef stretches for 1,250 miles (2,010 km.). Considered one of the natural wonders of the world, it is the world's largest coral reef, thought to be 30,000,000 years old and still growing. People come from all over the world to this part of Queensland to snorkel, skin dive, fish, or just relax. The reef provides an unending field of research for students of marine life. There are mollusks, from the giant clam, the world's largest shellfish, to the smallest cowries, which have a glaze like fine porcelain. Fish of striking color and pattern are everywhere. Great green turtles and many types of birds come to nest there. There is no comparable area on earth.

In contrast to the beauty of the coast, most of inland Queensland is a hot, dry region. In the north the extreme heat and humidity and the scarcity of water during the dry season make life rather difficult. Sugar, cotton, bananas, and other tropical fruit are grown along the coast.

In much of the interior of Queensland sheep and cattle roam on vast properties. In some areas water for the sheep and cattle comes from the deep underground Great Artesian Basin. Queensland is Australia's leading beef producer, and huge amounts of beef, veal, and dairy products are exported.

Above all, Queensland has great mineral wealth. The development of the state was enormously stimulated by the gold rush of 1867. The colony's most important goldfield was at Palmer River in the far north. Aborigines were hostile, the country was wild, conditions for travel and existence were very hard. The nature of the area is shown in some of the former place-names: Cannibal Creek, Revolver Point, Hell's Gate Pass, and Murdering Creek.

Gold continues to be mined but other metals are more important to Queensland today. Silver, lead, zinc, and copper are produced in large quantities in an area around the town of Mount Isa. A recently discovered phosphate deposit may be one of the world's largest. In addition, some oil and gas have been found in southern Queensland. And what is thought to be the world's largest bauxite deposit is located in the north on Cape York Peninsula.

Brisbane, Australia's third largest city, grew rapidly from a penal colony for the most hardened convicts to a city that is a great tourist attraction. Pleasantly situated on the Brisbane River, the city is surrounded by hills. It is the largest river port in Australia and is still growing. During

World War II many Americans were stationed in Brisbane, since it served as headquarters of the United States forces in Australia. A monument in the city commemorates the co-operation between the two countries.

Tropical flowers—bougainvillea, hibiscus, and frangipani—fill the numerous parks and gardens of the city. In the winter tourists come to take advantage of the pleasant tropical climate. But all year round Brisbane is a busy export center for the state's wool, wheat, minerals, and sugar products.

South Australia. The third largest state in Australia covers an area slightly larger than the countries of France, Italy, Belgium, and the Netherlands. South Australia has considerable agricultural wealth that is derived from its fertile and beautiful southeastern corner. It is there that almost all of the people of the state live.

The gentle Murray River flows westward and southward and provides water for the irrigated land where citrus fruits, plums, pears, peaches, apricots, and grapes are raised in the Mediterranean-like climate. The Barossa Valley is the largest wine-producing area of Australia. It was largely settled by emigrant German Lutherans in the mid-19th century. To the west are the rolling wheat fields of the Yorke Peninsula.

The entire northern part of South Australia consists of dry or desert land. Many of the sunken areas are really sand flats that flood only after an occasional heavy rain. At one time the isolated cattle stations and the aboriginal missions of this region received mail and supplies by camel caravan.

South Australia also has valuable minerals. Half the world's opals are found there. High-grade iron ore has been mined for years, and the state now has an important smelting industry. Recent discoveries of natural gas in the far northeastern corner of the state are further accelerating the industrial growth. South Australia is also a leader in shipbuilding, lumbering, and auto manufacturing. The urgent need for water all over the state is being met by a continuous development of reservoirs and pumping schemes.

Kangaroo Island, south of Adelaide, brings many visitors to the wildlife sanctuary and to the quiet beaches. It is also a gathering place for fishermen, since salmon and many other species are found in the water near the island.

Adelaide, the capital of South Australia, is internationally known for its Arts Festival. Every other year leaders in music, drama, and dance from many parts of the world join Australian artists for a fortnight of cultural events. They range from jazz to ballet and from poetry readings to art exhibitions.

The city of Adelaide owes much to its first surveyor-general, Colonel William Light, who laid out wide elegant streets and extensive parklands. The city actually includes three golf courses. Set between the Mount Lofty Ranges and the sea, Adelaide has a warm, lovely climate, and the quiet air of a relaxed, southern city.

Western Australia. The entire western part of the continent is one state equal in size to the combined European countries of France, Spain, Germany (East and West), Italy, Norway, and Sweden. For many years Western Australia was hardly settled, since most of the large area was dry, barren, and inhospitable. Today only about 1,000,000 people live in Western Australia.

The state has miles and miles of seemingly endless desert, yet it is also known as the wildflower state. There is fertile land in the southwest section where most of the people live. In the spring visitors take special tours to see acres ablaze with yellow, pink, and flame-colored blossoms, many of them found only in Australia.

In the southwest is the tall timber country, with forests of giant eucalyptus, jarrah, and karri trees. They are important for the strength and durability of their wood. There, too, are the wheat-producing areas, the dairy farms, and sheep ranches.

To the north are vast cattle stations of the Kimberley region, a semitropical land. Millions of dollars have been spent on building roads to bring cattle to market from this isolated interior. Near Kimberley is the cotton- and rice-growing district where irrigation from the Ord River makes it possible to grow these crops.

The late but dramatic development of Western Australia is well illustrated by the fact that the single-gauge railway system that spans the continent from Sydney on the east coast to Perth on the west coast was not completed until 1970. This railroad, which runs 2,461 miles (3,960 km.), includes a 300-mile (483 km.) stretch in Western Australia and South Australia without a curve. It is the longest straight railroad line in the world and crosses the Nullarbor Plain. This is a vast, empty, treeless, limestone plain, riddled with enormous underground caves and tunnels, which covers much of the southern part of the continent.

Western Australia also received its great boost from the discovery of gold. More recently, important deposits of nickel were found in the southern part of the state. In the 1960's tremendous high-grade iron ore finds were made in the northwest. Overnight, mining facilities, new towns, railroads, and ports were begun to aid in exploiting the Hamersley Range finds, one of the greatest iron ore reserves in the world.

Important oil fields on Barrow Island, which were first discovered in 1966, have attracted a great flood of foreign investment and added to the potential of enormous wealth in Western Australia.

More than half of Western Australia's population live in the capital city, **Perth**. A beautiful city, with an almost perfect climate, Perth is happily known as the City of Light, a name given to it following the historic flight of the American astronaut John Glenn in 1962. By arrangement, the people of Perth turned on all their lights to greet the astronaut as he sped across the dark sky in the Friendship 7, 150 miles (241 km.) above the earth.

Situated on the wide-spreading estuary of the Swan River, Perth is one of Australia's most attractive cities. The adjoining ocean port of Fremantle, on the same river, is the "western gateway" for Australia.

Perth is often compared to cities in southern California. People enjoy fine surfing and outdoor restaurants in Perth, as well as the open-air performances during the annual Perth Festival. The pride of Perth is King's Park, 1,000 acres (405 hectares) of natural bushland overlooking the city. Driving through King's Park a visitor gets a spectacular view of the city and the Swan River, the Indian Ocean beyond, and the distant mountain ranges.

Tasmania. In 1726, Jonathan Swift's hero Gulliver told of being driven by a violent storm to Van Diemen's Land. It seemed the remotest place imaginable at the time. Today Van Diemen's Land is called Tas-

Perth, the capital of Western Australia, is seen from King's Park.

mania, in honor of its discoverer, Abel Janszoon Tasman. It is an island state of the Commonwealth of Australia.

Tasmania is the smallest of the Australian states and the one that has the most vivid reminders of early colonial days. Grim ruins of its once famous penal colony can be seen at Port Arthur. There are also lovely old houses little changed from the mid-19th century.

A small area of great beauty, Tasmania includes lush farmlands, great lakes and waterfalls, quiet fishing ports, spectacular rain forests, and wild mountain ranges. This mountainous island is one part of Australia that usually has plenty of rainfall.

Because of the abundance of water in high country, the production of electricity by waterpower is relatively simple. Cheap and abundant hydroelectricity has attracted industries that might otherwise have been established on the mainland. They include a zinc refinery, aluminum smelters, and pulp and paper works.

Tasmania is also sharing in the mineral boom. High-grade iron ore at Savage River in the northwest is crushed and pumped through a pipeline to the coast, where it is made into pellets for shipment. Copper is mined on the island, but a great part of the mineral riches of Tasmania has barely been explored.

Much of Tasmania is given to the growing of apples, hops, pears, berries, and potatoes. The farmhouses are charming buildings of white stone or red brick. Tasmania is often called the Apple Isle. The first apple trees were planted in 1788 at Adventure Bay by Captain William Bligh, whose name will always be associated with the mutiny on his ship, the *Bounty,* in the Pacific Ocean. Captain Bligh stopped in Tasmania on his way to Tahiti. Apples are now exported from Tasmania to countries all over the world.

Hobart, the capital city of Tasmania, is Australia's second oldest city. Founded in 1804, it is said to be the home of Australia's oldest licensed inn and oldest theater. The oldest existing bridge in Australia, built by convicts in 1823, is situated at Richmond, in the Tasmanian midlands. The historic Bush Inn is at New Norfolk, up the winding Derwent River, near Hobart.

A modern city of some 148,000 people, Hobart is beautifully situated at the foot of Mount Wellington, which is often snowcapped, and has an extensive and deep harbor on the Derwent River estuary. In the city are elegant Georgian houses built by the owners of the sailing ships and whalers that first brought wealth to Tasmania. Outside the city are prosperous dairy farms and apple orchards.

Northern Territory. Not a state but a separate area administered by the federal government, the Northern Territory has been described as a land of hopes and heartbreaks. With an area three times the size of Sweden, it has only a little more than 70,000 people. The land ranges from relatively fertile, well-watered forests in the north to desert and semidesert areas in the center and south.

The raising of beef cattle is the main industry, and some of the properties are as large as some of the smaller states in the United States. These are the enormous cattle stations where the children attend school by radio and the flying doctors are on radio call.

The Territory's mineral wealth is considerable. Gold has been mined for many years, and recently reserves of manganese and iron ore have been opened up. Copper and uranium are also mined. One of the world's greatest deposits of bauxite has recently been opened up on Arnhem Land.

The southern area of the Northern Territory, in the heart of the continent, is the famed Red Centre of Australia, the name given to the center of the great outback. It consists of miles and miles of stark red-sand desert and mountain ranges.

At the center is Alice Springs, or The Alice as the city is called. It is a base for the Royal Flying Doctors Service, the School of the Air, and for tourists who come to see Ayers Rock, the most spectacular sight in this dramatic area. Ayers Rock is a monolith 5½ miles (9 km.) around its base, which rises 1,100 feet (335 m.) above the surrounding plain. In the caves around the base of Ayers Rock there are aboriginal paintings that relate the legends of this ancient people. The rock changes color with the time of day, from reds and oranges to purples and blues. At sunrise it looks like a gigantic burst of flames illuminating the desert landscape.

On the entire northern coast of Australia, **Darwin** is the only large settlement. Founded in 1872 and named for the British scientist Charles

Darwin, it is still much like a frontier town and often serves as the point of entrance and exit for people who come to tour the Northern Territory.

Safaris leave from Darwin to hunt with gun and camera for giant crocodiles, buffalo, and strange birds and rare animals. Game fishermen come and go from Darwin, and tours of the great outback often begin and end there. Many people come to learn about the aborigines and to see their art.

There are only two seasons of the year in this tropical city. The dry season, April to October, and the wet, November to March. It was during the wet season of 1974 that disaster struck Darwin. A cyclone destroyed 90 percent of the city and forced the evacuation of most of its inhabitants. A new Darwin is expected to rise on the site of the old city.

Canberra, Capital of Australia. Shortly before the federation of the Australian states in 1901, it was decided that the new nation should have a new seat of government. A federal territory, 939 square miles (2,333 sq. km.), was carved out of the state of New South Wales to be the site of the national capital. It is between Sydney and Melbourne, Australia's two major cities, not far from the Australian Alps.

"Canburry" is an aboriginal word for "meeting place," and from it came the name of the capital, Canberra. Today it is a meeting place for the nation and for many international conferences.

Canberra is located on a limestone plain, surrounded by hills, in an area of great natural beauty. Lovely old trees shade the city. The Molonglo River flows through it, and the blue Brindabella Ranges form a handsome backdrop. There is also much man-made beauty.

The Commonwealth government arranged an international competition for the design of the city. Of the 137 designs submitted, that of Walter Burley Griffin, an American, was awarded first prize. Griffin left his architectural practice in Chicago in 1913 to build Canberra, and in

Parliament House and the parade ground in Canberra, Australia's capital city.

1927 the Parliament House was opened by the Duke of York, later King George VI of England.

The government buildings of Canberra are surrounded by lawns and gardens and overlook the waters of a beautiful man-made lake, Lake Burley Griffin, in the center of the city. The parks and gardens glow with color, and in the autumn the entire city seems to lie in a golden haze from the trees that line the streets.

In addition to its government and commercial buildings, Canberra is a center for study. It is the site of the Australian National University, well-known for its postgraduate research facilities, and the Mount Stromlo solar observatory. Two space-tracking stations are nearby. Near the city are the embassies of all the countries having diplomatic ties with Australia, the official residence of the governor-general, and homes of the people who work in their capital.

Canberra is a busy, growing city of about 120,000 people. It has a clean, open, lovely aspect, a face that reflects a vigorous, young country.

HISTORY

Dutch Discovery.   The belief in a great south land was held in very early times. During the 2nd century, the Greek mathematician and geographer Ptolemy drew a map of the known world and sketched a huge unknown land to the south of Asia and a body of water we know now as the Indian Ocean. This land was called *terra australis incognita*, or "unknown southern land."

During the 17th century the Dutch, who regularly journeyed from the Netherlands to Java, their colony in southeast Asia, reached the continent, which they called New Holland. In 1606 Willem Jansz, commanding the Dutch vessel *Duyfken*, landed on the eastern side of the Gulf of Carpenteria, where the great bauxite deposits in Weipa are now being developed.

In the years that followed, other Dutch seamen and traders came along the western and southwestern coast of the continent. A number of Dutch navigators explored and named parts of the coast. Captain Tasman discovered the island that is now called Tasmania in his honor. On a later voyage he explored the north coast of Australia and gave Dutch names to a number of places. But the general reports of the great southern land were so discouraging that there was little enthusiasm for further exploration or settlement there.

The English Arrive.   The first Englishman to come to Australia was the adventurer William Dampier, who in 1688 brought his ship, the *Cygnet*, to King Sound, near the present town of Broome on the northwest coast. His report, *A New Voyage Around the World*, tells of this visit. He was the first Englishman to give an account of any part of the continent, but his description of the barren coast and its primitive inhabitants hardly encouraged other Englishmen to follow him.

It was not until three quarters of a century later, on April 20, 1770, that Captain James Cook of the Royal Navy, in command of H.M. bark *Endeavour*, sighted the continent near the eastern extremity of what is now Victoria. He had been sailing westward from New Zealand. On April 29th, the *Endeavour*, having sailed northward along the coast, dropped anchor in a large sheltered waterway, which was named Botany Bay for the variety of botanical specimens found there. Cook continued northward, passing by another sheltered harbor, which he named Port Jackson, later to become the site of Sydney. Finally he reached the tip of the continent, which he called Cape York in honor of the Duke of York, brother of King George II.

On August 22, the entire ship's company landed on a nearby island. Hoisting the British flag, Cook claimed possession of the entire eastern coast of the continent in the name of King George III of England and named it New South Wales. The scene of the ceremony was designated Possession Island.

At the end of Cook's voyage in 1770, all of the Australian coast had been mapped except the southern and southeastern parts. These blanks were filled in by Matthew Flinders, George Bass, and by a Frenchman, Nicolas Baudin, in their expeditions during the years 1798 to 1802. In 1802–3 Flinders circumnavigated the continent. He is believed to have been the first person to use the name Australia. Previously the continent had been known as New Holland, New South Wales, or Botany Bay. But nearly another century elapsed before all of Australia was explored.

Exploration.   The exploration and development of Australia was not an easy task. The life of the early colonists was extremely hard because farming with few tools was difficult on the poor land. Captain Phillip, the first governor of New South Wales, soon found better land and explored inland to the Blue Mountains.

Convict settlements were established at Hobart, on Tasmania; at Moreton Bay, which later became Brisbane; and in western Australia. Free settlers began to arrive in small numbers, and many convicts who became free stayed on in Australia. Other settlements were gradually established—at Perth, on the Swan River, at Adelaide, and at Melbourne.

For the first 25 years settlers around Port Jackson were largely hemmed in by the Blue Mountains, lying about 50 miles (80 km.) to the west. These mountains, though they never rise very high, are extremely rugged, especially on their eastern face. In 1813 a party of explorers found a way to the far side of the mountains and saw the fertile country beyond. The subsequent opening of the vast unknown land is a story of courage, hardship, and perseverance. For three quarters of a century explorers, convicts, seamen, surveyors, scientists, and settlers played important roles in this dramatic undertaking.

Among these men were Hamilton Hume and William Hovell, who in 1824 set out from near Sydney and made their way overland to Port Phillip Bay, now the site of Melbourne. On this journey Europeans crossed Australia's main river, the Murray, for the first time. Allan Cunningham, botanist and explorer, probed northward in the 1820's and discovered the fertile land at Darling Downs and a way from there to Moreton Bay. Charles Sturt in 1829–30 did much to trace the westward-flowing rivers of New South Wales.

In 1844–45 Ludwig Leichhardt, a German naturalist, traveled overland from near Brisbane to the northern coast where the city of Darwin now stands. On his third journey, attempting to cross the continent from east to west, Leichhardt and his party disappeared, and no trace of them has ever been found. Edward John Eyre is remembered for his epic excursion in 1841 around the Great Australian Bight.

The Australian continent was first crossed from south to north in

Between 1834 and 1853 Port Arthur, Tasmania, was Australia's main penal colony.

1860–61 by a party led by Robert O'Hara Burke and William John Wills. The leaders were successful in reaching the sea but lost their lives on the return journey. A third member of the group was cared for by the aborigines and finally returned to Melbourne.

By the end of the 19th century the opening of Australia was practically complete. The rugged country of southwest Tasmania has been investigated by air but is in many places still practically untouched by man. Most of the continent, however, has been visited by geographers, oil explorers, mineralogists, surveyors, anthropologists, and, of course, sheep breeders seeking possible pasturage.

GOVERNMENT

On January 1, 1901, the six self-governing colonies joined together to become states in a federation. The previous September, Queen Victoria had signed a proclamation that said, in part: "We, therefore, by and with the advice of our Privy Council, have thought fit to issue this Our Royal Proclamation and we do hereby declare that on and after the first day of January One thousand nine hundred and one, the people of New South Wales, Victoria, South Australia, Queensland, Tasmania, and Western Australia shall be united in a Federal Commonwealth under the name of the Commonwealth of Australia." Thus Australia became a new nation in a new century—the 20th.

Although they had delayed forming one nation, the Australian colonies had pioneered in developing features of democratic government that have been widely accepted. Australian voters were the first to use the secret ballot. Australia was also a leader in the drive for compulsory voting and for woman suffrage.

The Constitution that the Australian people enacted created a parliamentary democracy on the British model. The federal government of Australia retains control of foreign trade; defense; immigration; customs; external trade and commerce; postal, coinage, banking, and social services. It also controls the collection of income taxes and distributes grants to the states. The states control education, law enforcement, health services, and other important local functions.

Australia, although an independent nation, is a member of the British Commonwealth. Thus Queen Elizabeth II of Great Britain is Queen of Australia and is represented by a governor-general and six state governors. They have limited actual power since they ordinarily act only on the advice of the elected officials.

The Federal Parliament is made up of a 60-member senate (10 representatives from each state) and a 125-member house of representatives (elected on the basis of population). House members serve 3 years, senators, 6, and both are elected by voters under a special system that is designed to make sure that minority parties obtain representation in the government.

The actual head of the federal government is the prime minister. He is the leader of the political party holding a majority in the House of Representatives, and he is assisted by a cabinet. The prime minister holds office only so long as he and his party retain the confidence of a majority of the members of the House.

DEPENDENCIES

In addition to its two internal territories, the Northern Territory and the Australian Capital Territory, Australia is responsible for the administration of a number of islands and island groups that are designated as external territories. These are Norfolk Island in the South Pacific Ocean, and the Cocos (Keeling) Islands, Christmas Island, Heard Island, and the McDonald Islands, which are all located in the Indian Ocean. Papua New Guinea was an external territory of Australia before it achieved independence in 1975.

The Antarctic Division of the Department of Supply operates a number of permanent scientific research stations on the 2,400,000 square miles (6,116,000 sq. km.) that Australia claims as its Antarctic Territory.

As with other such claims, this is not recognized internationally. The un-inhabited Ashmore and Cartier Islands, in the Timor Sea, are administered as part of the Northern Territory.

AUSTRALIA, A WORLD POWER

Australian fighting men took a notable part in World War I and World War II. With soldiers from New Zealand, they formed the ANZAC (Australian and New Zealand Army Corps) and fought with great valor in Africa, in Europe, and in Asia.

In World War I Australian volunteers numbered over 400,000 out of a total population, at that time, of less than 5,000,000. They took part in many of the hardest fought actions at Gallipoli, in Flanders, and in Palestine. In World War II one in five of the male population, almost 700,000 men, enlisted, and more than half that number served overseas. Australians also served in Korea and in Vietnam.

Australia has taken an active role in the United Nations since its inception. The Australian Government has also contributed to many programs for the technical assistance of southeastern Asian and some African countries.

Australia is a member of many regional treaty organizations. The ANZUS Treaty provides that in the event of armed attack on any member in the Pacific, the three members—the United States, New Zealand, and Australia—would act together. Australia is a regional member of the United Nations Economic Commission for Asia and the Far East (ECAFE); a member of the Colombo Plan Consultative Committee, a co-operative plan by countries to promote economic and social development of South and Southeast Asia; and a member of the Asian and Pacific Council (ASPAC).

Association with Great Britain and co-operation with the British Commonwealth still remain important for Australia. But as can be seen from its treaty commitments, ties with the United States and regional associations with countries of Asia have greatly increased. Recently a senior Australian official said that 75 percent of his country's diplomatic activity is now concerned with Asian affairs, 20 percent devoted to relations with the United States, and only 5 percent left for the rest of the world.

Although Australia is still tied to the British Crown, there is a growing consciousness of an identity quite separate from the ancestral home in England. Many factors contributed to this awareness of Australian identity. The war efforts, the expanding economy, the increased trade with Asia, the integration into Australia of some 1,000,000 continental Europeans, and the developments in native culture and art all acted to give the Australians a heightened sense of nationality. The colorful celebrations in 1970 of the 200th anniversary of the landing of Captain Cook increased the feeling that Australia had indeed come of age.

This awareness of nationality has been shown in recent years by the use of Australian songs as alternatives to "God Save the Queen," although the latter is still used on royal occasions. "Waltzing Matilda," "Advance Australia Fair," and "Song of Australia" share honors as alternative national anthems. Also, under a recent law, the official designation "British subject" was replaced by "Australian citizen."

BRUCE W. PRATT, Editorial Director, *The Australian Encyclopedia*

Milford Sound, a South Island fiord.

NEW ZEALAND

New Zealand—two islands in the South Pacific Ocean—is 12,000 miles (19,300 kilometers) from Europe and 1,200 miles (1,930 km.) from Australia, its nearest neighbor. Distance and isolation have always been the central facts about New Zealand. Until the 20th century, these lonely "sea-girt islands" remained one of the least-known nations in the world. They were often dismissed, with Australia, as part of Down Under and considered of little importance. But today New Zealand has become a modern, flourishing nation with one of the highest standards of living in the world.

THE PEOPLE

Measured in terms of recorded history, New Zealand is a very young country. The story of its development was written by two different peoples—Maori and white men—both of whom played vital roles in making the nation what it is today.

Two Peoples—One Nation

The Maori, a brown-skinned Polynesian people, lived in New Zealand for some hundreds of years before the country was discovered by Europeans in the 17th century. Almost 200 years more passed before white men began to settle New Zealand in large numbers. Because the Maori were the first New Zealanders, their word for "white man" is *pakeha.* (It is believed to have originally meant "imaginary beings resembling men.") *Maori* itself means "normal," "usual," or "ordinary." In New Zealand today, the gap between the two peoples is steadily being closed.

At first, the arrival of Europeans brought the Maori more sophisticated methods of killing, as well as new diseases to which they had no resistance. As a result, during the 19th century the Maori population declined from about 250,000 to less than 40,000. But the 20th century brought a reversal of this trend, and the present Maori population is over 200,000.

The Maori have had full citizenship rights for more than 100 years. Today they are almost completely integrated with the *pakeha* New Zealanders, taking part in all aspects of life in New Zealand. Maori like Sir Peter Buck (Te Rangihiroa), a physician, anthropologist, and member of Parliament, have made important contributions to their country.

However, some problems remain. More and more Maori are leaving their old, close-knit communities and moving to the cities. The transition to an urban way of life creates tension, which is often compounded by lack of understanding shown by whites. Both government and private agencies have been set up to deal with these problems. Attempts are also being made to preserve the Maori's rich cultural heritage. Though some Maori feel the need for a "different but equal" status, intermarriage is increasing. In the words of an ancient Maori survival chant: ". . . has dawned the day/When must walk hand in hand/The brown race and the white. . . ." This is what most New Zealanders believe today.

Way of Life

Observers often conclude that many New Zealanders are "more English than the English." This may be due to the orderly flower gardens in front of most homes, to a certain manner of speech found in educated New Zealanders, and to the sentimental care with which many English institutions have been transplanted. But today the Englishness of New Zealand is largely an illusion. With increasing immigration, other European influences are becoming stronger, and they are all mingling in the blend that is the New Zealander.

Outdoor Life.   New Zealand's population is small in relation to its land area. One result is that most New Zealanders are house-owners rather than apartment-dwellers. Another is that every New Zealander has easy access to open country and that even city residents often manage to have a small cottage by the sea or in the mountains.

This means that outdoor activities and sports play an important part in New Zealand life. The New Zealander gives his leisure time with great energy to cricket, tennis, golf, soccer, outdoor bowling, skiing, hiking, mountain climbing, horse racing, swimming, sailing, and surfing. But however popular these are, none of them generates as much enthusiasm as does rugby football. Rugby is not a sport but a national cause in New Zealand. Every winter New Zealanders focus on it fanatically, as thousands cheer the teams that have won world renown at the game. New Zealand's position in international sports has also been maintained by its long-distance runners, including Murray Halberg and Olympic champion Peter Snell, and by racing drivers such as the late Bruce McLaren.

Since pioneering days are not long past, much admiration is still reserved for the "good keen man" who can rough it if necessary. New Zealanders also admire nonchalance in any situation, no matter how difficult. Thus they approved the first words of Sir Edmund Hillary (a New Zealander who learned to climb mountains in New Zealand's Southern Alps) after conquering Mount Everest. He was reported to have said casually that the world's highest peak had been "knocked . . . off."

The Arts.   Perhaps because so much energy had to be devoted to building up a young country, New Zealand has not produced many great artists. In the 19th century, there were the painters Charles Heaphy and W. M. Hodgkins, two of the first to capture the country's beauty on canvas; and poet-politician William Pember Reeves, who wrote lyrically about his homeland.

The 20th century, however, has seen the beginnings of a new trend. In the early 1920's, short-story writer Katherine Mansfield—a New Zealander who has been described as the "one peacock in our literary garden"—lived and worked in England. Today, more and more contemporary New Zealand artists are staying at home, drawing on the life around them for inspiration. Among the best-known are Dame Ngaio Marsh, mystery writer and playwright; Janet Frame, whose novels often deal sensitively with the mentally ill; and Sylvia Ashton-Warner, who has written both novels and nonfiction based on her experiences as a teacher in Maori schools.

CITIES

Wellington.   The country's capital and third largest city, Wellington was founded in 1840 by New Zealand Association immigrants. In 1825 the association had asked the Duke of Wellington—then in the British

Wellington, New Zealand's capital, is built around an excellent harbor.

FACTS AND FIGURES

NEW ZEALAND is the official name of the country.

CAPITAL: Wellington.

LOCATION: South Pacific Ocean. **Latitude**—34° 25′ S to 47° 17′ S. **Longitude**—166° 27′ E to 178° 35′ E.

AREA: 103,736 sq. mi. (268,675 sq. km.).

PHYSICAL FEATURES: Highest point—Mount Cook (12,349 ft.; 3,764 m.). **Lowest point**—sea level. **Chief rivers**—Waikato, Clutha. **Major lakes**—Taupo, Te Anau, Manapouri.

POPULATION: 2,800,000 (estimate).

LANGUAGE: English (official), Maori.

RELIGION: Protestant (mainly Anglican, Presbyterian, Methodist), Roman Catholic.

GOVERNMENT: Constitutional monarchy within Commonwealth of Nations. **Head of state**—British monarch represented by governor-general. **Head of government**—prime minister. **Legislature**—house of representatives. **International co-operation**—United Nations, Commonwealth of Nations, ANZUS Pact, Colombo Plan.

CHIEF CITIES: Auckland, Christchurch, Wellington, Dunedin.

ECONOMY: Chief minerals—coal, limestone, gold, iron sands. **Chief agricultural products**—meat, wool, butter, cheese, wheat, oats, barley, fruits, vegetables. **Industries and products**—frozen meat; dairy products; fishing; forestry, wood, and wood products; textiles and clothing; transportation equipment. **Chief exports**—meat, wool, butter, cheese, forest products, fishery products. **Chief imports**—machinery, iron and steel, petroleum products, cars, plastics.

MONETARY UNIT: New Zealand dollar.

NATIONAL HOLIDAY: 1st Monday in June, Queen's (official) Birthday.

NATIONAL ANTHEM: "God Save the Queen." (New Zealand's National Song is "God Defend New Zealand.")

NEW ZEALAND

Cabinet—for his support and patronage. Wellington scoffed at their plans for the new colony. Fifteen years later, however, they named the future capital for him in spite of his skepticism.

Located on the southern tip of the North Island, overlooking Cook Strait, Wellington has one of the deepest natural harbors in the world. The port area, with its miles of docks for oceangoing ships, also has ferry and hydrofoil connections to the South Island.

As the seat of government since 1865, Wellington is the site of New Zealand's parliament. Wellington's Dominion Museum has exhibits depicting New Zealand life and history. Wellington is the home of Victoria University, one of New Zealand's seven universities. The National Art Gallery has a large collection of paintings. In the Turnbull Library is the country's most important collection of material relating to the early history of New Zealand and the Pacific.

From the harbor, cable cars climb the steep green hills on which Wellington is built. From the highest, Mount Victoria, visitors and residents alike can enjoy a spectacular view of the harbor, where huge freighters and passenger ships and small white ferries ply the deep-blue waters.

Auckland. Near the northern end of the North Island is Auckland, New Zealand's largest city. With nearly 600,000 people, it is also the country's fastest-growing city, sprawling outward to keep pace with its rapidly increasing population.

Built on an isthmus between two harbors, Waitemata and Manukau, Auckland overlooks the protected waters of the Hauraki Gulf. A soaring bridge connects the city to its north shore suburbs. Auckland's location and excellent port facilities have made it a focal point for much of New Zealand's trade and shipping. It is also a major industrial center, with a wide variety of manufacturing plants and a newly established iron and steel industry based on local iron sands.

In recent years, Auckland has become the city with the largest Polynesian population in the world. In addition to the Maori who live there, new arrivals from all over the South Pacific make their home in the city. With two harbors, plus the country's largest international airport, it is not surprising that Auckland is also the point of entry and departure for more than half of all visitors to New Zealand. The busy, modern metropolis offers an exciting introduction to the country as a whole. The city's War Memorial Museum has one of the best collections of Maori artifacts in the world. The Auckland Art Gallery has paintings by New Zealand and European artists. Auckland University is known for its engineering, fine arts, and architecture schools. Every year the city stages an arts festival that attracts performers and artists from overseas.

Auckland's many quiet, green parks are peaceful oases amid the city's bustle. Mount Eden is one of several extinct volcanoes in the area. Only a few minutes away, miles of beautiful beaches afford excellent swimming and surfing. And of course, there is always sailing. Each year, on the last weekend in January, Auckland holds its Anniversary Regatta. The beautiful harbor is crowded with many hundreds of sailing craft, from oceangoing keelers to sailing dinghies crewed by small boys and girls.

New Zealand's largest city, Auckland, is a bustling industrial metropolis.

View of the campus of the University of Canterbury, in Christchurch.

Christchurch. Christchurch, set in the Canterbury Plains on the east coast of the South Island, is New Zealand's second largest city. It was founded in 1850 by a group of settlers, one of whom decided to name their new home for his old college in Oxford, England.

Through its port of Lyttelton, 7 miles (11 km.) away, Christchurch exports the wheat, wool, meat, and other products of the plains area. Hydroelectric power runs Christchurch's thriving industries—fertilizers, engineering, clothing, electrical equipment, and furniture.

The city's major landmark is the single spire of its Gothic-style cathedral. There are also many lovely parks and gardens, and the willow-bordered, winding Avon River—a namesake of the one that runs through Stratford, England—adds its special charm. Named for a school, Christchurch is proud of its excellent educational facilities—the University of Canterbury, Lincoln College, and Christ's College among them. In its origins and atmosphere, Christchurch is the most English of all New Zealand's cities.

Dunedin. On the Otago Peninsula about 200 miles (320 km.) down the coast from Christchurch is Dunedin. The name "Dunedin" is Gaelic for "Edinburgh." This southernmost of New Zealand's four main cities still has some of the solidity of the Scots who founded it in 1848.

In the 1860's Dunedin was a base for gold prospectors. Today the city is a manufacturing center, with iron and brass foundries and woolen mills. The products of these industries are shipped from Dunedin's harbor of Port Chalmers, as are the agricultural products of the Otago area. The University of Otago, located in Dunedin, is the oldest in New Zealand.

THE LAND

New Zealand's total area, which includes the two main islands and several smaller ones—Stewart, 20 miles (32 km.) off the southern tip of the South Island; the Chatham group, some 400 miles (640 km.) southeast of Wellington; and the Kermadecs, about 500 miles (800 km.) north-east of the North Island—is 103,736 square miles (268,675 square kilometers). It includes some of the most unusual, as well as some of the most beautiful, geographical features in the world. One New Zealand author, Sylvia Ashton-Warner, has written, "Nature in person lives in New Zealand."

The South Island

The South Island is the larger of the two main islands, and the one with the more varied landscape. It is dominated by the Southern Alps, a massive chain of mountains running roughly north–south down the island's west coast. Although only occasionally rising higher than 11,000 feet (3,300 meters), they provide magnificent Alpine scenery, with snowfields, crevasses, and glaciers that descend through rain forests almost to sea level. Soaring to 12,349 feet (3,764 m.), snowcapped Mount Cook is New Zealand's highest mountain—the Maori call it Aorangi, "the cloud piercer." Mount Cook is located in one of the country's many national parks, where visitors can enjoy mountain climbing, hunting, and hiking. The area has three vast glaciers—Tasman, Fox, and Franz Josef. Another attraction is the superb skiing. Airplanes take skiers to the heads of these glaciers for the long, smooth runs to the bottom, sometimes as much as 16 miles (26 km.) of uninterrupted skiing. This part of the South Island also abounds in beautiful, deep glacial lakes, such as Te Anau, Manapouri, and Wakatipu. All of them are popular resorts.

The southwestern coast, below the Alps, is deeply indented by isolated fiords, similar to those in far-off Norway. Fringed by mountains rising sheer from the sea, the fiords are among New Zealand's most dazzling beauties. The Fiordland, too, has been made a national park; one of its highlights is Sutherland Falls, the fourth highest waterfall in the world. Milford Sound, with its cone-shaped Mitre Peak, is the best-known fiord. Two of the others are Dusky Sound and Doubtful Sound.

Tasman Glacier, in New Zealand's Southern Alps—a sportsman's paradise.

Stretching from the foothills of the Alps to the east coast of the South Island are the vast open spaces of the region known as the Canterbury Plains. Farther south are the rolling hills and pleasant valleys of the Otago and Southland districts.

Across Foveaux Strait from the South Island is 600-square-mile (1,700 sq. km.) Stewart Island, known to the Maori as the Isle of the Glowing Sky. It is a peaceful spot with unspoiled beaches and forests that provide a haven for birds and other wildlife.

The North Island

Though the North Island's mountains may not soar as high as those in the South Island and though its contrasts may seem less stark, its landscape nevertheless has a special drama all its own.

Cutting across the center of the island is one of the most active volcanic complexes in the world. Tongariro National Park includes mounts Tongariro, Ngauruhoe, and Ruapehu. To the west of these three active peaks is the symmetrical cone of Mount Egmont, rising over 8,000 feet (2,400 m.) from the rich lands of the Taranaki district. Egmont, extinct for some 250 years, is often compared to Fujiyama in Japan. Today there are many fine ski trails in both the Tongariro and Egmont regions.

To the north of Mount Egmont are the Waitomo Caves, three huge limestone grottoes. Visitors marvel at the grotesque shapes of their stalactites and stalagmites, eerily lit by thousands of tiny glowworms. The caves seem to be a stage set by nature for some mysterious, long-forgotten ceremony.

Still farther north, centered around the old Maori town of Rotorua, is New Zealand's thermal region. Here, spouting geysers like Pohutu and the Prince of Wales Feathers put on spectacular performances, sending up plumes of hot water as high as 100 feet (30 m.). There are also boiling mud pools that plop and bubble, and fumaroles—where steam and gases rise into the air. Here, too, it is possible to catch a fish in one stream and cook it by plunging it into the next, for hot and cold springs exist practically side by side. The hot, mineral-rich springs are said to have beneficial effects for people suffering certain diseases, but they also give the area a distinct sulfurous odor that takes some getting used to.

There are several lovely lakes in the volcanic and thermal districts. Among them is New Zealand's largest, 238-square-mile (616 sq. km.) Lake Taupo, located almost in the center of the North Island. The lake is a fisherman's paradise, and anglers come there from many countries to catch some of the largest trout in the world.

Seaside resorts are strung out all along the east coast, from the Bay of Plenty to the Bay of Islands, and the area offers deep-sea fishing for marlin, shark, and tuna. In the North Auckland region there are protected stands of giant kauri trees, remnants of the kauri forests that once grew all over the North Island. Some of the trees have individual Maori names, and one of them is over 2,000 years old.

Climate

New Zealand's climate is strongly influenced by the sea that surrounds the islands. In the Fiordland, moisture-laden winds from the sea bring abundant rainfall, sometimes reaching as high as 250 inches (635 centimeters) a year. Westerly winds from the sea make New Zealand's

One of the many geysers in the North Island thermal region.

weather highly changeable, with bright blue skies one minute and gray, threatening ones the next. Temperatures are moderate, however, without great extremes of hot or cold.

Since New Zealand is below the equator, the seasons are the reverse of those in the Northern Hemisphere—a situation that occasionally causes problems for New Zealanders and foreigners alike. At the warmest time of year, New Zealanders eat a heavy Christmas dinner that faithfully follows its English original down to plum pudding with brandy sauce. Under his beard and red suit, Father Christmas perspires in the midsummer sun. Location in the Southern Hemisphere also means that temperatures tend to get colder toward the south, rather than the opposite.

Resources

The many rivers that drain its mountain ranges and lakes are New Zealand's most important natural resource. Often too short or too turbulent to be navigable for great lengths, many of them have been harnessed to provide cheap electricity for New Zealand's homes and industries. Of the large power stations in the South Island, the most important are Benmore, on the Waitaki River; Roxburgh, on the swift-flowing Clutha River, below Dunedin; and several on the Waimakariri river systems to the north. The country's largest hydroelectric plant is planned for Lake Manapouri, where over 1,000,000 kilowatts of electricity will be used to make aluminum from Australian bauxite ore.

The North Island's main power stations are on the Waikato River, which drains Lake Taupo. Since the North Island has the larger population and the smaller hydroelectric potential, cables under Cook Strait are also used to carry power north. At a large geothermal plant at Wairakei, just south of Rotorua, underground steam drives giant turbines, generating more electricity for the North Island. (The Wairakei installation is one of the rare ones in the world to use natural steam for this purpose; the only others are in Italy, Iceland, Mexico, and Japan.)

New Zealand's mineral resources include coal, most of which is mined in mountainous Westland in the South Island. Quarries in many parts of the country produce the limestone that, when crushed, is so vital to New Zealand's agriculture. The Otago was once the country's gold-rush center, but very little gold is mined there today. In the North Island there are iron sands, and natural gas has been discovered near Mount Egmont. Both are being developed. Oil is being sought in various parts of the country.

New Zealand's forests are another valuable natural resource. At first, both Maori and Europeans cut down and burned off trees to clear land for agriculture. Later, an active trade developed in New Zealand timber. Many of the islands' forests—including those of ancient kauri trees—were destroyed.

But eventually, under government reforestation programs, new trees replaced the old. In the 20th century, scientists introduced trees native to other countries, and today there are thriving forests of California pine, Japanese cedar, European larch, and Australian eucalyptus trees in New Zealand, side by side with the country's own rimu, matai, totara, tawa, and beech trees.

Forests cover about one quarter of New Zealand's land area, and their products—from pulp and paper to timber and furniture—are important exports.

ECONOMY

Endowed with mild temperatures and fertile soils, New Zealand was first settled as an agricultural country. Today it has one of the most efficient agricultural industries in the world, and in spite of the fact that over 60 percent of its people live in the cities, agricultural products are the country's most valuable exports.

There are over 60,000,000 sheep in New Zealand, raised wherever the land is suitable for grazing. They have made the country the world's leading exporter of mutton and lamb and one of the largest exporters of wool. One principal sheep-raising area is still the South Island's Canterbury Plains, the source of the first cargo of refrigerated meat shipped from New Zealand at the end of the 19th century.

Although sheep are the mainstay of New Zealand's agriculture, there are also some 8,000,000 cattle. About half of these are dairy herds, and New Zealand is the world's largest exporter of butter and cheese. Most of the dairy farms are located in the North Island, particularly in the Waikato and Taranaki regions. Beef production is also increasing steadily.

New Zealand's grain crops include wheat, barley, and oats, which are grown principally on the Canterbury Plains. A wide variety of vegetables are raised, and hay, grass, and clover are important as fodder crops for livestock.

The products of New Zealand's orchards—mainly in the Otago and

Sheep, raised for meat and wool, are a mainstay of New Zealand's economy.

Nelson areas of the South Island, and Hawke's Bay in the North Island—
include apples, peaches, pears, cherries, citrus fruits, and subtropical
fruits such as the tamarillo and Chinese gooseberry. Fresh or canned,
these are becoming more important exports. Berries are also grown, and
near Auckland and Napier, in Hawke's Bay, there are vineyards that produce fine wines. The grapes were introduced into New Zealand in the
early 19th century. Today experts compare some New Zealand wines
favorably with those of Europe.

Industries

Britain has always been New Zealand's most important trading partner, but today New Zealand is working to develop new markets in Asia—
particularly Japan—and the Western Hemisphere.

To meet this challenge, there has been a growing emphasis on the
development of manufacturing—including the production of automobiles, furniture, refrigerating equipment, and clothing. The government is
making massive efforts to attract outside investment and to expand New
Zealand's existing industries. Frozen crayfish, taken from New Zealand's
icy waters, and other fishery products are important exports.

Finally, as the increasing prevalence of jet travel makes the world
ever smaller, another industry is developing in New Zealand. Every year,
more and more tourists from all over the world are visiting the country
and delighting in its beauty.

Some New Zealand Rarities. Because of its geographical isolation,
New Zealand has some forms of plant and animal life not found anywhere else. These are one of the country's unique attractions for visitors.
The best-known is the flightless, long-beaked kiwi, which is New Zealand's national bird. Other birds native only to New Zealand are the tui

and the bellbird, both of which have lovely, melodic songs. There are also the kea, a species of parrot that preys on sheep; and the takahe, which was thought extinct until 1948.

Another New Zealand rarity is that the country has no land snakes. The tuatara lizard, New Zealand's lone reptile, is an accidental living fossil that should have been extinct 100,000,000 years ago. Besides its two sighted eyes, the tuatara has a sightless third eye—in the middle of its forehead.

In addition to the kauri and other New Zealand trees, there is the scarlet-flowered pohutukawa, sacred to the Maori; the nikau palm; and the rata, which twines around other plants, using them for support and decorating them with its bright, fragrant blossoms.

HISTORY

New Zealand's history began with the arrival of the Maori. Although there are many theories, the exact origins of this Polynesian people are lost in history and time. In 1947 the Norwegian explorer Thor Heyerdahl drifted across the Pacific from the coast of South America on the *Kon-Tiki* in an attempt to show they might have come that way. However, in spite of Heyerdahl's successful voyage, many scholars believe the Polynesians came from somewhere on the mainland of Southeast Asia. By feats of seamanship rivaling those of the Vikings (which took place in northern Europe at about the same time), the Polynesians eventually spread north to Hawaii, as far west toward South America as Easter Island, and south to New Zealand. In this way they populated the area now referred to as Polynesia.

Since the Maori had no written language, their history was passed from one generation to the next by word of mouth. According to this tradition, the Maori's ancient homeland is Hawaiki—a far-off place in central Polynesia that has never been satisfactorily identified. Hawaiki is also considered the "reunion place of spirits," to which all Maori return after death.

It has been estimated that the Maori came to New Zealand between A.D. 1000 and 1400, one important voyage of seven great canoes taking place about 1350. The canoes, made of hollowed logs, were either outriggers or double canoes. Sturdy and well-equipped, they carried men, women, and children; plants and seeds; and even some animals.

Early Life in New Zealand. Because the new arrivals first saw mountains shrouded with mist, they called their new home Aotearoa, "the land of the long white cloud." According to the mythology they developed, the country had been fished up from the ocean by a legendary Polynesian hero—"the man Maui, linked by his lineage with the gods." So the northern island came to be known as Te Ika a Maui ("Maui's fish") and the southern, Te Waka a Maui ("Maui's canoe").

At first the Maori depended for survival on hunting the moa, a large flightless bird that has long been extinct. Gradually, they developed skills in shelter building, weaving, agriculture, fishing, and bird snaring. As life became more comfortable, carving and other decorative arts reached a high level. Using tools made of hard New Zealand greenstone (a kind of nephrite, or jade), the Maori carved the arching prows of their canoes and the gateposts of their *pas,* or villages. They also carved intricate tikis, or fertility symbols, out of the greenstone itself. Every aspect of Maori life was regulated by ritual and a religion that has been described as mythology mixed with magic.

Recurrent wars were another important part of Maori life, and battles were ferocious. (An early visitor spoke of the earth trembling as a party of fighting men performed their pre-battle *haka,* or war dance.) As for the causes of war, one ancient Maori proverb says: *Ha wahine, he whenua i mate te tangata*—"Because of women and land, men die." Ritual played a significant role in war, too. Since many Maori were cannibals, the dead and captured were often eaten. Vanquished chiefs usually preferred this fate to its alternative—slavery. In spite of wars, however, by 1600 the Maori had become an agricultural people, and they had made the land their home.

The Coming of the White Man

By the end of the 16th century, European ships were ranging farther and farther from home. Gradually, reports of the existence of a southern continent came back to the ports of Europe. In 1642 the Dutch East India Company sent two ships commanded by its best navigator, Abel Janszoon Tasman, into the unknown south. His mission was "the discovery and exploration of the supposed rich southern and eastern land." In December of that year Tasman sighted "a large, high-lying land" and began mapping it. He called it Staaten Landt, but the Dutch later named it Nieuw Zeeland, after their own southwestern province. In a landing attempt, Tasman made contact with the Maori, who killed four of his crewmen. He sailed on, and Nieuw Zeeland was left to its belligerent inhabitants for more than a century.

Captain Cook. In 1769 New Zealand was visited by the greatest seaman and navigator of his time, the Yorkshireman Captain James Cook. He, too, was searching for the legendary southern continent. On a ship named *Endeavour,* Cook (and his crew of 90) sailed from Tahiti, where he had been observing a transit of the planet Venus for the British Admiralty and Royal Society. He circumnavigated the islands, charting them with remarkable accuracy. Sir Joseph Banks, a naturalist on board the *Endeavour,* recorded precise information on the islands' plant and animal life. Banks also noted that "the almost certainty of being eat as you come ashore adds not a little to the terrors of shipwreck." But after some hostile first encounters, Cook established reasonably good relations with the Maori.

Cook returned to New Zealand twice, in 1772 and 1777. There were other expeditions as well, including several led by Frenchmen. In 1781, after England's American colonies won their independence, England needed a new place to send its debtors and other "undesirables." There was some discussion in Parliament about using New Zealand as a kind of penal colony, but it was decided that the Maori were too fierce.

By the end of the 18th century, as Europeans learned more about New Zealand, the country slowly began to be opened up. At first the rich harvest of the southern seas attracted Australian, American, British, and French whalers and seal hunters. Busy whaling stations were built on the coasts of both of New Zealand's islands. Competition among them grew so fierce that at times it approached piracy, and soon several kinds of whales were almost exterminated. Trade in flax and timber also developed rapidly. The strong flax fiber, used for ropes, was bartered by the

Maori for imported goods. (Guns and rum were the preferred items.) The northern ports where these activities centered became colorful, frontier-type towns. As late as 1835, when the great English scientist Charles Darwin visited New Zealand, he described the people there as the "very refuse of society," and added ". . . we were all glad to leave New Zealand. It is not a very pleasant place."

Settlement, Expansion, and Development

But in spite of the black portrait painted by Darwin, other people were beginning to come to New Zealand. In the early 19th century, the economic crises that followed the Napoleonic Wars brought poverty to workingmen in Europe. Thousands emigrated—especially from England—in search of a better life. Missionaries came from England, too, and began converting the Maori to Christianity.

New Zealand Becomes an English Colony. By 1840, England had decided to annex New Zealand. In that year a treaty was signed with the Maori chiefs at Waitangi, in the North Island. The Maori recognized "Wikitoria" (Queen Victoria) as their sovereign, and they received guarantees of their property rights. They also agreed to sell land only to the English.

In 1837, Edward Gibbon Wakefield, a specialist in what he called "the art of colonisation," had set up the New Zealand Company to organize settlement of the new colony. He hoped to create an English society in the South Seas, its class structure based on laborers who might eventually become landowners. With assistance provided by the company, thousands of people immigrated to New Zealand. But Wakefield's plans were soon forgotten. The new settlers began to build a different society from the one he had envisioned.

All over New Zealand a period of development began. The hilly, open country of the South Island was ideal sheep-grazing land. Sheep farmers were soon exporting wool to Australia and England as well as supplying high-quality lamb and mutton for New Zealand's tables. Grain crops—mainly wheat and barley—were also raised in the South Island. In the 1860's gold was discovered in the south. This brought a rush of prospectors—some of whom had followed the hunt for gold around the world, from California to Australia to New Zealand—but the large finds were quickly worked out.

In the North Island there was progress too. The foundations of a thriving dairy industry were laid by English farmers who brought in Jersey and Guernsey cows. All kinds of vegetables were raised on farms similar to those in central England, and pear and apple orchards were planted as well. Since the North Island still depended on acquiring land from the Maori, farms and settlements tended to be small.

Self-Government. Yet on both islands, cities were being built, and in 1852, when there were some 50,000 European settlers, England granted New Zealand self-government. Six provinces—Auckland, Wellington, New Plymouth, Nelson, Canterbury, and Otago—were set up, and a central parliament was established at Wellington. The governor, Sir George Grey, planned for the peaceful development of New Zealand, with Maori and Europeans working together.

The Maori Wars. But trouble was brewing in the North Island. As more and more settlers arrived, pressure on the Maori increased. At the same time, Maori opposition to further land sales was growing. In 1860, in spite of the Waitangi treaty, land disputes flared into open war. Conflict continued throughout the 1860's. Various Maori religious cults kept up a fanatical resistance, but the settlers finally proved too strong. By 1872 the Maori wars were over.

With the return of peace, the government concentrated on attracting more settlers and on building up the economy. After a sizable loan from England, more than 100,000 new settlers came to New Zealand as assisted immigrants. The English loan was also used to build more roads, and a railroad network began to spread across both islands as well. Telegraph lines linked the more isolated settlements, and underwater cables made a stronger bond between the two islands. These improved communications eventually brought New Zealand closer to Europe, too.

The country's merchant marine was enlarged, and soon New Zealand's ships were calling at ports all over Asia and the South Pacific. But the greatest strides in the country's economic development began in 1882, when an experimental refrigerator ship took a cargo of meat on the long voyage to England. The *Dunedin* was a trailblazer. Vast new markets were now open to New Zealand.

The 20th Century

In this period of prosperity, with exports insuring a strong economy, New Zealand became one of the most progressive countries in the world. Herbert Henry Asquith, England's prime minister in the early 20th century, described New Zealand as "a laboratory in which political and social experiments are every day made for the information and instruction of the older countries in the world." Women were granted the vote in 1893. Land was redistributed, and an income-tax bill was passed. A welfare program, including unemployment insurance, old-age pensions, workmen's compensation, and family allowances, went into effect. The world's first system for government arbitration of industrial disputes was set up. By stages, education became "free, secular, and compulsory." During this time, too, the first steps were taken to bring the Maori more fully into the life of the country.

Many of these early measures were put through under the leadership of Prime Minister Richard J. Seddon. A former railroad worker and gold miner, Seddon—known as King Dick—became a personal symbol of his country's pioneering spirit. In 1907, after Seddon's death, New Zealand gained dominion status within the British Empire.

Two World Wars. The outbreak of World War I marked the real beginning of the country's participation in world events. New Zealand sent food and supplies to England and other Empire countries, and an expeditionary force to fight in France and the Middle East. These troops were respected for their bravery wherever they fought. At Gallipoli, in Turkey, in 1915, they and the Australians made the name ANZAC (Australian and New Zealand Army Corps) one to be reckoned with. New Zealand suffered tragic losses in the war: of a population of just over 1,000,000, nearly 17,000 were killed and 50,000 wounded.

World War II saw the nation's resources even more completely mobilized. Once more, vast quantities of food were shipped to England. Once more, New Zealanders saw action in many theaters of war—in Europe, in the Pacific, and in North Africa. They are still proud of their part in the German defeat at El Alamein, Egypt, in 1942, and in the bitter fighting at Cassino, Italy, in 1944.

GOVERNMENT

New Zealand's constitutional procedures are based on those of England. The British Crown is represented by a governor-general, and the basic legislative authority is a one-chamber parliament, usually referred to as the House of Representatives. (The upper house, which was called the Legislative Council, had no real legislative functions and was abolished in 1950.)

Parliament consists of representatives of 83 European and 4 Maori electoral districts. Elections are held every 3 years, and the leader of the party winning the largest number of districts becomes prime minister. He selects his Cabinet from the elected members of his party. Every resident of New Zealand over the age of 20 has the right to vote—and this includes all British subjects and even citizens of Ireland.

NEW ZEALAND AND THE WORLD

For some years after 1907, when New Zealand achieved dominion status within the British Empire, England was still home—an assured market and a protective parent. And, indeed, in many things New Zealand was a dutiful offspring. Not until the 1930's, with membership in the League of Nations, did the young country begin to take a more independent stand.

World War II, however, had a drastic effect on the country's foreign policy. The Empire evolved into the Commonwealth of Nations, and, as a member of the Commonwealth, New Zealand maintained its ties to the home country. But the war also brought a new awareness that New Zealand is close to—though not exactly a part of—Asia.

In 1951 the ANZUS Treaty was signed, with Australia, New Zealand, and the United States pledging mutual defense in case of attack. Three years later New Zealand became a member of SEATO (the Southeast Asia Treaty Organization).

As an important part of New Zealand's increasingly large role in Asian affairs, it has contributed generously to the Colombo Plan, set up by the Commonwealth to help raise living standards throughout South and Southeast Asia. In 1962 New Zealand granted independence to Western Samoa, which it had administered as a United Nations trusteeship. (An article on WESTERN SAMOA appears in this volume.) And in 1965, the neighboring Cook Islands were made a self-governing nation in free association with New Zealand. The Cook group, 15 main islands and several smaller ones, had been annexed under Prime Minister Seddon in 1901. Largely agricultural, they receive continuing assistance from New Zealand. New Zealand still administers Niue and the Tokelaus, both west of the Cooks. (See the article on OCEANIA in this volume.)

New Zealand is a small country that once seemed very far from the mainstream of life. But it has overcome all the obstacles imposed by its isolation. It has combined its two heritages—European and Maori—to become one of the most individualistic and advanced nations in the world.

JOHN MALE, Former Chief, Advisory Services
Division of Human Rights, United Nations

OCEANIA

Oceania is a vast region in the Pacific. Yet Oceania contains a total land area of only about 220,600 square miles (571,350 square kilometers). The islands that are scattered across Oceania are home to fewer than 4,500,000 people—less than half the population of Tokyo, the largest city of Japan.

The immense distances that separate the islands of Oceania from one another and the equally immense distances that separate Oceania from the industrially developed centers of Western Europe and North America have helped keep Oceania isolated from the mainstream of history until quite recent times. Until World War II there were still people in Oceania who had encountered only a handful of administrators, missionaries, and settlers from the outside world.

When the first Europeans visited the Pacific islands in the 16th, 17th, and 18th centuries, they returned home with descriptions of a region that sounded much like the Garden of Eden. Islands like Tahiti seemed to Europeans to be havens of beauty, innocence, and ease. The magnificent tropical trees and exotic flowers edging down to the blue waters of the Pacific provided a splendid background for the island people, who were handsome, clean, and usually friendly. The climate, although hot, was not uncomfortable. And since coconuts, breadfruit, and other fruit could simply be plucked from the trees and fish netted easily in the sea, there was no need to work hard in order to survive. To the

Aitutaki is a coral atoll made up of eight islets surrounding a lagoon.

"Tahitian Landscape" (1891?), an oil painting by Paul Gauguin. Minneapolis Institute of Art.

Outrigger canoes are built to move easily in the water near the islands.

visitors from Europe it seemed that there was little to do but live gently from one day to the next and dance, surf, swim, eat, and sleep when one felt like it. The region's flaws, such as violent storms, earthquakes, and local warfare, were not immediately apparent.

It is obvious why European sailors who had spent months aboard a cramped, dirty ship on a dangerous voyage should have found the islands appealing. Life in the islands seemed no less charming to the

---

ISLANDS ADMINISTERED BY NEW ZEALAND

COOK ISLANDS

LOCATION: Latitude—8° S to 23° S. **Longitude**—156° W to 167° W.

AREA: 90 sq. mi. (234 sq. km.).

POPULATION: 20,000 (estimate).

POLITICAL STATUS: Both Cook Islands and Niue have internal self-government.

SEAT OF GOVERNMENT: Avarua on Rarotonga.

CHIEF PRODUCTS: Citrus fruits, fruit juice, copra (dried coconut meat), pearl shell.

BACKGROUND FACTS: The group—Rarotonga, Mangaia, and six coral islands—is named for Captain Cook.

NIUE

LOCATION: Latitude—19° 2′ S. **Longitude**—169° 52′ W.

AREA: 100 sq. mi. (259 sq. km.).

POPULATION: 5,000 (estimate).

CHIEF PRODUCTS: Copra (dried coconut meat), bananas, basketware.

TOKELAU

LOCATION: Latitude—8° S to 10° S. **Longitude**—171° W to 173° W.

AREA: 4 sq. mi. (10 sq. km.).

POPULATION: 2,000 (estimate).

POLITICAL STATUS: Dependent territory of New Zealand.

SEAT OF GOVERNMENT: Administered by New Zealand High Commissioner, Apia, Western Samoa.

CHIEF PRODUCT: Copra (dried coconut meat).

BACKGROUND FACTS: Also called the Union Islands. The three atolls—Atafu, Nukunono, and Fakaofo—were discovered in the 18th century.

Tahitians fold their nets after bringing in a catch of fish. Fish is an important food source, along with the island's plentiful supply of fruit and garden products.

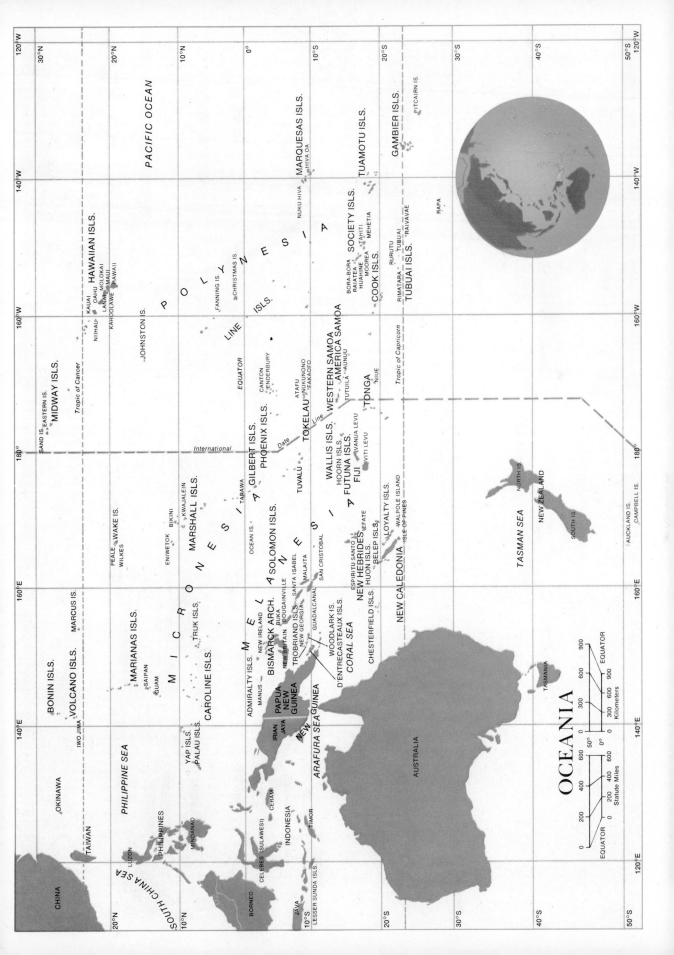

Europeans who remained at home. Chronicles of the voyages told of a last unspoiled corner of earth. Intellectuals such as Jean Jacques Rousseau saw in this untouched world the virtues of the unspoiled Noble Savage. The Noble Savage and his easy life provided a contrast to the hectic pace of life on the crowded continent of Europe.

The isolation of the people of Oceania dates back to the time when their ancestors came to the area. The arrival of explorers, missionaries, scientists, traders, and settlers between the 16th and 19th centuries began to break down that isolation. By the end of the 19th century almost every island group and island had been annexed by world powers.

But the end of Oceania's isolation did not become a reality until World War II. For 3 long years, from 1942 to 1945, hundreds of thousands of troops from Asia, Europe, and America fought one of the bloodiest wars this world has known in Oceania. At that time Oceania's strategic importance became apparent. It became impossible for the great powers and the nations that bordered the area ever to ignore it again.

It also became impossible for its people to avoid the outside world, whether they wished to or not. Thousands of islanders were thrown into close contact with outsiders for a long period of time. They fought with them in the jungles. They worked with them on the huge military bases. As a result the island people developed a whole range of new habits and new desires that could not be satisfied within their traditional economic and social systems.

Sometimes the clash between the islanders' way of life and the material wealth and power of the outside world produced strange results. Among the strangest were the cargo cults, forms of religious worship that grew up in parts of Oceania after World War II. In these cults "the good" was the coming of a new age, symbolized by a cargo of such things as cars, trucks, refrigerators, furniture, and canned goods. All these things the islanders had seen on foreign air bases during the war but had rarely been able to own. The people believed that if some local or foreign leader were properly prayed to or paid he would reveal the secret of how to obtain the cargo. Often the normal life of a village would come to a halt while the people built an airstrip and lit fires to guide in the expected cargo-carrying airplane.

Cargo cults reflect the islanders' sense of frustration when faced with the wealth and technology of the outside world. Present administrations in Oceania are trying to speed up political, economic, and educational development to give the people some of the benefits available in the industrialized countries.

THE LANDS OF OCEANIA

Oceania is situated in one of the most unstable areas of the earth's crust. Earthquakes and volcanic eruptions have occurred and continue to occur along both edges of the Pacific Ocean, in Asia and America. The earthquakes and volcanic eruptions that take place in the vast area in between are not as well-known, yet it is because of them that all of the smaller islands came into existence.

Continental Islands

The larger island groups, such as New Guinea, New Caledonia, the Bismarck Archipelago, and the Solomon Islands, are subject to volcanic

ISLANDS ADMINISTERED BY FRANCE

FRENCH POLYNESIA (Polynésie Française)

LOCATION: Latitude—about 7° S to 29° S. **Longitude**—131° W to 156° W.

AREA: 1,544 sq. mi. (4,000 sq. km.).

POPULATION: 94,500 (estimate).

POLITICAL STATUS: Overseas territory within the French Community.

SEAT OF GOVERNMENT: Papeete on Tahiti.

CHIEF PRODUCTS: Copra (dried coconut meat), vanilla, mother-of-pearl.

Windward Islands (Iles du Vent)

AREA: 463 sq. mi. (1,199 sq. km.).

POPULATION: 62,000 (estimate).

SEAT OF GOVERNMENT: Papeete.

BACKGROUND FACTS: Made up of Tahiti, Moorea, Mehetia, Tetiaroa, and Maiao, the Windward and Leeward groups together form the Society Archipelago or Society Islands.

Leeward Islands (Iles sous le Vent)

AREA: 183 sq. mi. (474 sq. km.).

POPULATION: 16,000 (estimate).

SEAT OF GOVERNMENT: Uturoa on island of Raiatea.

BACKGROUND FACTS: Consists of Huahine, Raiatea, Tahaa, Bora-Bora, Maupiti, and others.

Tuamotu Islands and Gambier Islands

AREA: 343 sq. mi. (888 sq. km.).

POPULATION: 7,000 (estimate).

SEAT OF GOVERNMENT: Apataki.

BACKGROUND FACTS: Tuamotu group is actually made up of two parallel ranges of islands, also called Low Archipelago.

Tubuai Islands

AREA: 63 sq. mi. (163 sq. km.).

POPULATION: 4,500 (estimate).

BACKGROUND FACTS: Rurutu, Tubuai, Raivavae, Rimatara, and Rapa are in this group. Also called Austral Islands.

Marquesas Islands

AREA: 492 sq. mi. (1,274 sq. km.).

POPULATION: 5,000 (estimate).

SEAT OF GOVERNMENT: Atuona on Hiva Oa.

BACKGROUND FACTS: Nuku Hiva and Hiva Oa are the largest islands in this group of 11, which was the setting of Herman Melville's novel *Typee*.

NEW CALEDONIA (Nouvelle Calédonie)

LOCATION (New Caledonia only): **Latitude**—20° 8' S to 22° 25' S. **Longitude**—162° 15' E to 164° 15' E.

AREA: 7,336 sq. mi. (19,000 sq. km.).

POPULATION: 96,000 (estimate).

POLITICAL STATUS: Overseas territory within the French Community.

SEAT OF GOVERNMENT: Nouméa.

CHIEF PRODUCTS: Nickel, iron, manganese, copra (dried coconut meat), coffee.

BACKGROUND FACTS: Discovered by Captain James Cook in 1774 and annexed by France in 1853. This territory has the following dependencies: Isle of Pines, Loyalty Islands, Huon Islands, Belep Islands, Chesterfield Islands, and Walpole Island.

WALLIS AND FUTUNA ISLANDS

LOCATION: Wallis: Latitude—13° 14' S. **Longitude** 176° 10' W. **Futuna: Latitude**—14° 15' S. **Longitude**—178° 05' W.

AREA: 77 sq. mi. (200 sq. km.).

POPULATION: 8,000 (estimate).

POLITICAL STATUS: Overseas territory within the French Community.

SEAT OF GOVERNMENT: Matautu.

CHIEF PRODUCTS: Copra (dried coconut meat), timber.

BACKGROUND FACTS: Made up of the Wallis and Hoorn islands. They became French outposts in 1842.

---

activity. However, they are not volcanic in origin. They are known as continental islands because they are made up of geologically ancient rock that owes its origin to the vast folding process that established the basic configurations of the Southeast Asian area. New Guinea is situated on the northern extremity of a continental shelf extending from the northern coast of Australia. There seems to be little doubt that New Guinea was once part of Australia. It may have been separated as a result of the rise in sea level following the melting of the ice cap.

Natural Resources. The range of natural resources likely to be present in these continental islands is considerably greater than in the non-continental islands. Continental origin implies a variety of types of rock and a variety of soils. The continental origin of these islands promises the possibility of the presence of a variety of minerals. Oil, gold, nickel, and copper are among the minerals that have already been discovered in these islands. Continental islands are also more likely to have better natural forests. This is true, for instance, in New Guinea, where the timber industry, although new, is already important.

High Islands and Low Atolls

The remaining islands of Oceania are either high volcanic islands or low coral atolls. Both the high islands and low atolls are of volcanic origin. The atolls are literally sitting on top of sunken volcanic islands.

High islands were formed as a result of volcanic activity and, consequently, are composed almost entirely of volcanic rocks and soils. They vary in size from large islands of over 4,000 square miles (10,000 sq. km.), such as Viti Levu in the Fiji group, to small ones of less than 1 square mile (2.6 sq. km.). They also vary in appearance. Some resemble the popular idea of a volcano, such as the small island of Kao in the Tonga group, which is an almost perfect volcanic cone. Many others have a sharp and broken outline, which may be familiar from tourist posters of some of the islands of the Societies and Samoan groups. Still others, like the main Fiji Islands, have been subjected to longer periods of erosion, which has broken up the volcanic rock and filled the valleys with the rich soil. There is also the possibility that minerals may be found, as was the case in the Fiji Islands, where gold and manganese deposits have been discovered.

The surface of low islands, or atolls, is composed entirely of coral sand. A typical atoll is never more than 20 or 30 feet (6–9 meters) above sea level and consists of a ring of long narrow islands and a reef enclosing a lagoon. Atolls are formed as a result of the sinking of high volcanic islands and the simultaneous upward growth of the coral reefs surrounding them. A coral reef is not made of rock, but of living and dead

Blowholes of gas rise from volcanic islands, such as these in the Tonga group. This phenomenon occurs between eruptions of active and dormant volcanoes.

organisms. Live coral is composed of billions of tiny creatures surrounded with lime. It is this that gives coral its solid appearance.

The creatures that make up coral reefs are known as polyps. They can exist only in water with a temperature of over 68 degrees Fahrenheit (20 degrees Celsius). As a result coral is found only in tropical waters. Polyps must also have something to which they can attach themselves, and they must have a good supply of food (they live on plankton), oxygen, and light. This means that they can live only along the shoreline near the surface of the water, up to a depth of about 150 feet (45 m.). When the rock or island onto which the coral is growing sinks, the coral grows upward to keep close to the surface, where life-giving oxygen and light are available. At the same time the lower layers of coral die and form a hard limestone base. A typical atoll is formed, finally, as the last bit of the island itself disappears beneath the water. All that is left is a lagoon with a fringing reef that gradually becomes partially or totally covered with coral sand thrown up by the action of the waves. The extent of the building up process by the coral organism varies greatly from atoll to atoll. At Eniwetok, for example, the depth of the calcareous deposits on top of the volcanic base reaches 4,630 feet (1,410 m.).

Natural Resources. Coral atolls have little in the way of natural resources except their incredible beauty. Their coral sand can support very few plants apart from the coconut. Minerals are non-existent, unless one considers guano a mineral. Guano, which is used as a fertilizer for crops, is the name given to seabird droppings that have been deposited over the centuries on roosting and nesting places and have solidified. Guano is found extensively throughout the islands of the Pacific, including the atolls.

The relatively vast and more highly concentrated deposits of phosphates on Nauru and Ocean Island are different from ordinary guano. It is believed that these deposits are made up of the droppings of huge prehistoric birds, now extinct. The droppings may have been submerged and compressed and then pushed up above sea level again.

No survey of Oceania's natural resources is complete without brief mention of the relatively untapped resources of the sea. For a long time the Pacific was well-known as a rich hunting ground for whales, although now their numbers have been sadly depleted. It is also well-known as a major source of bonito and especially tuna. But man is only beginning to be aware of the resources that may exist in the vast area of ocean and in the earth beneath. Perhaps one day the very sea itself and what it hides from view may provide the wealth this area now seems to lack.

Climate

Another important aspect of the physical environment that affects the lives of the people of Oceania is the climate in which they live. Since Oceania is situated almost completely in the tropics, it is subject to uniformly relatively high temperatures. It has more than adequate rainfall all year round.

Winds and currents in the Pacific are linked and follow a similar pattern. In general they flow in huge circles, clockwise in the Northern Hemisphere and counterclockwise in the south. The area in between the two wind systems is known as the doldrums. It is a highly unstable area,

Year-round high temperatures make a dip in the water especially refreshing.

where wind conditions can vary from nothing at all to the destructive hurricanes and typhoons that are so well-known to the Pacific islanders. The doldrums seem to follow the sun, reaching the Tropic of Cancer in June and the Tropic of Capricorn in December. As they move they interfere with the steady trade winds, bringing uncertainty and, quite often, devastating destruction to landsman and sailor alike.

The relatively high temperatures that are typical of the area have an important effect on agriculture. High temperatures raise the soil temperatures. Then when a large amount of rainfall is added, the soils are often partly destroyed for crop planting. As long as the soils are covered by thick rain forest and other natural vegetation, they are protected from the heavy rains, erosion, and loss of the valuable minerals that support plant life. But once the vegetation is removed and the soils are exposed to rain and intense sunlight, their valuable ingredients tend to leach out quickly.

The problem is not so great when the land is used for tree crops such as coffee and cacao, because these trees provide their own protection for the soil. But great care must be exercised in clearing and cultivating the land, for root crops in particular. For many years Europeans looked down on the farming methods used by the Pacific islanders. After the islanders had used a patch of land for a year or so they would move to another patch. The Europeans tried to introduce regular systems of crop rotation and to eliminate the "inefficiencies" of constantly leaving one patch and clearing a new one. Now there are many who believe that there is much to be learned from the "inefficient" system of the islanders, which helped to keep the tropical soils fertile.

THE PEOPLE

The peoples of the Pacific are usually divided into three groups: Polynesians, Micronesians, and Melanesians. These divisions were based on the observations of Europeans in the area. They believed that certain groups of islanders with common physical and linguistic characteristics, living in a definable geographical area, could be distinguished from other groups. In fact, only one of these groups—the Polynesians—meets these tests. Nevertheless, the terms do have a useful geographical meaning and it is in that sense that they are used in this article.

Polynesia (the name means "many islands") lies within the vast triangle formed by Hawaii, New Zealand, and Easter Island. The people within this area share a common basic language, social system, and religious beliefs.

Micronesia ("small islands") stretches westward from the borders of Polynesia, north of the equator, and includes the islands north of New Guinea to the borders of Oceania. The people within this area have little in common with one another. Some of them speak Polynesian languages, but most use non-Polynesian languages that are unrelated.

Melanesia ("black islands") encompasses the islands south of the equator and west of Polynesia and includes New Guinea. Apart from the fact that most of the people of this area have darker skins than the people in the two other areas, they have little in common with one another. Linguistic experts have been able to identify a Melanesian group of languages, but these form only a small proportion of the many hundreds of languages spoken in the Melanesian area. There are many physical types, social systems, and religious beliefs in Melanesia.

Where the People Came From

The mixed character of the Pacific islanders reflects the different places from which they came. There are, of course, many theories about the origin of these people. One of the best-known was set forth by the Norwegian Thor Heyerdahl. He tried to prove that the migrations came from the east by making his now celebrated voyage in the *Kon-Tiki*. Most scientists, however, believe that the migrations came from another direction, from Southeast Asia. A widely accepted theory is that the first people came into the area from the Southeast Asian peninsula at a time when New Guinea and Australia were still linked. These first inhabitants were nomads, who lived off what they could hunt or find. They moved out of Southeast Asia, across the narrow seas to New Guinea, and onto the Australian mainland. After the melting of the ice cap, and over a long period of time, peoples who had a slight knowledge of farming and of the other skills required to settle permanently moved into and through New Guinea and into the other islands of Melanesia.

Many years later other people with a more highly developed material culture came from Southeast Asia through what is now called Indonesia into Micronesia. To these were added groups from Asia and the

Students in American Samoa often watch educational television.

A young American Samoan climbs a coconut palm, an important food source.

Trobriand islanders perform a special dance to celebrate the end of the yam harvest.

Imports, such as these bicycles, are vital to the islands.

Philippines. Finally descendants of these predominantly Southeast Asian people moved out of Micronesia into Polynesia.

According to one of the theories—and it is hard to choose among the many that have been developed—the people we know as Polynesians gathered first in the Tonga-Samoa area in about A.D. 300. From there they moved to the surrounding islands. In time—probably about A.D. 1000—another center developed farther east in the Tahiti area. From there people moved to Hawaii in the north, Easter Island in the east, and New Zealand in the south.

No one is sure whether the voyages involved in reaching the centers such as Tonga-Samoa and Tahiti were planned. For many years the idea that they were deliberate voyages was almost never challenged. But recently some scientists have set out to show that the destinations reached by the voyagers were unplanned. Even granting the islanders' great skill as navigators, their ability to use the stars, and their knowledge of currents and winds, it is difficult to see how they could navigate so accurately over thousands of miles without any means of measuring longitude. A storm, for instance, could completely throw them off any system of dead reckoning they might have had.

Thus the peoples who settled in the Melanesian area were among the earliest arrivals into the area. Those who came into Micronesia and Polynesia entered the area much later and apparently by another route that skirted the Melanesian area. The peoples of Melanesia were denied the benefits of contact with the more highly developed cultures of the newcomers. The newcomers had themselves benefited from the developments that had occurred in the intervening centuries in Southeast Asia. This helps to explain the distinction made by the early explorers between the relatively advanced Polynesians and the less advanced New Hebrideans, Solomon Islanders, and New Guineans.

Differences and Similarities of the People

These differences in material culture were also reflected in political matters. It is probably true to say that everywhere in the Pacific the family was the basic and most important social unit and then became the basic political unit. In Polynesia there was great unity among family groups. This unity extended over whole islands and, in some cases, over groups of islands. This political unity was made possible by the existence of a common language and social system.

In Melanesia, on the other hand, there were literally hundreds of groups speaking completely dissimilar languages. In New Guinea today there are said to be between 600 and 700 different languages. Often only a few hundred persons made up a language group. Contacts with other groups in Melanesia mainly took the form of warfare or some limited trading activities. Added to this was the constant struggle for survival in an area with limited natural agricultural resources. As a result people lived in a state of insecurity and were constantly suspicious of their neighbors.

Using the Land.   To a Western European or an American, there were obvious similarities among the people. One example was their attitude toward land and work. All these peoples lived completely on what they could produce or take from the land. They hunted birds, wild pigs, and other animals. They gathered fruits, building materials, and firewood.

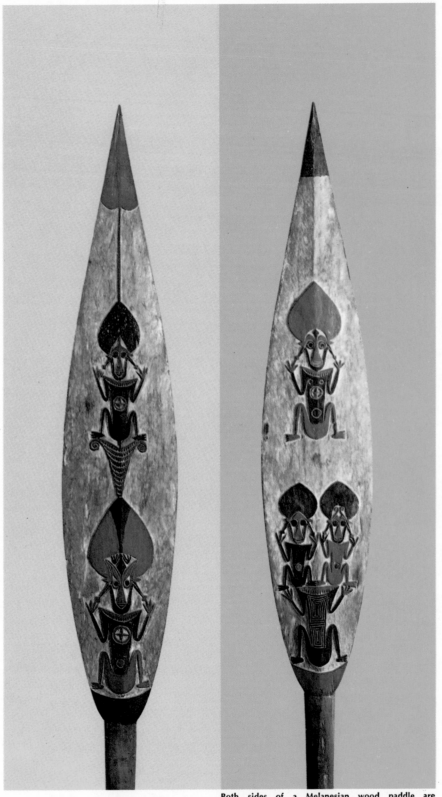

Both sides of a Melanesian wood paddle are richly decorated with carved and painted human figures in which the heads are especially emphasized.

They grew crops. Land meant life. As a result land was of prime importance in their social system as well as in their religious practices.

Among the people of Oceania control over the use of land was never the exclusive right of one man. The idea of exclusive individual ownership of land was unknown. The rights that an individual had to the land were related to those of the other members of the group to which he belonged. Individuals shared in the use of land with other members of their group for specific purposes. This went on as long as new needs did not arise making rearrangements necessary. Thus any one piece of land, while "belonging" to what we might call a political group, could be subject to all kinds of rights of use. These might include the right of some individuals to hunt, the right of others to collect firewood, and the right of still others to grow crops. Other land might be barred from use completely, for religious reasons.

It is easy to imagine the difficulties Asians and Europeans (as Caucasians in the Pacific are most often called) had when they tried to acquire land in Oceania. Even when the would-be buyer had the best of intentions (and of course, this was not always the case), the islanders never understood that by accepting money they were permanently transferring the land to someone else. The idea was foreign to them. Moreover, a European who had bought some land from a particular individual often found himself faced with many other unsatisfied "owners" who continued to trespass on his land or who also demanded payment.

The Islanders' Ideas About Work.   Another example of attitudes that were fairly common to all islanders was their attitude to work. Labor was not an end itself. One worked for socially desirable ends. Houses had to be built and maintained. Gardens had to be established and cared for. Food had to be prepared, communities had to be defended, tools and weapons had to be fashioned. These activities were carried out only when they were necessary. By and large these kinds of work were seasonal. And often they were linked with religious and social ceremonies. Few of these activities were carried out by individuals. Often the whole group would be concerned, or one family group would help another.

The contrasts between the islanders' ideas of work and those of a Western European are striking. But they were often ignored by newcomers. The idea of working regular hours each day, day after day, for no understandable purpose except to earn money, was quite strange to an islander. To him money was important only as a means of getting the goods the Europeans had to offer—steel axes, knives, and trinkets.

Once these desires were met, the islander's interest in work depended on how important money was to his way of life. This, in turn, depended upon what the European could offer and the extent to which the islander's life and that of his family became dependent on a money economy. More often than not, the limited range of goods available to him and the small amount of money he earned were not enough to make him want to work regularly for pay. This helps to explain why the Europeans often said that the islanders were slow and lazy, unreliable and stupid.

OCEANIA'S DISCOVERY BY THE WEST

The Explorers.   The first contacts between Europeans and the islanders began with the coming of the European explorers. It was these

Coffee, set out to dry here, is one of the chief products of New Caledonia.

Holidays in American Samoa are often celebrated with a parade.

explorers who did much to spread the somewhat exaggerated picture of the idyllic life led by the inhabitants of these exotically beautiful islands. These islands do rank among the most beautiful in the world. Their beauty is no myth. No one who has ever seen such islands as Bora-Bora, Moorea, and Tutuila could ever deny their beauty.

For the Western visitors of the 18th and 19th centuries there was a dramatic contrast between the beauty of the islands and the changes caused at home by the Industrial Revolution. These Europeans and Americans had seen quiet towns rapidly growing into overcrowded, filthy cities. Their countrysides were dotted with ugly factories, which even then were beginning to pollute the landscape. Life for many people in the West was becoming linked to the monotonous routine of the factory and the machine. Small wonder then that Oceania seemed to be a kind of paradise.

But what the explorers and their chroniclers overstated, sometimes more by implication than by so many words, were the conditions of life on the islands. It was the things they failed to report that left the total picture a somewhat false one and led to so much disillusionment.

Fruit was abundant, but meat was not. Island staples, such as taro and breadfruit, never succeeded in finding their way into the European diet. Little was said about rainstorms and hurricanes. The inadequacy of housing and the constant need for repairing the dwellings were not mentioned. Endemic malaria, other fevers, and intestinal parasites were scarcely mentioned. No hint was given that the open-hearted welcome the white man received and all the favors he was granted might be related to the axes, knives, and other useful tools he brought. The possibility that once his supply ran out he might not be so welcome was not even suggested.

Nothing was known about the islanders' moral and social systems, let alone their motivations. This lack of knowledge, and in many cases complete misinformation, were later to lead to unpleasant incidents. These led to Europeans' calling the islanders ungrateful and treacherous. But these were the harsh facts that became apparent only after one had lived in Oceania for longer than a few weeks. In many cases these facts were enough to explode the myth and lead the less strongly motivated Europeans to return home disillusioned. Of those who remained, many became "beachcombers"—a word coined to describe the easy, work-free life of sailors who left their ships to settle on the islands of the South Pacific.

The Missionaries. But most of the intruders into the Pacific who came to stay were hardier souls. The first in time, and perhaps in terms of their importance and influence, were the Christian missionaries. For the most part they represented the fundamentalist Protestant and Puritan churches of Western Europe and North America. These men and women were the products of the religious revivals of the late 18th and 19th centuries in Europe. They were, in general, endowed with great bravery and fortitude. Above all, they were driven by their fervor to convert the "heathen."

To the missionary, the legendary, idyllic work-free life of the Pacific islanders, with its reputed sexual freedom, represented the very essence of heathenism. It was a challenge to the missionaries' zeal. They went out to save souls for Christ by converting the islanders to Christianity. In

ISLANDS ADMINISTERED BY THE UNITED STATES

TRUST TERRITORY OF THE PACIFIC ISLANDS

LOCATION: Latitude—1° N to 20° N. Longitude—130° E to 172° E.

AREA: 687 sq. mi. (1,779 sq. km.).

POPULATION: 96,000 (estimate).

POLITICAL STATUS: United Nations Trust Territory administered by United States.

SEAT OF GOVERNMENT: Saipan island, in the Marianas, is the administrative center of the territory.

BACKGROUND FACTS: Covers area generally known as Micronesia. Consists of many islands held successively by Spain, by Germany, and by Japan as a mandated territory. Became UN trust territory after World War II.

Marshall Islands

LOCATION: Latitude—4° 30′ N to 14° 45′ N. Longitude—160° 50′ E to 172° E.

AREA: 70 sq. mi. (181 sq. km.).

POPULATION: 19,000 (estimate).

CHIEF PRODUCT: Copra (dried coconut meat).

BACKGROUND FACTS: The largest atoll in the Marshalls is Kwajalein; the best-known are Eniwetok and Bikini, sites of atom-bomb tests.

Caroline Islands

LOCATION: Latitude—3° 1′ to 5° 19′ N. Longitude—131° 11′ to 162° 59′ E.

AREA: 460 sq. mi. (1,191 sq. km.).

POPULATION: 36,000 (estimate).

CHIEF PRODUCTS: Copra (dried coconut meat), trochus shell.

BACKGROUND FACTS: Palau, Truk, and Yap are the most important groups within the Carolines. Truk was a Japanese naval base during World War II.

Marianas Islands (excluding Guam)

LOCATION: Latitude—13° 25′ N to 20° 32′ N. Longitude—144° 45′ E to 144° 54′ E.

AREA: 158 sq. mi. (409 sq. km.).

POPULATION: 29,700 (estimate).

PRINCIPAL TOWN: Agana on Guam.

CHIEF PRODUCTS: Copra (dried coconut meat), vegetables.

BACKGROUND FACTS: Islands discovered by Magellan in 1521. Guam became a U.S. possession in 1898; other islands became UN Trust Territory after World War II. Commonwealth status with the U.S. has been approved by the U.S. Congress.

GUAM

LOCATION: Latitude—13° 26′ N. Longitude—144° 43′ E.

AREA: 212 sq. mi. (549 sq. km.).

POPULATION: 450,000 (estimate).

POLITICAL STATUS: Unincorporated territory of the United States, administered by Department of the Interior.

PRINCIPAL TOWN: Agana.

CHIEF PRODUCTS: Fruits and vegetables, fish.

BACKGROUND FACTS: The largest and most populous of the Marianas, it is an important military base. Guamanians are full U.S. citizens.

AMERICAN SAMOA

LOCATION: All islands of Samoan group east of 171° west longitude are included in American Samoa.

AREA: 76 sq. mi. (197 sq. km.).

POPULATION: 31,000 (estimate).

POLITICAL STATUS: Unorganized, unincorporated territory of the United States.

PRINCIPAL TOWN: Pagopago.

CHIEF PRODUCTS: Copra (dried coconut meat), tuna, tarp, breadfruit, yams, bananas, coconuts, arrowroot.

BACKGROUND FACTS: Samoan group first seen by Europeans in 18th century. American Samoa became U.S. territory by 1899 treaty. The harbor at Pagopago on the island of Tutuila is known for its beauty and as the only port for large vessels in the Samoan islands. American Samoa includes:

Tutuila and Aunuu

AREA: 53 sq. mi. (137 sq. km.).

POPULATION: 22,500 (estimate).

Tau

AREA: 15 sq. mi. (39 sq. km.).

Ofu and Olosega

AREA: 3.5 sq. mi. (9 sq. km.).

POPULATION: 3,500 (estimate).

Swains Island

AREA: .9 sq. mi. (2.3 sq. km.).

POPULATION: 100 (estimate).

Rose Island

AREA: .4 sq. mi. (1 sq. km.).

POPULATION: 0.

WAKE ISLAND

LOCATION: Latitude—19° 18′ N. Longitude—166° 35′ E.

AREA: 3 sq. mi. (8 sq. km.).

POPULATION: 1,100 (estimate).

BACKGROUND FACTS: Atoll with three islets—Wake, Wilkes, and Peale. Site of attack by Japanese early in World War II.

JOHNSTON ISLAND

LOCATION: Latitude—16° 44′ N. Longitude—169° 17′ W.

AREA: .4 sq. mi. (1 sq. km.).

POPULATION: 150 (estimate).

BACKGROUND FACTS: Discovered by the British in 1807 and claimed by U.S. in 1858, it became a naval base during World War II.

MIDWAY ISLANDS

LOCATION: Latitude—28° 13′ N. Longitude—177° 22′ W.

AREA: 2 sq. mi. (5 sq. km.).

POPULATION: 2,500 (estimate).

BACKGROUND FACTS: It was discovered by Americans in 1859 and annexed in 1867. An air base and naval station. One of the important battles of World War II was fought here and is named for the group.

ISLANDS ADMINISTERED BY THE UNITED KINGDOM

GILBERT ISLANDS COLONY

SEAT OF GOVERNMENT: Tarawa is the administrative headquarters for the colony.

POLITICAL STATUS: Formerly called the Gilbert and Ellice Islands Colony. Became partially self-governing in 1971. In 1975 the Ellice Islands became the separate colony of Tuvalu.

CHIEF PRODUCT: Copra (dried coconut meat).

BACKGROUND FACTS: The Gilbert and Ellice Islands became a colony of the United Kingdom in 1915.

Gilbert Islands

LOCATION: Latitude—4° N to 3° S. **Longitude**—172° E to 177° E.

AREA: 102 sq. mi. (264 sq. km.).

POPULATION: 46,000 (estimate).

Phoenix Islands

LOCATION: Latitude—3° S to 5° S. **Longitude**—170° W to 175° W.

AREA: 11 sq. mi. (28 sq. km.).

POPULATION: 0.

POLITICAL STATUS: The islands of Canton and Enderbury in this group are under joint U.S.–U.K. administration for a period of 50 years that began in 1939.

BACKGROUND FACTS: Became a part of the Gilbert and Ellice Islands Colony in 1937, but have been abandoned since 1968 and are now uninhabited.

Line Islands

LOCATION: Latitude—2° N to 4° 40' N. **Longitude**—157° W to 160° 20' W.

AREA: 237 sq. mi. (614 sq. km.).

POPULATION: 1,100 (estimate).

SEAT OF GOVERNMENT: Christmas Island is headquarters for the Line Island district.

BACKGROUND FACTS: Sometimes called the Equatorial Islands. Fanning Island was discovered by Americans in 1798. Christmas Island is the largest island in the group. The third inhabited island is Washington Island. In 1972, 5 central and southern Line Islands became part of the colony.

Ocean Island

LOCATION: Latitude—0° 31' S to 0° 52' S. **Longitude**—169° 35' E.

AREA: 2 sq. mi. (5 sq. km.).

POPULATION: 2,500 (estimate).

BACKGROUND FACTS: Became a part of the protectorate (now a colony) in 1900.

TUVALU

LOCATION: Latitude—5° 30' S to 10° 30' S. **Longitude**—176° E to 179° 58' E.

AREA: 9.5 sq. mi. (24 sq. km.).

POPULATION: 6,500 (estimate).

SEAT OF GOVERNMENT: Funafuti.

SOLOMON ISLANDS

LOCATION: Latitude—5° S to 12° 30' S. **Longitude**—155° 30' E to 169° 45' E.

AREA: 11,500 sq. mi. (29,800 sq. km.).

POPULATION: 148,000 (estimate).

POLITICAL STATUS: Protectorate of the United Kingdom. Has internal self-government.

SEAT OF GOVERNMENT: Honiara on Guadalcanal.

BACKGROUND FACTS: The Solomon Islands were discovered in 1568 by the Spanish. Protectorate includes islands of Guadalcanal, Malaita, San Cristobal, New Georgia, Santa Isabel, Choiseul, Shortland, Mono, Vella Lavella, and several other islets.

NEW HEBRIDES CONDOMINIUM
(Nouvelles Hébrides)

LOCATION: Latitude—12° S to 21° S. **Longitude**—166° E to 171° E.

AREA: 5,700 sq. mi. (14,763 sq. km.).

POPULATION: 80,000 (estimate).

POLITICAL STATUS: Administered jointly by U.K. and France since 1887.

SEAT OF GOVERNMENT: Vila on Efate.

CHIEF PRODUCTS: Copra (dried coconut meat), frozen fish.

BACKGROUND FACTS: A 450-mile-long (720 km.) chain of about 80 volcanic islands. The largest is Espiritu Santo.

PITCAIRN ISLAND

LOCATION: Latitude—25° 4' S. **Longitude**—130° 6' W.

AREA: 1.9 sq. mi. (5 sq. km.).

POPULATION: 100 (estimate).

POLITICAL STATUS: British colony administered by a high commissioner.

SEAT OF GOVERNMENT: Auckland, New Zealand.

CHIEF PRODUCTS: Goods such as fruit, stamps, and trinkets for passing ships.

BACKGROUND FACTS: Discovered in 1767 but not inhabited until mutineers from H.M.S. *Bounty* came there in 1790 with a group of men and women from Tahiti.

addition they tried to impose a whole new moral and social order on the islanders. The missionaries wanted to build their own ideal society in the Pacific. It was to be a replica of the society they had striven for in Europe. It placed a premium on the virtues of uprightness, obedience, thrift, and hard work. It placed little emphasis on such notions as comfort and enjoyment. Many of the missionaries were narrow and bigoted people. Many had a limited capacity for appreciating art or music and a contempt for anything that was not European. They often condemned anything non-European, anything they called native.

For some years the missionaries made little or no headway. They suffered greatly, not only from the inevitable privations, but also at the hands of their hoped-for converts. But when the traders, planters, and settlers appeared on the scene, the island leaders felt overwhelmed. Needing an ally, the islanders turned to the missionaries as the only persons willing to support their interests. In return for the missionaries' support, the island leaders announced their conversion to Christianity. Mass conversions followed, mainly in the Polynesian areas, where the authority of the chiefs meant something. In this way Christianity—in name, at least—was established throughout Polynesia by the 1850's. True conversion to Christianity took much longer. The task was much more difficult in Melanesia, where the chief's power was less strong and the work of conversion had to proceed almost person by person. In fact, it still is going on today.

The Christian missionaries have done much for the islanders, in addition to bringing them Christianity. For many years it was the missionaries who provided the greater part of the education and the health services, for which the colonial governments had neither the interest nor the money.

On the negative side, however, the missionaries destroyed the islanders' native religious beliefs, ceremonies, music, art, and dancing. In this way they undermined the basis of the islanders' social systems and contributed to a complete collapse of their way of life. In Hawaii and a few other places where the islanders have been able to integrate themselves into the new Western way of life, the negative effects of change have not been as serious. But where such integration has not taken place, which is the case over almost all the rest of Oceania, serious problems remain.

The Settlers. The next decisive influence in the Pacific was the settlers. Their forerunners were the traders and whalers who came to Oceania for such products as sandalwood and oils. The traders established relations with the islanders and in this way introduced them to many Western influences, but the contacts were not lasting and had only a limited effect on the lives of the islanders.

The settlers first came to Oceania to cultivate coconuts. They exported to Europe the soft inside of the nut in the form of copra (dried coconut meat). In Europe copra oil was extracted for use in a variety of ways.

At first the traders arranged to collect coconuts at ports of call in the islands. Then they began to show the islanders how to dry the pulp. Finally, as the demand grew and the inadequacies of these arrangements became clear, Europeans decided to grow the coconuts themselves on the islands.

This decision meant that Europeans had to live with the islanders. Unlike the trader who could always cut and run when difficulties arose, the settler had to try to solve his problems. It meant acquiring land from the islanders on terms that would repay the considerable investment required in clearing and planting. It meant finding and maintaining a suitable supply of labor. It also often meant setting up schools and medical facilities. This led to the growth of a more varied European community. Finally, it meant that law and order had to be maintained in order to carry on economic activities.

Problems of all kinds followed wherever settlement took place. Trouble arose between the islanders and the settlers because of misunderstandings over land dealings. This often led to bloodshed and left behind bad feelings on both sides.

Often the island leaders could not meet the settlers' demands for law and order. This led to the settlers' taking the law into their own hands. This, in turn, involved the settlers in difficulties with their own governments. Eventually the settlers began to organize puppet governments on the islands, which they themselves controlled. They did this by supporting one of the leading chiefs and setting him up as king. This practice accounts in large part for the origins of the "kings" in the Pacific, for instance in Hawaii, Fiji, and Tahiti. The people of the islands participated only slightly in these European-run governments. The governments were further weakened by disagreements among the Europeans. Missionaries, planters, and town- or port-based commercial interests were involved in these squabbles.

By the last quarter of the 19th century chaotic and scandalous situations had developed in a number of island centers. In Fiji the King's pleas to Queen Victoria of Great Britain to assist him in controlling the activities of her subjects in his country went unheeded until 1874. In that year the British Government assumed responsibility for the Fiji Islands. An equally chaotic situation developed in Samoa—complicated by American, British, and German rivalries and opposing alliances with two indigenous chiefs—which led to the annexation and division of the Samoa group in 1899.

The Growing Influence of the World Powers.   This brings us to the third important influence in the historical development of Oceania—the intervention of the world powers of Europe and the United States. Throughout the greater part of the 19th century the world powers were reluctant to become involved in Oceania. This was due to the fact that the area was of no political or economic importance. Although the activities of the Pacific islanders often caused difficulties, these were minor irritations that could usually be handled by arrangements made on the spot by naval captains. However, during the first three quarters of the 19th century a few of the islands were acquired: Great Britain acquired New Zealand in 1840, and France acquired the Society Islands in 1842 and New Caledonia in 1853.

Between 1874 and 1901 almost every island group in the Pacific was acquired in one way or another either as a protectorate or a colony. Thus Germany acquired half of Samoa, the Carolines, and the Marshalls, and a third of New Guinea. France consolidated its rule in the remaining Polynesian islands near the Societies and shared the New Hebrides with Great Britain. The United States acquired Hawaii, half of Samoa, Guam, and the Philippines. Meanwhile, the British acquired Fiji, Tonga, one of the three parts of New Guinea, and the Solomon Islands; and they shared the New Hebrides with France.

Other changes took place that reflected the rise of new nations in the Pacific and the eclipse of others in Europe. Thus, at the beginning of the 20th century Great Britain transferred responsibility for some of its colonies and protectorates in the Pacific to its own former colonies of Australia and New Zealand. When Germany was defeated in World War I, it was forced to give up its Pacific colonies to be administered

The island of Serua, in Fiji, was once the home of mighty chiefs.

Silent reminders of World War II, when the islands were a battleground.

under the League of Nations mandate system by Australia, New Zealand, and another power in the Pacific, Japan.

As a result of these changes, Australia became responsible for the administration of Papua (the southeastern part of the island of New Guinea) and the Mandated Territory of New Guinea (the northeastern part of New Guinea). These territories were later united as Papua New Guinea. Australia also administered the Mandated Territory of Nauru. New Zealand became responsible for Niue Island, the Tokelau Islands, the Cook Islands, and the Mandated Territory of Western Samoa. Japan was granted direction over the Mandated Territory of the Pacific Islands (which comprised the Marianas, Marshalls, and Carolines).

After World War II and the defeat of Japan, Japan's mandated territory was placed under the United Nations International Trusteeship System to be administered by the United States. The other former mandated territories were also placed under the International Trusteeship System and continued to be administered by the same powers which had administered them before.

OCEANIA IN THE 20TH CENTURY

After these territories were acquired great emphasis was placed on establishing law and order, putting an end to intergroup fighting, and inducing people to settle their differences according to the legal codes introduced by the new administrators. This was seen as part of the Europeans' "civilizing mission" in the islands. A respect for European ideals of law and order seemed essential if Europeans were to live there, to grow crops there, or simply to trade there.

In the smaller islands this presented no great problem, but in the larger islands and island groups the process was more difficult and took much longer. In the Melanesian islands in particular, the task of imposing and enforcing laws was made more difficult by several factors—the rough and broken terrain, the fact that the people lived in small communities scattered over large areas, having little contact with one another, and the fact that they spoke many different languages.

In the Melanesian islands months of arduous travel on foot were necessary in order to make the first contacts with only a relatively few people. These initial contacts had then to be followed up over a period of years by further contacts before it could be claimed that the administration had established any degree of respect for its new system of law and order. In such places as New Guinea, the Solomon Islands, and the New Hebrides, this process took many years.

This process was necessary as a first step if the people of Oceania were to take their place in the modern world. However, it had the effect of undermining the old system of leadership and authority. It also undermined the existing systems of law and justice.

One of the main criticisms leveled at the colonial administrators of the 20th century has been that they failed to replace the old system with a new one with which the people felt they could identify. It seemed that for too long the new system remained something foreign that had simply been imposed from outside.

The establishment of law and order by the new administrators did have a positive side. For one thing it eliminated warfare. The elimination of warfare and the constant fear of death at the hands of one's enemies had an effect that is hard to overestimate. With the fighting ended, villages could be built at more convenient places—in valleys near a good source of water, instead of on an inconvenient ridge that was suitable only for military reasons.

The new administrators also brought health and educational services with them. They were assisted by the missionaries, who in many cases did more in providing these services than the governments. Indeed, one of the criticisms of the island administrators has been that they did too little to educate and to improve the health of the people whom they claimed to be civilizing.

The establishment of law and order by the new administrators also brought about change in economic life. In many of the islands there had once been a single economy that provided food and shelter for the community. Now there were two economies. The old one still existed in varying degrees. The new economy was run by the Europeans and to a lesser extent by Asians. The island people hardly took part in the European-run economy. This was an economy based on European-owned and -run plantations or mines. Asians participated as traders. The local people participated only as the providers of the land and sometimes as laborers.

Expanding Economies

The Europeans, in addition to growing and producing copra, began to experiment with other tropical products, such as rubber, sugar, coffee, cacao, and tea. The large-scale production for export of such tropical fruits as bananas and pineapples was also begun. But the planters met with many difficulties. There were problems of terrain, climate, and soils. Even if they were able to overcome these, there were the problems arising from the long distances the products had to be shipped to reach the world markets.

Another problem was the fluctuations in the prices that buyers were willing to pay for these commodities. Up until World War II production in most of the islands was limited to sugar and copra. The main exception was Hawaii, whose tropical products could be shipped relatively easily to markets in the United States.

The other main economic interest of the Europeans was in minerals. Gold was found in New Guinea and was the mainstay of the economy before World War II. Gold was also important in Fiji. In New Caledonia large deposits of nickel and chrome have been worked successfully since the early years of this century.

Both these economic activities in the islands and the commercial and trading activities to which they gave rise required skills or experience that the islanders did not have. The Europeans overcame this problem first by importing laborers from Asia—Chinese, Indians, Filipinos, and Indo-Chinese. When this practice became too expensive or when governments prohibited it, the Europeans were forced to turn to the local population and to develop jobs that even an untrained and unskilled person could readily perform.

At the same time the Europeans persuaded the governments to adopt measures that would induce the local population to take up steady employment. In this way the work that had to be done on a plantation was divided up into a series of minor and simple tasks that required a

Coffee can be grown on some islands such as New Caledonia.

Copra, or dried coconut meat, is a major product of Oceania.

relatively large number of laborers to carry out. As an inducement to work for money, governments imposed head taxes on every able-bodied male, which had to be paid in cash.

While these methods made it possible to use island labor, large numbers of laborers were not always available in the places where they were needed. It became necessary then to transport laborers to the plantations from other parts of the island or island group. In this way the contract or indenture system of plantation labor grew up.

This system has been the subject of much criticism over the years from many sides. Economists and labor experts pointed to the inefficient use of the manpower involved. They condemned the system for making it almost impossible for a worker to improve his skills and to become more productive. They attacked the low wages that were paid.

Even stronger criticisms were made of the inhumanity of the system of indentured labor. Critics pointed to the evil results of removing able-bodied males from their families and communities for years at a time and to the poverty that the practice caused. The indenture system has acted as a major barrier to the social, economic, and political advancement of the islanders.

Since World War II the pace of change in Oceania has increased. The people of the islands have become increasingly aware of their problems and have been pressing for changes. The administering countries have had to provide capital for services in the territories they administer. They have placed more of the responsibility for government in the hands of the islanders themselves.

The pressure of world public opinion has become a factor that cannot be ignored by the countries involved. Events that take place in Oceania are recorded in the world's newspapers. The situations in the territories of Oceania are regularly examined in the United Nations, which recommends new policies and courses of action.

As a result, the economies of many of the islands and groups have become more diversified. Per capita income has risen. New crops have been introduced. New industries have been established. Timber resources have been tapped. At the same time medical, health, and educational services have been expanded. Many hospitals, schools, and universities have been built and staffed.

On the political side there have also been developments. Territory-wide representative assemblies have been introduced. In a growing number of cases, local leaders participate in the executive branch of government. The Hawaiian Islands became a state of the United States of America in 1959. Western Samoa and Nauru achieved their independence in 1962 and 1968. Fiji and Tonga became independent in 1970, and Papua New Guinea in 1975. (Separate articles on FIJI, NAURU, PAPUA NEW GUINEA, TONGA, and WESTERN SAMOA, as well as on AUSTRALIA and NEW ZEALAND, appear in this volume. An article on HAWAII appears in Volume 5.) The Cook Islands and Niue now have full internal self-government in association with New Zealand, and the Solomon Islands have internal self-government under a British protectorate. The French territories are overseas departments of France. The people of the northern Marianas, part of the United States Trust Territory of the Pacific Islands, have voted to become a self-governing commonwealth of the United States.

OCEANIA'S FUTURE

Although these developments give hope for Oceania's future, there are still problems to solve.

Micronesia. In Micronesia the possibilities of economic development are not bright. The low coral atolls have generally poor soils and possess no mineral resources. Their economies depend almost entirely on copra. For the future they will probably have to rely on the development of fishing and tourism. In the Gilbert Islands the prospects for living are so grim that the possibility of moving the populations of some of the islands to other parts of Oceania is being explored. These problems are so fundamental that there is little hope for the future until they are solved. The fact that Micronesian territories are so scattered makes communications difficult and prevents the development of a sense of unity. This, in turn, delays political development.

Polynesia. In the Polynesian area the prospects are brighter. Copra is an important product here, but there are also others, such as sugar, tropical fruits, and vegetables. Tourism has also become an established and lucrative industry in such places as Hawaii, Samoa, and parts of French Polynesia.

This firm economic base in the islands of Polynesia is coupled with the fact that the population is concentrated on fewer islands that are closer together. This has helped the people in Polynesia to adapt their social and political life to the demands of the modern world more quickly. It was the Polynesian island groups such as Hawaii, Western Samoa, and the Cook Islands that were the first to achieve independence or self-government.

Melanesia. Melanesia presents a different picture. Fiji has a good potential economically. Sugar and coconut products are important, timber is abundant, and even though gold mining is a declining industry, there are still prospects for mineral development.

In the remaining parts of Melanesia—Papua New Guinea, New Caledonia, the Solomon Islands, and the New Hebrides—one of the problems facing the people is how to develop a sense of unity and nationhood among themselves. Even now the people of the various parts of Melanesia still often have only limited contact with one another. Most people in these areas still think of themselves as members of a village or family rather than as Solomon Islanders or New Caledonians. The formation of national legislatures to which representatives from all over the island group are elected has helped to introduce the idea of a wider unity even to villagers in remote regions.

All these island communities, whether they are in Micronesia, Polynesia, or Melanesia, face the problem of an old system that is crumbling or has almost disappeared and the development of a totally new political and economic system, which has only partially replaced the old. Every individual and family unit is required to adjust to the new way of life. How quickly change and the adjustment to change will take place, and whether there will be an atmosphere of peace and harmony between one group and another, between one community and another, between islanders and Europeans—all this depends on the efforts made by the people of Oceania. Progress and harmony also depend on the degree of assistance the islands receive from the world's wealthy nations.

JOHN MILES, Senior Political Affairs Officer, United Nations

PAPUA NEW GUINEA

Papua New Guinea, which lies northeast of Australia, is a country made up of part of one very large island plus many smaller islands and island groups. Most of its territory is located in the eastern part of the huge Indonesian archipelago (a chain of thousands of islands), which helps form a boundary between the Indian and Pacific oceans. Over past centuries these islands served as stepping-stones for numerous migrations of peoples from Asia. Papua New Guinea was formerly administered by Australia. It became an independent country in 1975.

THE LAND

Papua New Guinea has a total area of 178,260 square miles (461,691 square kilometers). About 85 percent of its land area lies in the eastern half of the island of New Guinea, the second largest island in the world. The western half of the island is called Irian Jaya and is a province of Indonesia. Farther east lie most of the island territories of Papua New Guinea. These include New Britain, New Ireland, the Admiralty Islands, and other islands of the Bismarck Archipelago; Bougainville and Buka in the Solomons; Woodlark; and the Trobriand, D'Entrecasteaux, and Louisiade island groups.

The most prominent geographical feature is a mountain system, the central cordillera, that extends the length of New Guinea and crosses the boundary between Papua New Guinea and Irian Jaya. Of the numerous mountain peaks, many reach considerable heights. High, steep valleys are located between the mountain ranges. Large, swift rivers rise in these mountains and valleys and flow north, east, and south to reach the ocean.

Port Moresby is Papua New Guinea's capital and largest city.

Luxuriant forest covers about 75 percent of the country. The soil, however, is generally shallow and infertile. There is very little highly productive agricultural land.

ECONOMY

Farming, forestry, and fishing are the chief occupations. Most of the people live on a subsistence level, which means that they produce their own food, clothing, and shelter. The chief food crops are sago (a starch obtained from a palm tree), taro (an edible root resembling a potato), yams, sweet potatoes, and bananas. This diet is supplemented by various vegetables, wild fruits, nuts, and fish. Meat from pigs, chickens, and wild animals is available in limited supply in some villages.

Plantations supply copra (dried coconut meat), cacao, coffee, tea, rubber, palm oil, and cattle for export. Mining plays an important role in the economy. Bougainville island is rich in copper, and gold and silver are also mined.

THE PEOPLE

The people of Papua New Guinea are a mixture of related and unrelated ethnic and tribal groups. Although there are about a thousand tribes living in thousands of villages and speaking hundreds of different languages, the people can be placed in three main divisions. They are the pygmies, the Papuans, and the Melanesians. To these groups may be added a relatively small number of whites, who are mainly from Australia.

The distribution of the population is extremely uneven. There are vast swamplands, mountainous areas, and infertile regions that are virtually uninhabited. By contrast, some coastal areas and some interior valleys of greater fertility have high population densities.

About half of the people are Christian, including both Roman Catholics and Protestants. The non-Christian half of the population maintains its traditional religious beliefs, which include the worship of ancestors and spirits.

The official language of the country is English, and instruction in the public schools and missionary schools receiving aid from the government is in English. However, with hundreds of languages spoken, communication remains a problem. A language called Melanesian Pidgin is widely used in the northern part of the country and is rapidly being adopted in other sections. Pidgin is a modified and simplified form of English with some words from other languages. It may someday become the national language of Papua New Guinea.

The way of life of most of the people centers around the village. There is usually a longhouse in the village, or one serving several villages, which is reserved for men only. Upon reaching a certain age boys are initiated into the men's secret cult. The symbols of spirits and ancestors are kept in the longhouse, and religious rites are performed there. Each village also has a rectangular-shaped park, which provides a meeting place for feasts, dances, and other social activities.

Although only a small minority of the people live in cities, in recent years the rate of population increase has been greater in the cities than in the countryside. Many people have migrated to the cities in search of better opportunities. The capital and largest city is Port Moresby. It has most of the manufacturing and service industries of the country. Lae and

FACTS AND FIGURES

PAPUA NEW GUINEA is the official name of the country.
CAPITAL: Port Moresby.
LOCATION: Pacific Ocean north of Australia.
AREA: 178,260 sq. mi. (461,691 sq. km.).
POPULATION: 2,700,000 (estimate).
LANGUAGE: English, Melanesian Pidgin, many other languages and dialects.
RELIGION: Christianity, animistic beliefs.
GOVERNMENT: Constitutional monarchy. **Head of state**—British monarch represented by a governor-general. **Head of government**—prime minister. **Legislature**—House of Assembly. **International co-operation**—United Nations, Commonwealth of Nations.
CHIEF CITIES: Port Moresby, Lae, Rabaul, Wewak.
ECONOMY: Chief agricultural products—copra, coffee, tea, cacao, rubber, palm oil, yams, sago, taro, bananas. **Industries and products**—mining, lumbering, fishing. **Chief minerals**—copper, gold, silver, manganese. **Chief exports**—copra, coffee, copper, rubber. **Chief imports**—foodstuffs, livestock, machinery, manufactured goods, drugs.
MONETARY UNIT: Kina.
NATIONAL HOLIDAY: September 16, Independence Day.

PAPUA NEW GUINEA

Rabaul rank second and third in size. The largest cities are located on the coast and are major ports.

HISTORY AND GOVERNMENT

There is little information about the early, unwritten history of Papua New Guinea. However, it is known that the highlands were inhabited at least as early as 8000 B.C. Some scholars believe that there were several migrations from Asia by way of the islands of the Indonesian archipelago. The first people were fishermen, hunters, and food gatherers. Later arrivals were farmers who introduced various fruits and vegetables as well as domesticated animals such as dogs and pigs.

The first Europeans to visit New Guinea were probably Spanish and Portuguese navigators in the 16th century. But little was known about the land or its people until the 19th century. In the early 19th century the Dutch colonized western New Guinea. Later in the century the Germans established themselves in the northeastern part of the island and the British in the southeast. In 1906 Australia took over British New Guinea and governed it as the Territory of Papua. The Australians occupied German New Guinea at the beginning of World War I. Later they administered it as the Territory of New Guinea, first under a mandate from the League of Nations and then as a United Nations Trust Territory. In 1949 the administration of the two territories was unified.

Papua New Guinea achieved self-government in 1973 and complete independence on September 16, 1975. It is a member of the United Nations and the Commonwealth of Nations.

Papua New Guinea is a parliamentary democracy under a constitutional monarch. The head of state is the British monarch, who is represented in Papua New Guinea by a governor-general. The legislature is the unicameral (single-chamber) House of Assembly. The prime minister, the head of the government, is the leader of the majority party in the House of Assembly, and chooses his cabinet from among its members. The judiciary consists of a supreme court and a national court.

THOMAS FRANK BARTON, Indiana University

FIJI

The people of Fiji originally called their home Viti, but this presented spelling and pronunciation problems for outsiders. "Fichi," "Feejee," and finally "Fiji" were common misspellings. The last one became so widespread that it is now used to describe the entire island group. Today Fiji is recognized as the most important island group in Oceania north of New Zealand. The lovely, tropical islands are a center of communications and transportation in the southwest Pacific.

THE PEOPLE

Except for Hawaii, Fiji, with over 500,000 inhabitants, is the most populous island group in Oceania. The population is growing rapidly, having almost doubled between the end of World War II and 1970. Slightly more than half of the people are descendants of settlers from India. Most of the Fijian Indians are descended from laborers who were brought to the islands to work on the sugar and pineapple plantations in the 1880's. The next largest group are of Fijian origin. Europeans, part-Europeans, Chinese, and Pacific islanders make up smaller groups within the population.

Almost three fourths of Fiji's population live on Viti Levu, the larger of the two main islands in the group. The most heavily settled areas are the coastal towns and the river valleys where the land is suitable for farming. The capital and leading city of the islands is Suva, on the east coast of Viti Levu, where over 10 percent of the people make their home.

Away from the modern, bustling city of Suva, Fijians continue to live much as their ancestors did, although schools and radio are bringing

A view of Suva harbor on Fiji's principal island, Viti Levu.

Fijian women net fishing in the waters near Suva.

new ideas to even the remotest villages. Fijian villages are usually made up of a cluster of about 20 houses called *mbures*. A *mbure* is a framework of logs or bamboo with mats woven of coconut leaves or reeds that can be let down when it rains to keep the house dry. Taro, cassava, yams, bananas, and breadfruit grown in or near the villages are the staples of the Fijian diet. For a *meke* ("feast") these foods are served with crabs, crayfish, and a delicacy called *kokoda,* a kind of pickled fish. Even on non-feast days it is unlikely that anyone will go hungry because, according to custom, food as well as farm tools must be shared with those who ask for them.

THE LAND

The more than 800 islands and islets of Fiji occupy some 7,055 square miles (18,272 square kilometers) of land scattered over some 250,000 square miles (647,500 sq. km.) of ocean. About 105 of the islands are inhabited. The island of Rotuma, which is about 240 miles (386 kilometers) northwest of Fiji, is included in the group.

Fiji's two main islands—Viti Levu and Vanua Levu—are volcanic in origin. They are generally rugged, with high, sharp peaks reaching over 4,000 feet (1,200 meters) above sea level. Unlike many of the Pacific islands, the two main islands of Fiji have relatively large areas of flatland that have been built up by the action of rivers. Many of the smaller islands are low coral atolls with sandy beaches and stately palms.

The climate of Fiji is tropical with an average year-round temperature of 80 degrees Fahrenheit (27 degrees Celsius). The southwest trade winds bring heavy rains to the east side of Viti Levu and Vanua Levu. Suva, on the rainier side of Viti Levu, receives over 120 inches (300 centimeters) of rain a year, while the drier north coast has only about 70 inches (178 cm.) annually. As a result the vegetation on the rainy side is dense and tropical. On the dry side grasses and shrubs grow.

THE ECONOMY

For years Fiji imported many of the things it needed and paid for them with money from exports of sugar, copra (dried coconut meat), and gold. But that pattern has changed. Fiji is becoming increasingly self-reliant. The development of small manufacturing industries, the introduction of new crops, and the expansion of livestock and dairy industries have helped keep pace with the rapidly growing population. This development has reduced Fiji's dependence on the outside world for basic items of food and clothing. At the same time, its exports have increased. Sugar and copra are still important products, and although gold mining is declining, other minerals being mined, such as bauxite, phosphate, and oil, have begun to take its place in the economy. The most important developments, however, are the enormous expansion of the tourist industry since 1960 and the efforts being made to exploit the extensive timber resources. Timber production and related industries are expected eventually to become the country's most important economic activities. At present, the sugar industry is still the most important. Sugar exports represent more than half of the value of total exports, while the industry provides employment directly for almost one quarter of the population and indirectly for an even greater proportion. Sugar is produced on small holdings mainly by Fijians of Indian origin.

As in all the Pacific islands, copra has always been important to the economy. Fiji's copra exports are valued at over $5,000,000 per year. Copra used to be exported in its raw form, but is now exported in treated form as oil or meal.

The tourist industry is already earning about half as much as the whole of the sugar industry. First-class hotels have been built and related industries have been developed to provide for the some 100,000 people who visit Fiji annually.

Sugarcane is the basis of commercial farming in Fiji.

FACTS AND FIGURES

FIJI is the official name of the country.
CAPITAL: Suva (on the island of Viti Levu).
LOCATION: Southwest Pacific. **Latitude**—15° S to 22° S. **Longitude**—175° E to 177° W.
AREA: 7,055 sq. mi. (18,272 sq. km.).
PHYSICAL FEATURES: Highest point—Mount Victoria (4,341 ft.; 1,323 m.). **Lowest point**—sea level.
POPULATION: 519,000 (estimate).
LANGUAGE: English.
RELIGION: Christian, Hindu, Muslim.
GOVERNMENT: Constitutional monarchy. **Head of state**—British monarch represented by governor-general. **Head of government**—prime minister. **Legislature**—parliament. **International co-operation**—United Nations, Commonwealth of Nations.
ECONOMY: Chief mineral—gold. **Chief agricultural products**—sugarcane, coconuts, rice, cacao, pineapples. **Industries and products**—sugar, coconut products, gold.
MONETARY UNIT: Fiji dollar.
NATIONAL HOLIDAY: Oct. 10, Independence Day.
NATIONAL ANTHEM: "God Bless Fiji."

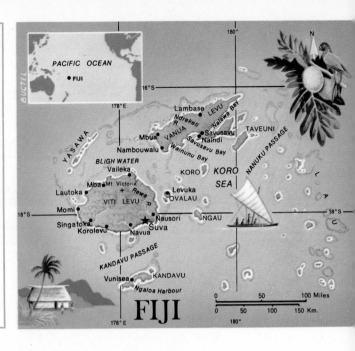

FIJI

HISTORY

It is believed that the Fijians' ancestors came from somewhere in Southeast Asia and, over a long period of time, made their way to Fiji through New Guinea, the Solomons, and the New Hebrides. The first European known to have reached the islands was the Dutch explorer Abel Tasman, who sailed through the Fiji group in 1643.

There were considerable hazards for the early visitors to the islands because the Fijians of that time were cannibals who considered "long pig" a delicacy. Nevertheless, Americans and Europeans came to plunder the sandalwood resources, which they exhausted entirely in a period of about 10 years. Whalers and other ships called there for supplies. Crew members who deserted or survived shipwrecks became beachcombers. Traders came to sail their schooners through the islands, and some settled there. At about the same time, missionaries arrived and began to persuade the leading chiefs to abandon cannibalism and become Christians.

There were incidents of violence—attacks by Fijians and attacks by Europeans. To gain favor, missionaries and traders alike sold and often gave arms and ammunition to native leaders. This had a disastrous effect on the relations among Fijians. It also made it possible for one of the leaders, Cakobau ("chief of Bau"), to become ruler over a large portion of his group. Faced with defeat in 1854, Cakobau embraced Christianity and gained the support of the missionaries. However, Cakobau was never able to rule all the groups in his realm and in 1874, after years of struggle, Fiji became a British crown colony. It became independent on October 10, 1970—96 years to the day after it had been ceded to Queen Victoria.

Government. Fiji now is an independent state within the Commonwealth of Nations. The government is led by a prime minister who presides over a form of cabinet government. The parliament is composed of the House of Representatives, whose members are elected, and the Senate, whose members are appointed. Although Fiji is a young nation, it is developing rapidly as the crossroads of the South Pacific.

JOHN MILES, Senior Political Affairs Officer, United Nations

NAURU

The Republic of Nauru, which has an area of 8.2 square miles (21 square kilometers), has the distinction of being the smallest independent nation in the Pacific. Nauru is also one of the most isolated islands in the southwest Pacific. Situated just south of the equator, Nauru has as its nearest neighbor Ocean Island, which is about 190 miles (300 kilometers) to the east and is, like Nauru itself, covered almost entirely by phosphate deposits. In fact, phosphate, which is used as a fertilizer, is the most important topic in all descriptions and discussions of Nauru.

Nauru's phosphate deposits account for its present comparative wealth and for the fact that it became an independent country. The phosphate deposits, which cover approximately four fifths of the tiny island, are among the highest quality in the world. But the phosphate is being worked out rapidly. It is expected that at the present rate of extraction the deposits will be exhausted by the year 2000. When that day arrives the Nauruans' present source of income will have disappeared. In addition, the island will have been made uninhabitable because it will have neither soil nor other resources, unless they are brought to the island by ship.

The People

The Nauruans are believed to be a mixture of Pacific peoples—Polynesian, Micronesian, and Melanesian. Nothing is known of their origin or how their ancestors came to Nauru. Their language does not seem to be related to any of the other languages of the Pacific islands. German missionaries translated the Bible into one of the most widely used dialects, providing the basis for a standard Nauruan language. Education, which is free and compulsory for Nauruan children between the ages of 6 and 17, has also helped to unify the people of the island. In addition to state-run schools, there are schools provided by the Roman Catholic mission to Nauru. Young Nauruans who want to continue their education in college or technical school generally go to Australia.

The prospering phosphate industry provides most of the working population with jobs. Earnings from phosphate also help to provide welfare benefits and social services for those who need them. Since the end of World War II an ambitious housing program has been under way and a large number of European-style houses have been built. Each of the new houses is provided with a 10,000-gallon freshwater tank. The tank, or the rainwater catchment on the roofs of older houses, is a vital necessity on an island that has no rivers or streams and receives only a small amount of rain annually. In especially dry years water is imported.

Cultural and local political life on the island centers on the Domaneab—which is the Nauruan for "meeting place of the people." The legislature is located at Yaren.

Perhaps the most unusual feature of island life is the widespread hobby of capturing and taming seabirds such as noddies, frigate birds, and man-o'-war hawks. Special perches are built for the birds on the beach and the islanders save their pets the trouble of hunting for their food by bringing them bits of raw fish to eat.

FACTS AND FIGURES

REPUBLIC OF NAURU is the official name of the country.

CAPITAL: No official capital.

LOCATION: Southwest Pacific Ocean, 33 miles (53 km.) south of equator. **Latitude**—0° 31′ S. **Longitude**—165° 56′ E.

AREA: 8.2 sq. mi. (21 sq. km.).

POPULATION: 6,500 (estimate).

LANGUAGES: Nauruan, English.

RELIGION: Nauruan Protestant Church, Roman Catholic.

GOVERNMENT: Republic. **Head of government**—president. **Legislature**—legislative assembly. **International co-operation**—Commonwealth of Nations, South Pacific Commission, International Telecommunications Union of the United Nations.

ECONOMY: **Chief mineral**—phosphate. **Chief agricultural product**—none. **Industry and product**—phosphate mining. **Chief export**—phosphate. **Chief imports**—food supplies, building materials, machinery.

MONETARY UNIT: Australian dollar.

NATIONAL HOLIDAY: January 31, Deliverance Day (1946) and Independence Day (1968).

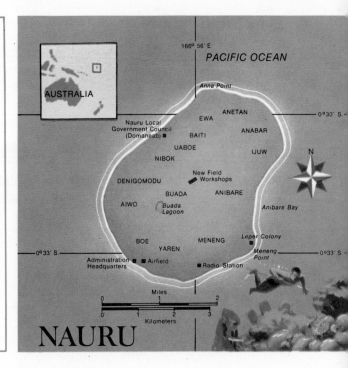

NAURU

History

Europeans first learned about Nauru in 1798 when a whaling captain, John Fearn, visited it. Because of the favorable impression it made on him, he called it Pleasant Island. In the years that followed, Nauru received its share of traders and beachcombers. As in many other Pacific islands, one effect of their presence was to create conflict among the various groups on the island and to cause the death rate to rise sharply as a result of fighting and the introduction of previously unknown diseases.

Nauru became a pawn in the game of international politics, and in 1886, as a result of the Anglo-German Convention, it was allocated to the German sphere of influence in the Pacific. German administration went into effect in 1888 when a commissioner arrived. It was not, however, until the turn of the century, when an Englishman discovered the rich phosphate deposits, that Nauru began to have economic value to Europeans. An English company acquired the rights for mining the phosphate in Nauru and Ocean Island. Mining operations began on Nauru in 1906.

In 1920 Nauru became a League of Nations mandated territory. During World War II, Nauru was occupied by Japanese forces, and in 1943, 1,200 Nauruans were taken to Truk in the Caroline Islands. At the end of the war, 743 survivors were returned to Nauru, where they arrived on January 31, 1946. This day, remembered in Nauru as Deliverance Day, was the date chosen as Nauru's Independence Day in 1968. At the end of World War II, Nauru was placed under the International Trusteeship System of the United Nations and was jointly administered by Australia, New Zealand, and Great Britain. Again, as under the League, Australia administered the territory on behalf of the other two nations. It continued to do so until Nauru achieved independence.

Economy

At the end of World War I, the assets of the phosphate company were acquired by the three governments—Australia, Great Britain, and New Zealand. The management of the industry in both Nauru and Ocean Island was placed in the hands of the British Phosphate Commissioners. It was their function to direct the extraction of the phosphate and supply it to each of the three governments, on a nonprofit basis, for the costs of production.

Included in the costs of production were royalties that were paid to the Nauruans. These included not only royalties paid to landowners but royalties paid into various trust funds to be used for public purposes or to be held for use by the Nauruans when the supply of phosphate ran out. In addition, the commissioners paid royalties to meet the costs of administration and other special public expenses.

The Nauruans' feeling that they could get better prices for their phosphate if they were in charge of the industry themselves was one of the principal reasons for their insistence that control should revert to them as part of the independence settlement. Eventually the three partner governments agreed to relinquish their control of the industry and to sell it to the Nauruan Government.

Nauru's bid for independence was a move that took most observers by surprise, and when originally suggested it was met with great derision in many quarters. For years, the Australian Government, as trustee, with the support of the United Nations, had thought of the future of Nauru in terms of finding a new home for its people. Although a number of possible sites were found, they were unsatisfactory to the Nauruans because each involved settlement in another country and meant that the Nauruans would lose their identity as a people. Finally, the Nauruans defied all the rules relating to international sovereignty. They demanded and received their independence, instead of accepting association with Australia.

Government

Nauru has a representative legislature and a cabinet composed of ministers presided over by a chief minister drawn from the legislature. The chief minister is also the president and head of state, and he is elected by the legislature, thus thoroughly mixing the usually separated executive and legislative functions of government.

The Future

The Nauruans are optimistic about their future. Their young people are well educated since many have received a higher education and technical training in Australia. Their leaders are energetic and, as they have already shown, are willing to make imaginative decisions. The knowledge that the island shall one day be unable to support them does not seem to daunt the Nauruans. Their job is to discover whether it is feasible to fill the worked-out phosphate areas with soil. The government has also purchased a number of cargo ships as the nucleus of its fleet, and it has hopes of attracting tourists and business enterprises. These are some of the ways the Nauruans seek to guarantee their young nation's future on the island.

JOHN MILES, Senior Political Affairs Officer, United Nations

Tongan women working on a ceremonial rug.

TONGA

The Kingdom of Tonga became a fully independent nation in June, 1970, after 70 years as a British protected state. The kingdom, which is made up of about 150 islands and islets, is situated southeast of the Fiji islands and southwest of the Samoan islands. It has a total area of only 270 square miles (700 square kilometers).

Tonga is divided into three main groups: Vava'u in the north; Ha'apai in the center; and Tongatapu in the south. Geologically, however, Tonga consists of two parallel, narrowly separated formations. Both are volcanic in origin, but the eastern chain has sunk, so that the islands are low-lying and coral covered. The islands in the other chain are high islands with richer volcanic soils.

The People and Their History

Most of the approximately 83,000 Tongans are Polynesian. The Tongans' ancestors arrived in the islands early in the Christian Era. They probably formed part of the first group of migrating Polynesians.

Tonga is one of the oldest kingdoms in the Pacific. According to tradition, *tu'i tongas,* the spiritual kings of Tonga, can be traced back to the 10th century. The influence of the *tu'is* was quite widespread and sometimes seems to have extended to other areas of Polynesia.

Tonga was first discovered by Europeans in 1643 when the Dutch explorer Abel Tasman landed on Tongatapu. English and Spanish sea captains followed. One of them, the 18th-century Englishman, Captain James Cook, was so impressed with the hospitality of the islanders that he called their group the Friendly Islands, a name that is still sometimes used. Another well-known visitor was Captain Bligh of the H.M.S. *Bounty.* The famous mutiny took place while the *Bounty* was in Tongan waters.

Christian missionaries made two unsuccessful attempts to gain a foothold in the islands, first in 1797, then in 1822. Both times they were forced to leave. Finally, in the early 1830's, an alliance was formed between the missionaries and one of the chiefs. The chief received the support of the missionaries, who supplied him with useful European goods, including arms and ammunition. For his part, the chief announced his conversion to Christianity and saw to it that his subjects were similarly converted. This chief became the acknowledged leader, first in his own Ha'apai group, then in the Vava'u group, and finally in the Tongatapu group. In 1845 he became the ruler of the whole of Tonga as King George Tupou I.

King George ruled until his death in 1893. During the last 10 years of his reign his island kingdom was torn by religious dissension. Aided by the Reverend Shirley Baker, one of the leading missionaries, King George seceded from the Wesleyan, or Methodist, Church and formed his own separate and independent church, the Wesleyan Free Church of Tonga.

Soon, with Baker ruling as premier, stories of religious persecution in the islands began to reach the outside world. There were complaints that people were being unjustly imprisoned and exiled and that their lands were being confiscated because they refused to join the King's new church. The problem was solved when, after investigation by British officials in Fiji, King George was persuaded to dismiss Baker as his premier.

Tonga's early encounter with Christian missionaries is reflected in its social and cultural life today. Christian values and practices have a profound effect on all Tongans. Churches are regularly well-attended. The Sunday Sabbath is observed with great strictness. According to Tonga's Constitution it is not lawful to work, play games, or trade on Sunday. The influence of the Christian churches also extends to such areas as education. The churches run 50 of the country's 129 primary schools and 41 of its 44 post-primary schools.

Government

Tonga's government is a mixture of Polynesian and European elements. The king rules his people on the basis of a constitution that com-

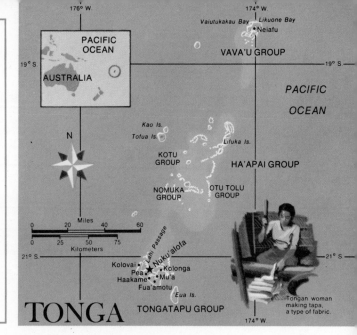

FACTS AND FIGURES

KINGDOM OF TONGA is the official name of the country.

CAPITAL: Nuku'alofa.

LOCATION: South Pacific Ocean. **Latitude**—15° S to 23° 30' S. **Longitude**—173° W to 177° W.

AREA: 270 sq. mi. (700 sq. km.).

PHYSICAL FEATURES: Highest point—On island of Kao, about 3,380 feet (1,030 m.). **Lowest point**—sea level.

POPULATION: 83,000 (estimate).

LANGUAGE: Tongan.

RELIGION: Wesleyan Free Church of Tonga.

GOVERNMENT: Monarchy. **Head of state**—king. **Head of government**—premier. **Legislature**—legislative assembly.

CHIEF CITIES: Nuku'alofa, Neiafu.

ECONOMY: Chief minerals—phosphate, oil. **Chief agricultural products**—copra, bananas. **Chief exports**—copra, bananas. **Chief imports**—textiles, foodstuffs, hardware.

MONETARY UNIT: Pa'anga.

NATIONAL HOLIDAY: June 4, Independence Day.

Tongan woman making tapa, a type of fabric.

bines democracy with the traditional aristocracy of the Polynesians. The principal executive body is the Privy Council. The king presides over the Council, which is appointed by him. The function of the Council is to advise the king. Tonga's Legislative Assembly is presided over by a speaker appointed by the king. The Legislative Assembly consists of the seven privy councillors, seven Tongan nobles, and seven elected representatives of the people.

Economy

Most Tongans get their livelihood from the land. Every Tongan male, when he reaches the age of 16, is entitled to receive 8¼ acres (3.4 hectares) of land for cultivation and a small allotment in town for his house.

Tonga has two important export products—copra and bananas. However, the country has always imported more than it has exported. In an effort to provide the kind of economic base that would help Tonga to overcome this unfavorable balance of trade, a development plan to increase exports was begun in 1965. Emphasis was also placed on long-term projects in education, health, communications, and agriculture.

Another way that the government hopes to promote the small kingdom's development is through the establishment of a tourist industry. A modern hotel has been built at Nuku'alofa. Cruise ships already call there, and the possibility of building a jet airport is being considered. Small seepages of oil were discovered on Tongatapu in 1968 and are being investigated as another source of income.

The Future

Tonga is a small country. Its population is growing so rapidly that it is now difficult to provide land allotments for the young men, and it may soon become impossible. To add to this problem, there are few other employment opportunities at present. But the quality of Tonga's leadership, together with the apparently successful blending of traditional Polynesian customs and institutions with those of the Western world, offers hope for the future of this small Pacific kingdom.

JOHN MILES, Senior Political Affairs Officer, United Nations

Coconut, a staple food of the islands.

WESTERN SAMOA

Western Samoa, which gained its independence in 1962, is made up of the large islands of Savai'i and Upolu, as well as seven smaller islands and islets. The remaining islands of the Samoa group make up American Samoa, a dependency of the United States. The beauty of the islands has attracted many visitors. Perhaps the best-known of these was the English poet Robert Louis Stevenson, who spent the last years of his life on Upolu. Known to the Samoans as Tusitala—Teller of Tales—Stevenson was buried on a mountain overlooking the town of Apia in 1894. His tomb is inscribed with his own words:

> Here he lies where he long'd to be,
> Home is the sailor, home from the sea,
> And the hunter home from the hill.

The Land

The islands of Western Samoa are mountainous and of volcanic origin. Mount Silisili on Savai'i rises to over 6,000 feet (1,830 meters) and Mount Fito on Upolu reaches over 3,000 feet (910 m.) above sea level. Many of the mountains in the interior are covered by dense tropical rain forests. The lower mountain slopes and valleys have fertile soils, but Western Samoa lacks mineral resources. Its climate is tropical. Rainfall is abundant, but there are no large rivers because the water sinks into the porous lava soil. The village people usually get their water from nearby springs. Rainwater is also captured and stored in tanks.

Of the two main islands, Upolu is the most densely populated. Apia, the largest town in the group, is on the northern coast of Upolu and has grown into an important commercial and transportation center.

The People

Although the people of Western Samoa are trying to join their economy to that of the more developed nations, they continue to live simply and with few luxuries. Their homes, called *fales,* are well-suited to the climate. The roof is made of thatched leaves, and the sides of the house are left open. For privacy and protection from winds and rain there are blinds made of coconut palm leaves. The earth floor of the house is usually covered with pebbles. Mats are piled up to make seats and beds. Most Samoans wear plain clothes suited to the warm climate. The lavalava, a piece of cloth wrapped around the waist, is worn by men. Women wear a two-piece dress called a *puletasi.*

Tropical fruits, such as bananas, coconuts, avocados, papayas, and pineapples, as well as poultry, fish, and boar are the basis of the Samoan diet. Foods are still often prepared in a separate cookhouse over an open fire. Distinguished visitors are honored at the ava ceremony. *Ava* is a traditional drink made by pounding the roots of a tree of the pepper family and adding water. The beverage is served in a wooden bowl by the *taupo*—the daughter of the chief. According to custom a little of the drink is spilled on the ground for the gods and then the toast is given, "*Ia manuia* [May you be blessed]."

The Economy.   Western Samoans have been blessed with a climate and soil good for growing certain kinds of foods, but the lack of natural resources for industry presents serious problems. The economy is still geared mainly to producing enough food to live on. The three main export crops are copra (dried coconut meat), cacao, and bananas. The export of taro is growing in importance. And efforts are being made to develop a coffee industry and to investigate the timber resources.

At present Western Samoa's ability to pay for the goods and services it imports from the outside world depends on the three agricultural products. Even in the best years, the value of goods imported into Western Samoa considerably exceeds the value of goods exported. The serious problems faced by the economy have been underlined in recent years by the fluctuations in the world prices for copra and cacao, as well as by damage caused by an insect that attacks coconut trees, and by the diseases that have affected cacao trees and banana palms.

Education.   One hope for the future lies in the growth of educational opportunities for young Samoans. Education is provided by the central government and, as in most Pacific islands, by the Christian missions. The most pressing problem is to provide enough schools to raise the level of education to the point where the country is able to produce trained manpower for its economic and political life.

History and Government

As with most other areas of the Pacific, the early history of Samoa is a mixture of legend and educated guesses. The first European to sight the island group was the Dutch explorer Jacob Roggeveen, in 1722. Other explorers followed. The first lasting contact took place when an English missionary arrived in 1830. Within a few years Christianity was well established in Samoa.

The importance of Samoa to the great Western powers made it the object of conflicting claims among the Americans, British, and Germans for much of the latter part of the 19th century. In 1900 a treaty was

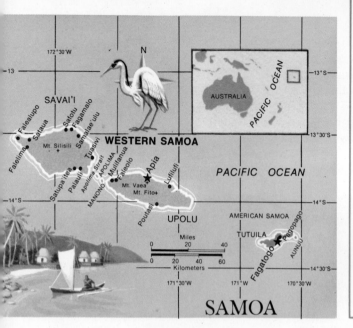

SAMOA

FACTS AND FIGURES

WESTERN SAMOA—Samoa i Sisifo—is the official name of the country.

CAPITAL: Apia.

LOCATION: South Pacific Ocean. **Latitude**—12° 53' S to 14° 07' S. **Longitude**—171° 24' W to 172° 48' W.

AREA: 1,097 sq. mi. (2,842 sq. km.).

PHYSICAL FEATURES: Highest point—6,094 ft. (1,857 m.) on Savai'i. **Lowest point**—sea level.

POPULATION: 141,000 (estimate).

LANGUAGE: Samoan, English.

RELIGION: Protestant, Roman Catholic.

GOVERNMENT: Republic. **Head of state**—Ao o le Malo. **Head of government**—prime minister. **Legislature**—legislative assembly. **International co-operation**—United Nations Economic Commission for Asia and the Far East, World Health Organization (WHO), South Pacific Commission.

CHIEF CITY: Apia.

ECONOMY: Chief agricultural products—taro, coconuts, bananas, cacao, tropical fruits, coffee. **Industries and products**—copra, clothing, food processing. **Chief exports**—copra, bananas, cacao. **Chief imports**—meat, sugar, cotton goods, motor vehicles.

MONETARY UNIT: Tala.

NATIONAL HOLIDAY: January 1, Independence Day.

NATIONAL ANTHEM: "The Flag of Freedom."

signed making Western Samoa a colony of Germany and giving the United States control over Eastern Samoa. German rule of Western Samoa lasted until World War I, at the end of which the colony was placed under the control of New Zealand by the mandate system of the League of Nations. From 1945 to 1962 Western Samoa was a United Nations trust territory, administered by New Zealand.

The 1962 Constitution is a mixture of traditional Polynesian and Western elements. The head of state is elected for a 5-year term by the 47-member Legislative Assembly. The head of government is the prime minister, who is appointed by the head of state from among members of the Legislative Assembly. The Legislative Assembly is made up of 45 Samoans and two Europeans. The Europeans are elected by all adult Europeans. The Samoan members must be matai, as the traditional chiefs are called, and only matai have the right to vote for them.

A matai is chosen after discussion by members of the family and thereafter has the right to call upon the services of members of his group, as well as to control the family lands and often their products. He usually holds his position for life, although in some cases it is possible for the family to remove him. To most Western Samoans the system is democratic, since any Samoan can hope to become a matai.

The Future

Western Samoa has adopted a realistic policy concerning its development and its relations with the outside world. A board has been formed to develop new aspects of the economy including the livestock industry and tourism. Soon after its independence, Western Samoa concluded a Treaty of Friendship with its former administrator, New Zealand. By this treaty, New Zealand will assist Western Samoa in foreign relations and to consider requests for technical and other assistance. Western Samoa also takes part in the work of the South Pacific Commission.

JOHN MILES, Senior Political Affairs Officer, United Nations

ILLUSTRATION CREDITS

The following list credits, according to page, the sources of illustrations used in volume 2 of LANDS AND PEOPLES. The credits are listed illustration by illustration—top to bottom, left to right. Where necessary, the name of the photographer or artist has been listed with the source, the two separated by a dash. If two or more illustrations appear on the same page, their credits are separated by semicolons.

4– Jere Donovan
5
7 Luis Villota
8 Emil Muench—Lenstour Photos
9 F. & N. Schwitter Library
10 Editorial Photocolor Archives, N.Y.
11 Vroom
12– Jeppeson Maps—The H. M. Gousha Co.
13
15 Margaret Durrance
16 George Buctel
18 Luis Villota; Victor Englebert
19 Torben Huss; Luis Villota
20 Editorial Photocolor Archives, N.Y.; Japan National Tourist Organization
21 Dankwart Von Knobloch—Lenstour Photos; Eric L. Ergenbright—Lenstour Photos
23 B. G. Silberstein—Alpha KK
24 Joshua Tree Productions; Joshua Eyal
25 Peter Turner
26 Editorial Photocolor Archives, N.Y.; J. Alex Langley—DPI
27 Fujihira—Monkmeyer Press Photo Service
28 Monkmeyer Press Photo Service
29 Michael Charles; Banyan Productions, Singapore
32 Annan Photo Features; Wesley B. McKeown
33 Annan Photo Features; Wesley B. McKeown
34 George & Rhoda Sidney
36 Shostal Associates; Brian Brake—Rapho Guillumette Pictures
39 Editorial Photocolor Archives, N.Y.
40 Michael Charles
41 Center of Asian Art and Culture, The Avery Brundage Collection, San Francisco
42 Charles Shapp
43 Marilyn Silverstone—Magnum Photos
48 Frank Schwarz—Lee Ames Studio
49 De Wys, Inc.; William Mares—Monkmeyer Press Photo Service
51 Chris Kutschera
53 Editorial Photocolor Archives, N.Y.; Margaret Durrance
54 Gosta Glase—Ostman Agency
56 Harrison Forman
58 Warren Slater—Monkmeyer Press Photo Service
59 Owen Franken
60 Mulvey-Crump Associates, Inc.
62 Editorial Photocolor Archives, N.Y.; Doranne Jacobsen—Editorial Photocolor Archives, N.Y.
63 Luis Villota
64 Owen Franken
65 Editorial Photocolor Archives, N.Y.; Doranne Jacobsen—Editorial Photocolor Archives, N.Y.
66 Doranne Jacobsen—Editorial Photocolor Archives, N.Y.
67 Anne Bringsjord
68 Howard Koslow
70 Grolier, Athens
71 International Foto File
72 International Foto File
73 George Buctel
74 International Foto File
76 J. Johansson—De Wys, Inc.
77 Lanks—Monkmeyer Press Photo Service
79 George Buctel
80 Harrison Forman
81 Inge Morath—Magnum Photos
83 Harrison Forman
84 F. & N. Schwitter Library
85 George Buctel
86 Harrison Forman
87 Herbert Fristedt

88 A. Earle Harrington; F. & N. Schwitter Library
90 Herbert Fristedt
92 Herbert Fristedt; Charles Shapp
93 George Buctel
94 Eric Brown—Monkmeyer Press Photo Service
95 Multi-Media Photography, Inc.; Herbert Fristedt—Ostman Agency
97 Kay Honkanen—Ostman Agency
99 Charles Shapp
101 Herbert Fristedt
102 George Buctel
103 Editorial Photocolor Archives, N.Y.; Herbert Fristedt
104 Editorial Photocolor Archives, N.Y.
105 Jordan Penkower—Editorial Photocolor Archives, N.Y.
106 Jordan Penkower—Editorial Photocolor Archives, N.Y.
107 George Buctel
108 Editorial Photocolor Archives, N.Y.
110 Editorial Photocolor Archives, N.Y.
111 William Mares—Monkmeyer Press Photo Service
112 J. Alex Langley—DPI
113 Frank Schwarz—Lee Ames Studio
115 Aramco; S. E. Hedin
117 Brent Brolin
118 Brent Brolin
119 Walter Hortens
121 Alan Band Associates
122 Walter Hortens
123 Brian Brake—Rapho Guillumette Pictures
124 George Buctel
125 Frank Schwarz—Lee Ames Studio
126 Frank Schwarz—Lee Ames Studio
127 Frank Schwarz—Lee Ames Studio
128 Omar Bessim—Monkmeyer Press Photo Service
129 Mulvey-Crump Associates, Inc.
130 Omar Bessim—Monkmeyer Press Photo Service
131 Editorial Photocolor Archives, N.Y.
132 Gosta Glase—Ostman Agency
133 Gosta Glase—Ostman Agency
134 Emil Muench—Ostman Agency
135 George Buctel
137 Emil Muench—Ostman Agency
139 Wesley McKeown
140 Bernard Siberstein—Rapho Guillumette Pictures
142 Owen Franken
143 George Buctel
144 Owen Franken
145 Harrison Forman
147 Owen Franken; Harrison Forman
148 Owen Franken
151 Emil Muench—Ostman Agency
153 Emil Muench—Ostman Agency; S. E. Hedin
154 F. & N. Schwitter Library
155 George Buctel
157 J. Alex Langley—DPI
158 J. Alex Langley—DPI
159 Mulvey-Crump Associates, Inc.
160 Warren Slater—Monkmeyer Press Photo Service; Herbert Lanks—Monkmeyer Press Photo Service
162 Bill Hubbel—FPG
163 J. Alex Langley—DPI
164 J. Alex Langley—DPI
165 Esther Gerling—FPG
166 Fujihira—Monkmeyer Press Photo Service
167 Esther Gerling—FPG
168 Fujihira—Monkmeyer Press Photo Service; J. Alex Langley—DPI
169 Esther Gerling—FPG

172 Wesley McKeown
177 Edward J. McCabe
178 Jere Donovan
180 Bergman-Sucksdorf—Ostman Agency
181 F. & N. Schwitter Library
182 F & N. Schwitter Library
183 Philcarol—Monkmeyer Press Photo Service
184 Rosemary Russo Levin
185 F. & N. Schwitter Library
186 F. & N. Schwitter Library
187 F. & N. Schwitter Library
188 Emil Muench—Ostman Agency
189 Editorial Photocolor Archives, N.Y.
190 S. E. Hedin
191 F. & N. Schwitter Library
192 S. E. Hedin—Ostman Agency
193 F. & N. Schwitter Library
194 S. E. Hedin—Ostman Agency
195 S. E. Hedin—Ostman Agency
196 Carl Purcell
197 Monkmeyer Press Photo Service
199 Robert W. Young—Lenstour Photo Service
200 Editorial Photocolor Archives, N.Y.
201 F. & N. Schwitter Library
203 S. E. Hedin—Ostman Agency
204 Monkmeyer Press Photo Service
205 Emil Muench—Ostman Agency
206 Fujihira—Monkmeyer Press Photo Service
207 S. E. Hedin
208 A. Earle Harrington
209 S. E. Hedin
210 F. & N. Schwitter Library
211 S. E. Hedin—Ostman Agency
212 S. E. Hedin—Ostman Agency
213 Editorial Photocolor Archives, N.Y.
214 Fujihira—Monkmeyer Press Photo Service
215 Wesley McKeown
216 Frank Schwarz—Lee Ames Studio
217 Rosemary Russo Levin
219 Editorial Photocolor Archives, N.Y.
220 Margaret Durrance
221 George Buctel
223 Frank Schwarz—Lee Ames Studio
224 Alan Band
225 George Buctel
228 J. Alex Langley—DPI
229 J. Alex Langley—DPI
230 George Buctel
232 Emil Muench—Ostman Agency
233 Fujihira—Monkmeyer Press Photo Service
234 A. Earle Harrington
235 George Buctel
236 A. Earle Harrington
238 Randolph King
239 George Buctel
240 Frank Schwarz—Lee Ames Studio
242 J. H. Pickerell—Alan Band Associates
244 Editorial Photocolor Archives, N.Y.; Victor Englebert
246 Herbert Fristedt; Peter Turner
247 Editorial Photocolor Archives, N.Y.; Jack Fields—Photo Researchers
248 Banyan Productions, Singapore
249 George & Rhoda Sidney
251 Per Reppen—Ostman Agency
252 Per Reppen—Ostman Agency
253 Fujihira—Monkmeyer Press Photo Service
255 George Buctel
257 S. E. Hedin; Fujihira—Monkmeyer Press Photo Service
259 Rosemary Russo Levin
261 George Buctel
262 Jules Bucher—Photo Researchers
263 S. E. Hedin—Ostman Agency
264 S. Erikson—De Wys, Inc.
267 S. Erikson—De Wys, Inc.